Studying Russia and Its Wars

Studying Russia and Its Wars

Academic Stocktaking in Times of Insecurity

Edited by

Yuliia Kurnyshova
Andrey Makarychev

BRILL

LEIDEN | BOSTON

Cover image: © Pawel Czerwinski. Available for use via Unsplash.

This book received funding from the Estonian Academy of Science, 2024.

The Library of Congress Cataloging-in-Publication Data is available online at https://catalog.loc.gov

Typeface for the Latin, Greek, and Cyrillic scripts: "Brill". See and download: brill.com/brill-typeface.

ISBN 978-90-04-74735-7 (hardback)
ISBN 978-90-04-74736-4 (e-book)
DOI 10.1163/9789004747364

This book is printed on acid-free paper and produced in a sustainable manner.

Contents

Acknowledgements

This edited volume is the main academic result of the project titled "New Russia Studies: a Plan for Teamwork", generously supported by the Estonian Academy of Science in 2023–2024. Within the framework of this project, we have convened two academic workshops—"Studying Russia's War: Methodological and Conceptual Innovations" (October 6, 2023, hosted by the Johan Skytte Institute of Political Studies, University of Tartu) and "Russia and Its War(s): Scholarly Audit and Political Implications" (June 7, 2024, Tallinn). Project's findings were summarized, presented and discussed at two panels—"Studying Russia and Its Wars: an Academic Audit" at the Tartu Conference on East European and Eurasian Studies (June 12, 2025), and "The Russo-Ukrainian War: Academic Lessons and Policy Learning" at the Aleksanteri Conference (Helsinki, October 23, 2025). We are thankful to all colleagues involved in the project as contributors, discussants, moderators, panel chairs and peer reviewers.

The Editors
October 2025, Helsinki and Tartu

Notes on Contributors

Francoise Dauce
is Director, Centre of Russian, Caucasian and Central European Studies (CERCEC) in Paris. Her works focus on forms of political domination, both coercive and liberal, which are exercised in different areas: the army, civil society and the media.

Dinissa Duvanova
is Professor and Chair of International Relations Department, Lehigh University. She is political economist specializing on governance, bureaucratic politics and technology-enabled forms of political participation in Russia, Eastern Europe, and Central Asia. Duvanova's current empirical research focuses on protests and repressions in Russia, Ukraine, and Kazakhstan.

Vladimir Gel'man
is Professor of Russian and Eurasian Studies at the Aleksanteri Institute, University of Helsinki, Finland, and Research Associate at the Yerevan Center for International Education, Armenia. His research interests include Russian and post-Soviet politics and governance in a theoretical and comparative perspective with a special emphasis on political regime dynamics, political institutions, policymaking, electoral and party politics, regional and local government.

Ivan Gomza
is Professor at the Graduate Department of Social Sciences and Humanities, Kyiv School of Economics. His scholarly interests comprise warfare, democratization, authoritarian regimes, nationalism, contentious politics, and good governance.

Sanshiro Hosaka
is research fellow at the International Center for Defense and Security in Tallinn. He specializes in Soviet/Russian propaganda and covert actions; Western academia and agents of influence; Soviet/Russian intelligence; historical memory of Ukrainians

Aliaksei Kazharski
is associate professor at Charles University (Prague). His research interests include Central and Eastern Europe, regionalism, national identities, and social constructivist analysis of international politics

Ivan Klyszcz
is a Research Fellow at the International center for Defense and Security in Tallinn. Dr Klyszcz's research is centred on issues of international security, namely, great power competition, the grand strategy of major powers, and Russian foreign policy.

Yuliia Kurnyshova
is Researcher at the Aleksanteri institute, University of Helsinki and affiliated researcher at the University of Tartu. Her research project explores the political and security implications of Russia's war against Ukraine, with a focus on agency, security discourse and theory of securitization.

Andrey Makarychev
is Professor of Regional Political Studies at the University of Tartu. His areas of expertise include biopolitics, visual analysis of international relations and the Russo Ukraine war.

Katalin Miklóssy
is Docent of political history, and Head of Eastern European studies, works at the Aleksanteri institute, University of Helsinki. She focuses on the evolution of the rule of law with special interest in the impact of regional security conceptions on nationalism and conservatism.

Anselm Schmidt
is grant consultant at University of Tartu.

Kirill Shamiev
is a postdoctoral researcher at the Institute of International Relations and Political Science of Vilnius University. He studies Eurasian domestic politics, civil-military relations and foreign & security affairs.

CHAPTER 1

Studying Russia's Wars:

An Analytical Introduction

Yuliia Kurnyshova and Andrey Makarychev

Russia's military intervention in Ukraine put an end to many core elements of the post-Cold War security order, including NATO—Russia enhanced and prioritized relations, EU–Russia strategic partnership, and Russia's 'soft power' projections (from the Sochi Olympics and FIFA World Cup to the 'Russian world' doctrine). Russia's withdrawal from the Council of the Baltic Sea States and expulsion from the G8 exemplified Moscow's institutional disengagements with the Euro-Atlantic international society. Yet despite these evident ontological changes, the academic community still runs short of new scholarly approaches and interpretative tools for properly examining and contextualising these transformations.

Apparently, other turning points of a similar force and long-term consequences in the recent history have left strong imprints in various academic domains. Thus, the end of the Second World War triggered a long debate among German intellectuals on collective guilt and responsibility. In the meanwhile, in the USA the post-war years laid institutional foundation for new disciplinary fields, such as area studies which became an intellectual component of American leadership in non-Western world. The 1968 events gave a strong boost to the blossoming of the French intellectual left, including Michel Foucault, Jacques Ranciere, Alain Badiou, Jean-Luc Nancy and other philosophers. The end of the Cold War paved the way to the proliferation of peace research, critical geopolitics and border studies, the securitisation—desecuritisation paradigm, and, of course, academically institutionalised post-Soviet studies. In the aftermath of the 9/11 attack, new scholarly conceptualisations of the state of exception appeared; and terrorism studies established itself as a core discipline within the domain of security studies. Against this background, the question still to be answered is: what would be the impact of Russia's aggression against Ukraine on academic disciplines of international relations, comparative politics and foreign and security policy analysis?

This book is an invitation to think of Russia anew to overcome the imperfections, misjudgements, and fallacies of what might be dubbed old Russia studies. In the old paradigm, Russia was axiomatically perceived as an inherent part of

 | DOI:10.1163/9789004747364_002

a 'wider Europe' with a bunch of justifications, from references to history and culture to contemporary patterns of consumption and urban life in Moscow and St. Petersburg. Therefore, European concepts of civil society, governance, federalism, and others were considered applicable to Russia, a country that was seen as part of a broader—not unproblematic yet manageable—transition from communism to democracy. This transformation implied that "bad" governance is to be eradicated, and "good civil society" to be supported. Since transition was patronized by Western institutions (through material incentives, knowledge transfer and political support of reforms), the West was believed to share (at least moral) responsibility for its outcomes.

This edited volume discusses how the Russo–Ukrainian war makes the international scholarly community reconsider some of the previous assumptions concerning Russia's domestic and foreign / security policies, and think of novel approaches to study Russia afresh, which is impossible beyond the context of the war in Ukraine. The collection of chapters offers an audit of the academic legacy that has prevented most scholars and researchers from not only predicting but also considering as a serious possibility the full-scale war that Russia has ultimately unleashed against Ukraine, which continued a series of previous Russian military interventions in Chechnya, Georgia and Syria. The military interference of Russia in Ukraine took by surprise not only politicians in many countries, but also experts who, for many years after the end of the Cold War, were confident of the growing mutual dependence between Russian and European economies, and of the changing function of borders—from dividing to connecting. Many international relations specialists misunderstood Russia's role in the "near abroad" (for example, by referring to a "civil war" in Ukraine, a "conflict", a "crisis", etc.), and have long considered the so-called post-Soviet integration processes under the auspices of Russia as a peculiar analogue of European integration. Now, using the example of Belarus, we see that the Union State is, above all, a military project that allows Lukashenka's regime to be drawn into the war against Ukraine and to place nuclear weapons even closer to potential targets in Europe.

Most scholars have not only overlooked the aggressive potential inherent in the seemingly innocent calls for multipolarity emanating from the Russian leadership. The ability of the Ukrainian society and the Ukrainian Armed Forces to effectively resist aggression was also a surprise for many. All this points to significant gaps in research optics, which necessitates substantial rethinking of many previous assumptions.

The need for a fresh look at Russia is fostered by new challenges that became visible long before the war. Russian mainstream discourse has re-signified many Western concepts and infused different meanings into them, which made

Western narratives of Russia's normative and institutional convergence and integration with Europe more problematic and less explanatory. Playing these 'language games', Russia transformed integration into spheres of influence, human rights into protection of Russian speakers abroad, soft power into interference in elections in the West, cultural diplomacy into propaganda and disinformation, civilisation into self-sufficient autarchy and impunity (see Yuliia Kurnyshova's chapter in this book), and regionalism into a policy of keeping "extra-regional powers" at a distance from Russian borders.

A re-examination of Russia studies is needed nowadays due to new problems and challenges we face in the academia. From a domestic perspective, Russia's rejection of democracy and liberalism was much more radical than the bulk of academics predicted and expected. From an international viewpoint, this rejection dramatically extended the spectrum of possible options and scenarios for the future, some of which being considered antiquated and outdated only a few years ago, such as territorial reshuffling of borders by force or a threat of a nuclear attack. These unfortunate transformations require a new political vocabulary since such generalised notions as authoritarianism or illiberalism don't seem to be sufficient anymore to properly understand the evolution of Putin's regime whose ideational foundations gradually but steadily shifted towards embracing narratives of esoteric spirituality, "historical rights" and Russian exceptionalism.

New critical junctures cropped up in the war-incited debate. We are witnessing a new cycle of the old realism—liberalism debate on interests and norms, and the emergence of new concepts such as "Westsplaining". We see a new polarisation in the Russia debate along the lines of "Putin's war" vs "Russia's war", as well as disintegration versus consolidation of the regime, mostly articulated in a policy relevant rather than an academic language. The interest in the whole bunch of issues pertaining to postcolonialism and decolonisation, with the ensuing "de-centering" of Russian studies (see Aliaksei Kazharski's chapter in this book), is underway, paralleled by a growing gap between "Western" and "non-Western" IR (see Ivan Klyszcz's chapter). New facets of the structure—agency debate appear: to what extent was Russia's aggression facilitated or shaped by structural factors of a global scale? Why are some countries eager to finalise their post-Soviet transition with full integration with the West, while others prefer an opposite route? What prevented most of the analysts to take seriously the prospect of Russia's invasion in Ukraine? Why many experts underestimated Ukraine's ability to effectively resist the aggression? Why security repercussions of the Moscow–Minsk alliance were not duly considered?

The crisis of interpretation and explanation affected almost all major schools of political theory. Several common problems of theorising Russia can be identified. First, many theories suffered from an excessively high level of generalisation, and therefore absolutised their core arguments. This is certainly the case of liberalism with its "end-of-history" paradigmatic vision of the future, as well as the adjacent modernisation theory with its heavy emphasis on interdependence which ultimately transformed into political sloganeering (from '*Wandel durch Handel*' to 'No European security without Russia'). The idea of Russia's belonging to Europe correlates with the liberal idea of inclusiveness and faith in the ability of liberal democracies to transform their autocratic and violent neighbours; however, different interpretations of this approach took a dogmatic and teleological turn, preventing many policy experts and public intellectuals from thinking beyond the paradigm of inclusion.

A similar reproach can be extended to IR constructivism (see Ivan Klyszcz's chapter in this volume) which theorized intersubjectivity as an indispensable characteristic of EU–Russia relations. Comparing European and Eurasian integration projects as allegedly grounded in similar principles depicted by the neofunctionalist theory also looks problematic. By the same token, the post-modernist interpretation of marginality and liminality as a source of positive / productive practices and experiences (for instance, in border regions) showed its limitations too. Deconstruction of dichotomies and multiple truths inherent in post-modernist theorizing does not look unproblematic under the conditions of war and therefore might require revision. So does (neo)realism whose major tenet of spheres of influence, as the case of Russia's revolt against the Euro-Atlantic security order has proven, can be effectuated only through wars. Besides, due to the war all attempts of great power management look either dysfunctional, or morally compromised.

Secondly, most of the theories built their arguments on (the expectation of) success stories aimed to confirm their validity. Yet after February 24, 2022, it became obvious that old approaches should be radically revised, including theories that presumed Russia's inevitable integration into international and global structures. Neither the economic benefits of trade with the West nor the EU's status of a strategic partner prevented the Kremlin from launching the war against Ukraine. Accordingly, approaches to Russia based on the Western understanding of political rationality and the role of material or political incentives to cooperate are unlikely to remain relevant.

From a practical viewpoint, the liberal international society could not delegitimise practices of authoritarianism, which exposed limitations of normative power and democracy promotion as foreign policy tools. By the same token, the growing post-political trend (simply speaking, a managerial approach to administering foreign policy) might be beneficial for non-democratic regimes. De-securitising momentum in liberal foreign policies of many Western countries was often conducive to de-securitising (neglecting threats coming from) Russia, being accompanied by US post-hegemonic disengagement and decay of liberal practices within the West itself.

The war has transformed and modified many assumptions pertinent to specific disciplines such as, for instance, border studies. Russia's invasion of Ukraine made clear that the function and operation of borders between Europe and Russia has radically changed—from experimenting with different types of connectivity and inclusiveness to erecting fences closing check points, and militarising border areas. This U-turn is a challenge to the whole paradigm of border geopolitics at EU's eastern margins which since the end of the Cold War was based on the presumption of borderlands as connectors, meeting points and "bridges" (Makarychev, Stoklosa, Yarovoi 2024). Given this dynamic, one may wonder how different border-related spatialities can be comparatively approached from methodological and theoretical perspectives, and to what extent was the war a turning point in the academic debates about borders, boundaries and frontiers in a broader sense?

From the viewpoint of political geography, the war was a huge challenge to the post-Cold War practices of "inclusive regionalism" presumed on the integrative function of regions-in-the-making. At the same time, Russia's invasion raised new questions: can Russia and Belarus be considered parts of Eastern (but still) Europe? If not, then what happens with the concepts of Eastern and Central Europe? (see: Kazharski and Makarychev 2024).

Let us look at post-colonial studies, a discipline with an accusatory anti-Western legacy, which does not appear to be helpful in case of Russia's war in Ukraine, since the Euro-Atlantic West is a target of Russia's incursions, not a culprit. In some cases, post-colonial solidarity with Ukraine does work (as for instance in the Baltic states), yet in many other instances it does not (in the global South) (Kurnyshova 2024). This attests to a rather limited space for translating post-colonial theory into practical de-colonisation.

Against this backdrop, the degree of applicability of many Western theories for understanding non-Western illiberal regimes is an open question. In the academia we often make a mistake similar to the one made by many political leaders—many of us look at Russia through the prism of Western (mostly liberal) lens. Some researchers continue to believe that the available academic tools make it possible—provided they are used correctly—to explain the causes of the war and the cruelty of the aggressor. Their logic boils down to the assumption that any behaviour can be explained, and therefore epistemologically normalized. Therefore, deviant behaviours do not exist—there are simply different logics, rationalities and intentionalities (Makarychev and Nizhnikau 2023). Nevertheless, it might be expedient to include the category of deviant behaviour in our political lexicon. Psychologists actively use it, but political scientists mostly hesitate, which often leads to normalisation of barbarism (see Yuliia Kurnyshova's chapter in this volume).

Against this backdrop, this book has two goals. One is to decipher the "mental maps" that guide the instigators of the war in the Kremlin, and another is to rethink the previous approaches that became obsolete, to better understand or explain Russian affairs, and provide methodological suggestions for future research. Some chapters focus on the workings of Russian affairs, in particular—on domestic or foreign policy developments which have been misunderstood or misrepresented by previous scholarship. On this count, several chapters highlight weaknesses in existing academic literature.

Let us briefly touch upon the first goal that might be subdivided into several points. Firstly, we need to abandon the search for a long-term—strategic or ideological—plan in the Kremlin's foreign policy. The era of "big texts," as postmodernists long ago told us, is over, and Putin illustrates this well. Its power is performative, that is, based not on deeply held ideas of identity or stable political models, but on specific acts of demonstration of force aimed at specific audiences, which ultimately illustrates the immanent level of uncertainty and ambiguity that is directly inscribed in the very power structure and the political fabric of the Putin regime.

Secondly, of huge importance is that the Russian leadership has never acted within the logic of a nation state in the established sense of the word. Cultural nationalism, which is necessary for the construction of a nation-state, in Russia was transformed into imperialism on the grounds that millions of Russian speakers live outside of the boundaries of national jurisdiction. As a result, Russia has become a country without borders in the sense that it does not fully control the annexed Ukrainian territories, and this annexation is not recognized by the international community.

Thirdly, we need to re-study and better understand the nature of Russian sovereign power (see Andrey Makarychev's chapter in this volume). From a stabiliser of the political system in the early days of Putin's rule, sovereignty has become a tool through which the Kremlin coerces and controls rather than it administers, manages and governs. This is how the Russian state appropriates time (through the politics of memory and glorification of history), space (through the creation of "grey zones" with ambiguous status such as Abkhazia, South Ossetia, Transnistria, LNR and DNR) and, finally, the bodies of people (through their sending to the front or, in case of refusal, to prison).

Our second goal is to rethink some of the previous assumptions that were central for Western scholarship on Russia. One of them relates to boosting academic efforts to decolonise our knowledge of the former Soviet Union. This is necessary to avoid making judgments about countries of Central Asia or Southern Caucasus through the prism of Russia's views or through the eyes of Russia specialists. Unfortunately, during the three decades after the collapse of the USSR, this happened quite often, which led to a distorted perception of the current war and the responsibility of the Russian state for it. De-exceptionalisation of Russia through debunking its claims for an irreducible uniqueness (see Aliaksei Kazharski's chapter in this volume) is a meaningful contribution to the de-colonisation agenda.

However, it is at this juncture that an epistemic controversy pops up: by depriving Russia of its mythological exceptionality, either as a deviation from the internationally accepted norms or as a civilisational peculiarity or specificity, we face a challenge of implicitly normalising Russia's political behaviour through including it in a broader pattern of democratic backsliding and militant nationalism on a global scale. Given the fact that, according to recent research, only 8 per cent of the population in the world live under democratic rule (Democracy Index 2024), this opens new facets of the long-standing discussion on norms and digressions. Arguably, the war has dissociated Russia from the liberal core of the Euro-Atlantic international society, but at the same time Russia remained part of illiberal policy practices of a global scope.

These points are quite general, but they can be used as the basis for new approaches to the study of Russia. Using diverse research techniques, we see a very different Russia, much less comparable in most respects to European countries. We cannot give an exact answer to the question of why Russia chose first to imitate Europe, and then to distance itself from it instead of partnership and integration. But we can show and describe how Russia went through this trajectory, which many preferred not to see and which the war against Ukraine highlighted with all clarity.

1 The Structure of the Book

The book starts with the deconstruction of Russia's mythologised claims for a great power role: in his chapter *Ivan Gomza* comes to the conclusion that attempts to raise its international status through military means only diminish Russia's resources and potential, and ultimately limit its global ambitions. However, in the next chapter *Dinissa Duvanova* argues that those experts who anticipated the fast collapse of Putin's regime failed to spot the growing state capacity and institutional efficiency in achieving the economic goals of the political regime. At the same time, she criticises those scholars who explain the resilience of Russian authoritarianism through a genuine popular support.

The juxtaposition of these two arguments implies that Russian political system is efficient in achieving Putin's goals, yet at the same time is not supported by the population.

Vladimir Gelman in his contribution places the Russian Studies debate in an institutional framework. He characterizes of the invasion of Ukraine as an "exogenous shock" for Russian academic community, which illustrates an attempt of some Russian scholars to distance from the war by defining it as an "external" event, rather than accepting it as an intrinsic component of the Russian political regime to which a significant part of the academic community gravitates institutionally, politically and ideologically.

Andrey Makarychev in his chapter offers an alternative to the institutional paradigm and seeks to explain Putin's model of sovereign power through an intricate—and often indistinguishable—conceptions of biopower and necropolitics as two pillars of what the author dubs Russian Corporeal Order. Engaging with the biopolitical avenue of research related to Russia today, the author shows that the body can be approached as a prism through which this country under its current regime can be better understood.

In the next chapter *Yuliia Kurnyshova* looks for a place of civilisational discourse—and the ensuing concept of barbarity—in academic debates about Russia's war in Ukraine. She identifies the dehumanising and othering effects inherent in concepts of barbarity and civilisation, and discusses Russia's wilful subscription to "barbaric" practices as a core element of the invasion of Ukraine.

Other authors deploy the transformations within the Russian Studies in cross-country and comparative perspectives. *Aliaksei Kazharski* juxtaposes the trajectory of Putin's political regime with Balkan countries' experiences of nationalism and authoritarianism. This outlook is instrumental for understanding the ongoing and future changes in the Russian Studies from the viewpoint of comparative analysis, opening this discipline to dialogues with other fields of research. Two other contributors look at how the war in Ukraine changed European academic landscapes (in Hungary, Poland and Finland in *Katalin Miklossy*'s chapter), or might be approached from the vantage point of the French school of pragmatic sociology (in *Francoise Dauce*'s chapter).

This opens the book content to a broader set of disciplinary perspectives on the challenges faced by the Russian Studies. In this vein, *Kirill Shamiev* engages with security studies, discussing civil-military relations against the backdrop of the failures of Russian defence reform. In the next chapter *Ivan Klyszcz* discusses how the Russian discipline of International Relations might resonate within the global IR literature. In their co-authored chapter *Sanshiro Hosaka* and *Anselm Schmidt* propose to use more extensively "novel methodologies from beyond the ivory tower of classical academia" for challenging Russian mainstream discourses and digging deeper into the Kremlin-sponsored industry of political technologies. Their contribution is an answer to the ongoing methodological debate on a much-discussed shortage of reliable empirical data to study Russia, given the limited reliability of official sources of information and even greater limitations for professional field research in Russia.

Taken together, these chapters may be read as an invitation to think beyond traditional and established political categories and explore the cognitive value of critical thinking for a better unpacking of Russian foreign policies. Of course, the topics covered and the issues raised in this book may overlap: for example, *Klyszcz* discussed constructivist trends in Russian IR referring to postcolonial turn in IR, while *Kazharski* discusses parallels and analogies between Russia and the Balkan countries in direct references to civilisational discourses analysed by *Kurnyshova* and—again—constructivist theories of actors' role identities. These overlays stem from the complexity of the challenges faced by Russia Studies, and attest to the necessity of rethinking and re-signifying the existing explanatory frameworks inherent in this discipline.

References

Democracy Index: conflict and polarisation drive a new low for global democracy. 2024. Economic Intelligence EU, February 15, https://www.eiu.com/n/democracy-index-conflict-and-polarisation-drive-a-new-low-for-global-democracy/

Kazharski, Aliaksei and Andrey Makarychev, 2024, "Eastern Europe" at War: Introduction to the Special Issue, *Alternatives: Global, Local, Political* 49 (2), 47–51.

Kurnyshova, Yuliia. 2024. Ukraine and the global South: Ambiguities of a post-colonial perspective, *New Perspectives* 32 (3), 211–220.

Makarychev, Andrey and Ryhor Nizhnikau. 2023. Normalize and Rationalize: Intellectuals of statecraft and Russia's war in Ukraine, *Journal of International Relations and Development* 26, 632–642.

Makarychev, Andrey; Katarzyna Stoklosa and Gleb Yarovoi. 2024. Introduction. Europe-Russia Borderlands During the War in Ukraine: New Practices and Conceptualizations, *European Foreign Affairs Review* 29, 1–8.

CHAPTER 2

New Approaches to Study Russia's Foreign Policy

Role of Great Power Ideology and Nested Asymmetric Relations

Ivan Gomza

> Une idée fausse, mais claire et précise, aura toujours plus de puissance dans le monde qu'une idée vraie, mais complexe.
>
> ALEXIS DE TOCQUEVILLE

1 Introduction

In a seminal volume on late Imperial Russia, historian Hans Rogger has contrasted two opposing views on Russian foreign policy objectives and capabilities. While Karl Marx professed world domination as the goal of Russian policy, the Foreign minister under Alexander III, Nikolay Giers, surmised that Russia was too weak to acquire the Turkish Straits (Rogger 1983, 163). The two opposing interpretations of Russia's capacity have hardly changed since the 19th century: scholars, especially after the Russian invasion of Ukraine in 2022, vehemently debate the motivations, goals, and possible repercussions of the aggression. In particular, some argue that Russia pursues limited goals of asserting itself in the traditional sphere of influence, whereas others see the relentless quest for global domination.

Much of this discussion hinges upon the assumption of Russia's preponderant role in Eurasia and globally, namely the interpretation of Russia as a Great power. The notion of Great or major power has been in the limelight of political science and international relations scholarship for decades. A succinct definition considers a state a Great power when it satisfies three criteria: (1) unusual capabilities with which to pursue its interests in interstate relations, (2) broad and expansive foreign policies beyond its immediate neighbourhood or region, and (3) strategy to influence the course of international affairs relatively independently of other major powers (Volgy et al. 2011, 6). Additionally, Great powers establish, maintain, and, if necessary, modify the extant system

 | DOI:10.1163/9789004747364_003

of international political order (Mastanduno 2009, 124). However, the notion of major power describes not only the state's capacities but also its status in the international realm. Status is a category that relates to images and perceptions; it reflects 'collective beliefs about a given state's ranking on valued attributes [... which] manifests itself in two distinct but related ways: as membership in a defined club of actors, and as relative standing within such a club' (Larson et al. 2014, 7). A major power belongs both to formal international institutions and informal networks of states, which make a difference in international affairs. Besides, its opinion on the situation is greatly valued and sways others. Being a Great power grants considerable prestige, altering other actors' perceptions of what a major power—as opposed to a small one—can and is allowed to do (Gilpin 1981, 31). Therefore, the image of Great power is a critical asset in international relations.

As the first paragraph of this paper suggests, Russia's standing is ambivalent. On the one hand, Russia's Great power status is often taken for granted. In academia, a veritable cotton mill industry peddles the idea (Rosefielde 2004; Hedenskog et al. 2005; Kanet 2007; Rangsimaporn 2009; Anno 2019; Radvanyi and Laruelle 2023; Gioe and Smith 2024). In a typical statement, Ellman observes, 'Russia was one of the Great powers for much of the 18th, all of the 19th and almost all of the 20th centuries. Its victorious armies entered Berlin in the 18th century, Paris in the 19th century and Berlin again in the 20th century' (Ellman 2023, 172). Similarly, Mearsheimer acknowledges Russia's superpower status as a given fact since 2001 (Mearsheimer 2023). Policymakers readily espoused the view. Senator McCain, for instance, declared that the American nation 'faces the most diverse and complex array of crises since the end of World War II—Great power competition with Russia and China' (McCain 2016), the Congressional Research Service informed that 'the current situation of GPC [Great Power Competition] is a three-power situation involving the United States, China, and Russia' (CRS Report 2024), both 2017 and 2022 US National Security Strategy admit Russia's participation in Great power rivalry (White House 2017; White House 2022), while NATO 2022 Strategic Concept declared 'The Russian Federation [to be] the most significant and direct threat to Allies' security and to peace and stability in the Euro-Atlantic area' (NATO 2022).

The vision of Great power status is shared in Russia. Baev identified two dominant opinions proper to most Russian policy-makers: (1) stipulation that by the merits of history and geography, Russia is entitled to Great power rank, (2) preference for a world order controlled by a group of major powers to a unipolar world (Baev 2010, 117). Clunan concurs: 'All political elites viewed Russia's past status as a great power positively and as a core facet of Russian identity' (Clunan 2009, 55). Given the crystallisation of the Great power identity, it is hardly surprising that Russian official statements and scholarly texts in foreign policy analysis outlets fully-heartedly promote it. To cite just a few, Yeltsin averred that 'Russia was and continues to be a great world power' (cited in: Adomeit 1995, 45), and Putin, in his 1999 address, agreed, 'Russia was and will remain a Great power. It is preconditioned by the inseparable characteristics of its geopolitical, economic, and cultural existence. They determined the mentality of Russians and the policy of the government throughout the history of Russia, and they cannot but do so at present' (Путин 1999). Likewise, a paper published by the Russian Ministry of Foreign Affairs unequivocally stipulates that 'according to objective indices and its real steps on the world scene, it [Russia] is a Great power' (Kokoshin 2002), another text, published twenty years later, clarifies fundamentals of Russia as a superpower (Хасанова 2023), and yet another lionises the Sino-Russian cooperation, professing amazement with 'economic, political, and military capacity of the two superpowers' (Яровая and Сидяков 2020, 532). In addition, Russian media outlets are filled with numerous texts declaring Russia an 'energy superpower',[1] a 'corn superpower' (Medetsky 2016), or a 'helium superpower' (Уралов 2021). Given the official framing, it is hardly surprising that ordinary Russians tend to perceive their country correspondingly. The panel data on public opinion collected by *Levada-Tsentr* (Левада-Центр 2020) since 1999 (see Fig. 2.1) points to an established trend.

Overall, the interpretation of Russia as a Great power remains entrenched in foreign policy, academia, and laypeople's perceptions. This has practical implications for the Russo-Ukrainian war because, as numerous IR students—especially from the structural realist school—argue, if Russia is a superpower, then it has a legitimate sphere of influence, including Ukraine, and it vehemently and violently opposes the Western encroachment (Bock, Henneberg, and Plank 2015, 101–109; Götz 216, 301–302; Posen and Van Evera 2023).

However, some observers are not ready to acknowledge Russia's Great power status without consistent scrutiny (Adomeit 1995; Petersson 2014, 30–34; Smith 2014; Šćepanović 2023). The growing consensus of this school was aptly grasped by Alexander Hill, who stipulated that 'Russia is arguably some way from being

1 The notion was officially advanced by Putin at 2005 Russia Security Council meeting (*Министерство иностранных дел Российской Федерации* 2005) and aptly referred to both in Russian press (Гривач and Денисов 2008; Марцинкевич 2020) and Western academia (Goldman 2008, 14). For a critical review of the "energy superpower status" after the 2022 invasion consult *The Insider* (2023).

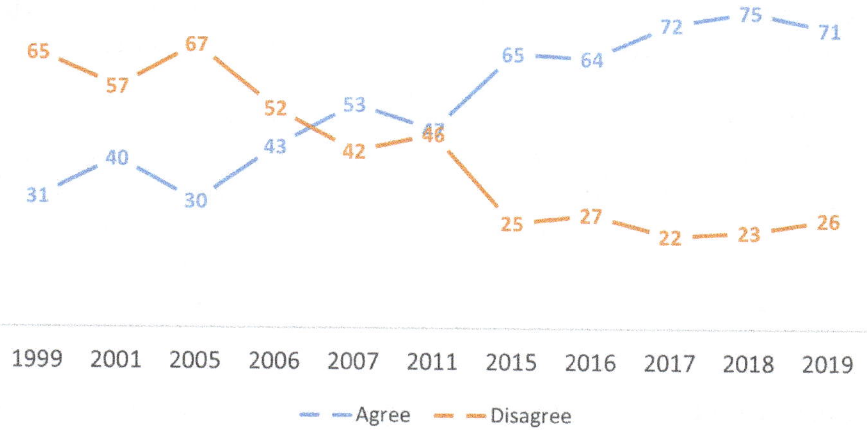

FIGURE 2.1 Is Russia a superpower today?
ЛЕВАДА—ЦЕНТР. 2020. "ВЕЛИКАЯ ДЕРЖАВА." [LEVADA-TSENTR, "THE GREAT POWER"]. EFFECTIVE APRIL 24, 2025 HTTPS://WWW.LEVADA.RU/2020/01/28/VELIKAYA-DERZHAVA/

able to claim to be a Great power' (Hill 2023, 320). However, the exact measurement of how much Russia lacks in terms of being considered a Great power is challenging. For some, Russia is "a Great power with qualifications" because although it 'has the military might and will to be a Great power, its economic foundation is weak' (Hancock 2007, 72) or, as Obama's White House Press Secretary scathingly remarked, 'They are now the 15th-largest economy in the world, and they rank somewhere behind Spain' (Obama 2015). Others advocated an even more radical reconceptualisation of Russia's status, effectively arguing that it is nothing more than a second-tier power. Some members of the US expert community suggested: 'Russia should indeed hardly be more than a regional power […] influential in Eastern Europe, in its borderlands, but not on a global scale' (Helm, Smeltz, and Burakovsky 2020). The disagreement between the two groups is not of essence but of scale. Both extrapolate long-term economic and demographic trends to signal Russia's diminishing capacities, but the latter believes the country has already reached the threshold of the status modifier, while the former estimates it is still heading toward it.

Yet in Spring 2024, a seasoned US official and foreign affairs expert opined that 'Russia cannot be wished away. At some point, it will again become a Great power […] The West, beginning with the United States, will have to deal with this fact' (Hunter 2024). Reiteration of Russia's Great power status brings to life a pernicious conclusion that this power should be accommodated one way or another for peace in Europe, which in turn brings to life an unfulfilled expectation that, given Russia's structural economic weaknesses, such an accommodation is within reach. Twenty years ago, two other seasoned foreign affairs experts observed Russia's circumstances and asked rhetorically:

> Will Russia join Iran and North Korea? This is hardly an attractive option for a country whose elite aspires to world leadership more than anything else and looks down on Iran and North Korea as backward, Third World regimes [...] It is difficult to imagine how a rational, even selfish, assessment of Russian interests would lead Russia to conclude that it would be best served by undermining the United States (Rumer and Wallander 2003, 70).

Given the Moscow-Teheran cooperation in UAV and missile technologies since 2022, the 2024 Putin's visit to Pyongyang to deepen military and economic cooperation with North Korea, including ludicrous children's tourist visitations to North Korea (The Moscow Times 2024), and Kremlin's unveiled animosity toward the US, the rhetoric did not quite hit the mark. Admittedly, much water has run under the bridge since 2003, and the IR context has changed significantly, which made the prognosis difficult. The problem, however, is that Russia's assertive anti-Western policy manifested itself in 2007 at the Munich Security Conference, in 2008 war against Georgia, in 2014 annexation of Crimea, in covert operations in the east of Ukraine since 2015, in 2016 support for Bashar al-Asad, in 2023 military advising in Central African Republic and Mali—not to forget the elephant in the room, the 2022–202[...] Russo-Ukrainian war. In other words, Russia's Great power status, interpreted in the anti-West framework, is a part of the problem, not that of a solution.

As this chapter will argue, the source of misleading predictions, failed policies, and human tragedies might be partially explained by a mistaken interpretation of Russia's status. To put it bluntly, Russia is not—and has never been—a Great power; rather, it has always been a regional power with extended geography and inflated ambitions. As a regional power, it was entangled in nested asymmetric relations (the notion will be clarified below) and skilfully used the distorted measurements of its capacities to obtain more than it could typically manage.

The chapter is organised as follows. First, it considers Russia's status in the international system since 1721 and illustrates that despite various institutional trappings (the Romanov empire, the Soviet state, and contemporary Russia), it has rarely been considered a consequential actor, especially when acted unilaterally. Second, it introduces the notion of Great Power Ideology (GPI) to highlight Russia's continuous efforts to be accepted as a major power. Next, it elaborates on the notion of asymmetry in international relations and introduces the concept of nested asymmetric relations, which explains Russia's foreign policy, including the recent eruption of the Russo-Ukrainian War. Finally, the chapter concludes by suggesting possible avenues for scholarship and policy implications.

2 A Feeble Giant: A Historical Overview of Russia's Status in the International Relations

When Muscovy emerged from Mongol domination (circa 1480), it was an obscure principality outside what was considered 'proper Europe' (see Wolf 1994, where he dissects this trend as fully established by the 1700s). The perception was amplified by policy espoused by several generations of tsars who spent most of the 16th century fighting other Slavonic statelets and most of the 17th expanding eastward (up to the Kamchatka peninsula charted by Russians in 1699). The single most important Russia's adversary throughout history was the Ottoman Porta: Russia fought it in twelve separate wars spanning 350 years. As a result, although due to travelogues by Richard Hakluyt, Samuel Collins, and Jacques Margeret, reconnaissance operations by Conrad Bussow, Philippe Avril, and Peter Petreius, and official reports by Adam Olearius, Sigismund von Herberstein, and Charles Whitworth, the Russian tsardom was not an unknown entity, for most Europeans, including statemen, it remained an exotic curiosity of little practical relevance akin to the Great Qing or the Ahom kingdom. For instance, a study exploring the "Special Relationship" between Stuarts and Romanovs concludes they were focused on trade (Dukes, Herd, and Kotilaine 2009, 208), just like English relations with other distant kingdoms in Asia.

The situation changed only slightly during the 18th century. Although by the 1720s, tsars achieved several essential victories and territorial gains from Sweden and the Polish Commonwealth, which eventually inspired Peter I to stylise his state as an empire, Russia's international status in European affairs remained unexceptional. The key word is *European*. In the early 1700s, Russia pivoted to the West, deepening its participation in regional politics. Russian historians, traditionally inclined to aggrandise Russia's role, claim that 'by 1749 Russia solidified its position in Baltics and became a regional superpower [региональной сверхдержавой]' (Анисимов 2020, 27). Contrary to such exaggerations, Russia joined the European concert as an auxiliary performer. A fitting illustration is the 1748 Aix-la-Chapelle Congress, which settled the War of the Austrian Succession. This conflict is crucial since it represented one of the first episodes of Russia's involvement in European power politics when Russia projected its power as far as the Rhine. However, this achievement was of external making: as the exhausting 8-year war dragged on, the principal belligerents (France and Prussia on one side, Great Britain and the Habsburg Monarchy on the other) were eager to engage Russia as a possible ally (Lodge 1926, 64). Austrians played their cards better, and 36,000 Russian soldiers were deployed from Livonia in 1748 to support the Austrian cause. However, because of lengthy deliberations and faulty logistics, the troops arrived in Germany after the peace talks had been initiated and played no military role. The most significant events occurred around the Aix-la-Chapelle congress: Russia was purposely excluded from deliberations because of France's official demand, which 'incensed and humiliated [Tsarina] Elizabeth and her court who had hoped that Russia would finally come of diplomatic age by being invited to a European peace congress' (Nester 2014, 47). The circumstances which prompted Paris not to accept St. Petersburg as a participant sheds additional light on Russia's secondary role. In 1743, Russian police had intercepted the personal correspondence of the French Ambassador to St. Petersburg, Marquis de La Chétardie. The French envoy articulated disdain toward the tsarina's intellectual and moral features in his letters, alleging she was characterised by:

> d'une paresse d'esprit et d'une légèreté qui la mettaient à la discrétion de ses entours; effrayée et effarouchée de la moindre affaire et du plus petit effort intellectuel; n'ayant de goût que pour les fêtes, la toilette, les intrigues de sa domesticité (Rambaud 1895, 3)
>
> [laziness of mind and a lightness that placed her at the mercy of her surroundings; frightened and startled by the slightest matter and the smallest intellectual effort; having a taste only for parties, grooming, and the intrigues of her domestic life]

Besides, La Chétardie might have been involved in a conspiracy, which was a severe matter because, in 1741, he organised a coup that put Elizabeth on the throne (Миних 1891, 200–201; Naumov 2015, 73–74). When La Chétardie was sent with disgrace back to Paris, Louis XV got vexed—and repaid with interest in 1748.

Several features of Russia's participation in the War of the Austrian Succession highlight its secondary status: there was a regime change partially orchestrated from abroad and strove to modify St. Petersburg's foreign policy in accordance with French designs; Russia's military was considered 'a hired gun' invited to play an auxiliary role or, as one historian later acerbically put it, 'everyone knew what the Tsaritsa Elizabeth was; all that remained was to fix the price' (Showalter 1996, 106); and it suffered a diplomatic disgrace being excluded from the peace congress. One also must not forget the demographic factor: in the 1750s, Russia's population was still less than France's (see Table 2.1). This combination hardly warrants the apologetic interpretation of Russian historians who allege that 'Russia's participation in the War of the Austrian Succession proved that from then on, any settlement of European political issues was impossible without Russia' (Анисимов 2020, 32). It is noteworthy that the statement echoes the official appeal by the leading courtier under Elizabeth, Bestujev, who, addressing the Empress, sycophantically exclaimed:

> Tant que Sa Majesté Impériale eut manifesté sa volonté d'intervenir avec une force imposante, aussitôt toute la situation des affaires européennes avait pris un aspect bien différent [...] Le corps d'observation russe ne s'est pas avancé plus loin qu'il ne fallait pour propager dans l'Europe entière la gloire des armes impériales, conquérir à la souveraine le titre flatteur de pacificatrice du continent, et en outre lui rapporter des sommes considérables, tant par le versement des subsides que par le soulagement que l'absence des troupes procurait au Trésor (Rambaud 1895, 3).
>
> [As soon as Your Imperial Majesty had manifested her will to intervene with an imposing force, the entire situation of European affairs had taken on a very different aspect [...] The Russian observation corps did not advance any further than it was necessary to spread throughout Europe the glory of the imperial arms, to conquer for the sovereign the flattering title of pacificator of the continent, and in addition to bringing her considerable sums, both by the payment of subsidies and by the relief that the absence of troops provided to the Treasury.]

The next great European war, the Seven Years' War, improved Russia's status—but only to a point. As a reminder, Russia fought alongside Austria and France against Britain and Prussia due to the famous *Renversement des alliances*. A standard account of the Seven Years War's result stipulates that although 'Russia entered the Seven Years' War somewhat as an outsider [...] she began to be considered as a significant European power' (Marston 2001, 91) and 'acquired capabilities that placed her among the Great powers' (Baugh 2011, 662) first and foremost due to its military performance because 'almost alone among the belligerents, Russia at the end of the war gave the impression of total invulnerability and the sense of having something mighty in reserve' (Duffy 1981. 124). Yet although victories at Kunersdorf (1759) and Kolberg (1761) proved the valour of Russia's troops, they did little to further Russia's goals throughout the war, which Western historians believe to be destruction of Prussia as a major power (Weigley 1991, 178), and Russian specialists explain to be 'the elevation of the

Russian empire as the dominant state in Europe—not a great one, for Russia has already become such, but the most prominent among the great' (Анисимов 2020, 751).

The net outcome of the Seven Years' War was far from it for two reasons. First, Russia failed to meet either of its objectives. Prussia remained a formidable power, and its king, Frederick II, was widely recognised as a military genius. Politically, Prussia retained Silesia, the bone of contention between Berlin and Vienna. Although conquering Berlin and East Prussia, Russia eventually relinquished all territorial gains. Moreover, St. Petersburg startled the partners with a brusque U-turn when the successor signed a separate peace treaty and an offensive-defensive alliance with Prussia against former allies after Empress Elizabeth's death (Weigley 1991, 192). Therefore, other European powers remained mindful of the fickleness and volatility of the Romanov dynasty, especially pronounced during the succession crises. The number of succession crises is not to be ignored: out of 20 monarchs in the dynasty, six experienced irregular power transitions (i.e., were murdered), while successors often adopted unforeseeable foreign policy changes. Second, Russian troops' prowess notwithstanding, the Franco-Austro-Russian alliance suffered from the worst instances of coalitional warfare, with agendas and objectives so different that 'despite the wealth of manpower that Austria and Russia could bring directly to bear against Frederick, they, more than he, operated against the backdrop of a ticking financial clock' (Schumann and Schweizer 2012, 228–229). Eventually, no international congress was organised, and belligerent individual treaties were signed. Adding an insult to injury, any possible participation of Russia in such a congress was forestalled by Vienna and Paris, the two allies enraged by St. Petersburg's abandonment, and Berlin and Vienna opposed Russia's intrusion into "German" affairs (Чечулин 1896, 103). By the mid-1760s, there was a hardly satisfying tendency of constant sidelining for a state eager to be recognised by peers.

Admittedly, Russia's foreign policy since the 1760s was far from a total blunder. Under Catherine II, it renewed the commercial treaty with Great Britain, defeated the Ottomans, partitioned Poland, and forced Prussia to take a more accommodating stance (LeDonne 2004, 176). In 1779, St. Petersburg even adroitly manoeuvred into the role of mediator between Prussia and Austria and, consequently, assumed some sway in the Holy Roman Empire's internal politics, which it mainly accomplished to gain further international prestige (Aretin 1986, 142).

The apogee of Russia's influence and status was, undoubtedly, at the Congress of Vienna, when after the wars of 1792–1815, it became 'the dominant European power [and] appearance of [Emperor] Alexander's armies in western

Europe made Russia's vast manpower and military might feared for a generation' (McKay and Scott 2012, 342). The Russian Emperor headed to Vienna with 'a very definite plan as to what Russia was to receive as a reward for her actions in safeguarding and liberating Europe' (Strakhovsky 1947, 159). However, other participants begged to differ about the exact amount of Russia's reward. The most divisive issue was the so-called 'Polish-Saxony question', which, in a nutshell, revolved over (1) Russia's desires to incorporate additional Polish territories and (2) the compensation Prussia should get for them (Nicolson 1964, 148–149). France, however, used the contradictions between Russia, on the one hand, and Great Britain, Austria, and Prussia, on the other, to effectively split the four powers and block the fulfilment of Alexander's agenda. Alexander's self-assurance played a considerable role in this outcome, for it morphed into what other members of the Congress perceived as overbearing behaviour. At some point, Russia's emperor addressed Metternich 'with such haughtiness and violence of language such as might have appeared extraordinary even if applied to one of his own servants' (de Talleyrand 1881, 52). It is needless to explain that such a behaviour alienated Austrian and British representatives, while 'the memory of Russian victories nurtured a perception of the new empire's awesomeness and the conviction that any challenge would be met with overwhelming retribution' (LeDonne 2004, 425).

In the 1830s–1850s, Nicholas I, who continued his brother's assertive policies, focused on the 'Eastern Question'. In a number of small wars against Ottomans, Iranians, and tribes of the Caucasus, he achieved control over the Danube mouth and the North Caucasus. In Europe, Russia fixed challenges during the two waves of popular revolutions in 1830 and 1848, predominantly because one of the main foci of insurgency was Poland but also because it cherished the conservative status quo. For a cursory observer, the Romanovs remained invincible. Eventually, it was the 'Eastern Question', which proved otherwise. When, in 1853, St. Petersburg acted as if the Ottoman Porta was a puppet state and occupied Moldavia and Wallachia, Paris and London opposed the move militarily, and Austria diplomatically. Next, it turned out that the Russian might was a mirage. As F. Kagan commented, 'the system of serfdom rendered virtually impossible the maintenance of any sort of trained reserve upon which to call in wartime [...] Russia had to maintain a standing army almost three times as large as that of the next largest in Europe. The problem was, even that large an army was not large enough' (Kagan 2002, 130). To illustrate, out of 800 thousand soldiers officially under arms, Russia could master only 1\8 to protect its naval bases in Crimea (Stone 2006, 120–121). Besides, it turned out that logistics to Crimea were so faulty that it was easier for the French and British to supply their troops than for the Russians. However, the most significant miscalculation was diplomatic: St. Petersburg failed to foresee (let alone prevent) the Franco-British alliance, dismissive neutrality of Berlin, and a hostile move by Vienna.

Nicholas I misestimated Russia's weight, believing it could act unilaterally. The retribution was exact: the new congress Russia did participate in was that of Paris (1856), and it imposed the integrity of Ottoman Porta, the demilitarisation of the Black Sea, and Russia's loss of the delta of Danube and Kars. The losses were not considerable in material terms, but the damage to status was enormous. As Talyor commented, 'No such compulsory disarmament had ever been imposed upon a Great power [...] and the allies would not have presented such terms to any Power whom they regarded as truly European. At bottom [they] regarded Russia as a semi-Asiatic state, not much above the level of Turkey and not at all above the level of China' (Talyor 1954, 85). This was a disgrace for the successor of an Emperor who, 40 years earlier, came to Paris and Vienna to decide the continent's fate. The most severe blow was dealt to the elite's self-esteem, as Grand Duke Constantine bitterly averred: 'we cannot deceive ourselves any longer; we must say that we are both weaker and poorer than the first-class powers' (Lieven 1992, 6).

After the 1856 defeat, the imperial court realised that meaningful changes were in order. Under Alexander II, there were numerous reforms in the administration and the military. The foreign policy course was re-aligned because domestic policies assumed primacy and dictated cautious behaviour in the international arena. Besides, the balance of powers in Europe was especially precarious due to seismic shifts provoked by the rivalry between Berlin and Vienna and the erratic foreign policy of Napoleon III. As a result, the amicable relations between Bismarck and Gorchakov, established during Bismarck's ambassadorial tenure in St. Petersburg, where he became aware of the 'Crimean syndrome' suffered by Russian elites and acted accordingly (Kolb 1990, 156), the mutual esteem between Alexander II and William I of Prussia (Wetzel 2012, 58), and the renewed cooperation with Vienna notwithstanding, St. Peterburg passively observed the turmoil of the Second Schleswig War (1864), the Austro-Prussian War (1866), and the Franco-Prussian War (1870–1871), forced to accept the outcomes as *faits accomplis* lamenting, as one Russian diplomat did, 'a hard way that showed us we are powerless in Europe' (Beyrau 1987, 74). A kind of consolation came in October 1870, when Russia did manage to impose an irreversible accomplishment of its own: capitalising on the French defeat at Sedan, St. Petersburg denounced the Paris Peace Treaty's Article 11 and remilitarised the Black Sea. Notably, Gorchakov took the step only after securing approval from Bismarck (Wawro 2003, 238).

The Russo-Turkish War of 1877–1878, 'the first trial of Russia's new and 'modern' war planning system' (Rich 1998, 115), proved the military reforms yielded some sweet fruits because Russian troops managed to crush the Ottomans and reach the outskirts of Constantinople. There were bitter complications: the Russians were exhausted by the thrust to Constantinople (Talyor 1954, 246), so when the San-Stefano treaty, which granted substantial territorial gains to Russia and its satellites, was questioned in London, St. Petersburg had no resources to insist. Moreover, Russia's success alarmed Great Britain and Austro-Hungary, now supported by Bismarck for diplomatic reasons. It constituted 'a decisively anti-Russian act [and thus undermined] the friendship between Russia and Prussia, which had been the most stable element in European politics for more than a hundred years, was shaken for the first time' (Talyor 1954, 248). Russian historians refer to this shift in a tale-telling self-aggrandising manner: 'Bismarck stopped cosying up [заискивать] to Russia' (Троицкий 1997, 364). Whatever the interpretation, the Congress of Berlin provoked tangible disappointment in Russia by forcing it to relinquish territorial gains and revealing, once again, its weakness vis-à-vis Great powers.

Forced to tread carefully in European affairs, Russia renewed its thrust into Asia and incorporated parts of Central Asia and the Pacific rim. Military efforts, combined with peasant resettlement into the Far East and, later into the century, the construction of the Trans-Siberian Railway, produced a hope to use Siberia's potential for Russia's advantage. The eastward expansion was not a goal but rather a method. Mikhail Petrashevskii encapsulated the general mood while rhapsodising: "Our present position in Asia, its strengthening or weakening may be considered an indication of the level of development of our society [...] Siberia is destined to achieve for us a diploma with the title of a truly European nation" (Bassin 2004, 203).

However, the eastern path to glory was no less elusive because Russia clashed with its traditional adversaries, Great Britain, and an unknown contender, Japan. The deeper Russia penetrated Manchuria and Korea, the more anxious Tokyo grew, for 'Russia occupied an anomalous position in Japanese strategic thought. On the one hand, it was still feared as a territorial threat. Yet a decade of commercial relations between Japan and half a dozen Western nations had revealed Russia to be an economic pygmy' (Auslin 2005, 15). Russia, however, dismissed Japan as a non-entity and pursued a policy of brinksmanship. Eventually, having exhausted all conciliatory options, namely the delimitation of spheres of influence, with Korea going to Japan and Manchuria to Russia (Malozemoff 1958, 106–109; Miyoshi Jager 2023, 284–289), and having concluded a maritime alliance with Great Britain (Nish 1966), Tokyo delivered St. Petersburg a devastating blow in the war of 1904–1905. Nothing highlights the magnitude of the disaster better than the fact that throughout the conflict, Russia failed to win a single significant encounter (Стейнберг 2004, 233). The road from Mukden and Tsushima through the 1905 revolution to the eventual demise of imperial Russia was taken.

Even the Soviet Union, having inherited Russia's peculiarities and handicaps, was a superpower with adjectives. When established by a militant group in 1922, the USSR remained an isolated pariah state, wreaked by a decade of warfare, civil strife, and ensuing colossal societal dislocations. Therefore, it could not even pretend to be a consequential IR actor for the first two decades. In contrast, the spike in the USSR's status occurred after (and due to) WW2. However, as Moshe Lewin points out, 'in 1945 Soviet Russia was a mighty state, but in reality a shaky one. It was hungry, devastated, exhausted, terrorised, ruled by a decaying power complex: *a needy, ailing superpower*' (Lewin 2005, 242). Even when the country liberated its productive forces after 1953, everyone, including the Soviet leadership, 'understood that the Soviet Union was the *underdog* of the Cold War. It was playing perpetual catch-up with America but [...] the Soviet system was simply not delivering the kinds of breakthroughs that would speak to its supposed superiority' (Radchenko 2024, 595). The shift to *détente* was caused precisely by the apprehension that neither human sacrifices (like those under Stalin) nor efforts to galvanise internal and external support (a strategy applied by Khrushchev) would help to overcome the US. Hence, Brezhnev's effort to build the USSR in the global and European economy and co-exist with the US. So, even at the apogee of its status, the USSR was a 'junior superpower [...] able to keep up with the theatrical drama until the 1980s' (Rich 2009, 288). When *détente* deteriorated, the Soviet Union failed to re-enter the race with the US because of the 'deterioration of the Soviet economy, ecology, and quality of everyday life [...] as well as deep and growing problems of a multinational state, contrasted dramatically with the spectacular upsurge of the United States and Western Europe in the 1980s. By 1985, the USSR was a superpower *only in the military sense*' (Zubok 2007, 307). Since the military complex required vast expenditures, Gorbachev initiated the reduction of the arsenals. *New Thinking* was a desperate attempt to bolster soft power and achieve Great power prestige by other means (Larson and Shevchenko 2003), which failed despite Gorbachev's successful image-building abroad. Therefore, throughout its brief history as a superpower, the USSR remained apprehensive and insecure, both of its prominence and ability to meaningfully impact global policies. For this reason, Freire's conclusion that the Soviet Union exemplified the status-inconsistent overachiever, an actor who enjoys Great power status without having the capabilities of a Great power (Raquel Freire 2011, 74), is pertinent.

3 A Trapped Bear: Russia's Great Power Ideology and Its Contingency Upon Nested Asymmetric Relations

The historical overview of Imperial and Soviet Russia's status in European affairs suggests that contrary to commonly held assumptions, Russia has never been a proper Great power because of its role in the IR structure and how other players perceived it. As Iver Neumann concludes, 'Russia's quest for recognition as a Great power has not been a successful one' (Neumann 2008, 129). Rather, throughout its history, Russia was a second-tier actor endowed with a considerable amount of material and human resources, thus nurturing a consequential preference to focus on hard (as opposed to both soft and smart) power. This combination enabled Russia to play a more significant role than its economic, cultural, and political capacities granted.

The fact that Russia lacked the critical components to become a genuine Great power explains why, instead of being it, Russia elaborated a *Great Power ideology* (GPI)—a complex of practices designed to sustain and promote its major power image at home and abroad. I use the notion of ideology deliberately, following the insight by Althusser that although ideological constructs constitute an illusion, they allude to reality (Althusser 1971, 163); that is, they can be helpful to conjure the reality behind the illusion. Different adjectives, often applied to Russia's status (e.g., a "military" or "energy" superpower), are indicators: it is not a *proper* Great power but a pretender. Yet the allusion to being a Great power serves as a foundational cornerstone of Russia's foreign policy. Russian elites are keen on imitating symbolic and material trappings of Great powers, e.g., participation in international congresses, membership in international institutions, a dedicated apparatus to promote its narratives, a colossal army ostensible armed with the newest equipment, and a pretence to possess a universal culture.

The GPI is useful for manipulating three audiences: the domestic population, the global international community, and, most essentially, a restricted club of great powers. *Domestically-oriented GPI* ensures popular compliance by making efforts and sacrifices meaningful. Typical roles of 'the protector of Slavs people' under Romanovs, of 'the first proletarian state that cannot help assisting other revolutionary nations' under the Soviets, and of 'the traditionalist bulwark against liberal values' under Putin not only convey an image of a mighty force able to provide the lofty promises but also rationalises the relative backwardness and the need to double the efforts. Lieven's observation that Imperial Russian 'resources did not match her military commitments [so ...] in its effort to meet these commitments the state at times came close to bankrupting the treasury and undermining the Russian economy' (Lieven 1983, 12) serves as a fitting forebear for both Soviet sinking costs to support socialism in Central Europe and tropical Africa (Frank 1994), which in 1980 consumed $23 billion or 1.5% of Soviet GPD (National Foreign Assessment Center 1981), and Putin's *Gazprom* sponsoring promotion of traditional values (Gazprom Annual Report 2019), which contributed to $12.9 billion of financial losses reported in 2024 (Reuters 2025). The notion of GPI, therefore, can explain the conundrum highlighted by Dinissa Duvanova (Chapter 3, this volume); namely, GPI establishes ideational sources for the Russian state's often surprising resilience since it motivates the populace to live up to the higher standards of effort and not flinch from sacrifices a Great power is ostensibly expected to make. By effectively leveraging domestically-oriented GPI, the narrow elites in St. Petersburg and Moscow improve their grasp on power, for 'any class striving for power must [...] present its interest as the general one' (Marx and Engels 1969, 34). GPI, therefore, is partially responsible for the tragic situation of common Russians: lured by the elite-initiated quest for status, they 'serve the autocrat who effectively and efficiently suppresses freedoms, plunders national resources, prosecutes the regime's opponents, and wages war'.

Globally-oriented GPI promotes visions of Russia's exceptionalism and 'messianism' (Taras 2024). In particular, it represents Russia as a warlike though benevolent nation that establishes just order, brings balance, shields the weak, mediates between different civilisations, and offers alternatives to the West (Curanović 2021, Chapter 4). Although these tropes are potent even when consumed domestically, they grow especially useful for gaining prestige and admiration among world nations. During the Imperial Era, the globally-oriented GPI promoted a positive view of Russia among fellow Orthodox (e.g., Rumanians) or Slavs (Czechs), persuading them they could count on assistance from a big brother. In Soviet times, the Kremlin peddled the idea of an anti-colonial revolutionary Messianic state that would liberate subjugated people all over the globe. Arguably, Putin's Russia offers a mixture of two images, simultaneously promoting 'Russki Mir' and charming the segments of 'the Global South' dissatisfied with the West-centered world order.

The most essential task, however, is prescribed to *superior-oriented GPI* because it manipulates local social status. Status is ubiquitous in social relations. On a personal level, humans can detect status information and conjecture their rank within a hierarchy, with some researchers suggesting that 3-year-old children are able 'to draw inferences about how dominance influences social interactions and to predict how dominant and subordinate individuals are likely to differ on a range of characteristics' (Charafeddine et al. 2015, 591). Status is competed for. From sports to academia, professional activities are arenas where contenders compete for status and eventually establish some kind of ranking. High status grants more prestige and material benefits. It is hardly surprising that the international community operates as a status-oriented competition.

Two status features are essential for adequately interpreting Russia's GPI. First, status is local; second, although status acquisition is possible both via bottom-up and top-down pathways, the latter is more gratifying. Localisation of status means that most significant competitions for status occur between peers: a top league team does not play with amateurs, and a fresh PhD student does not compare their Scopus index with that of a professor. States also coalesce into status communities, being 'hypersensitive to position with respect to peer competitors, and more or less oblivious to the fights over rank that take place among and between all other nations' (Renshon 2017, 43). And just as a gifted musician graciously accepts praises from the audience but relishes positive acclaim from another virtuoso, a state welcomes status-related behaviour by lower-standing states, but it strives for recognition by higher-ranking members of the international community.

Russia's *superior-oriented GPI* is designed to convince the status community Russia historically dreamed of entering, the Great Powers Club, that Russia is worthy of being accepted as a peer. As the previous section suggested, the request has not been granted because Great powers deemed Russia 'Oriental', 'backward', or 'corrupt'. Given the permanently poor state of the economy, entrenched bad governance, and persistent patrimonial practices, the verdict is mainly correct and is of Russia's own making. Blaming, however, is beyond the point of this paper; it is essential that historically, the Russian political class experienced acute dissatisfaction with their position in the status community.

Russia's claim for greatness was based on abundant natural and human resources. This, however, does not constitute a sufficient condition to enter the Great Power Club: the Democratic Republic of Congo is blessed with mineral resources, but historically, the country suffered from predation by outsiders and neighbours. Luckily for Russian elites, there was another convincing tool to display and improve Russia's status: warfare.

In order to elucidate the way warfare contributed to the reification of Russia's superior-oriented GPI, I suggest applying the analytical framework of nested asymmetric relations, which offers novel solutions for those studying Russian and international affairs. Combined with the status dissatisfaction model, nested asymmetric theory offers additional insights into Russia's GPI, the origins of the Russo-Ukrainian war, and possible approaches to managing the conflict.

Due to a historic-geographical fluke, many of Russia's wars took the form of asymmetric conflicts. In this paper, I do not use the notion to describe 'fourth-generation irregular warfare (with irregular attacks on vulnerable military units, population, infrastructure, culture, and institutions)' (Smith, Corbin, and Hellman 2001, 72) that pits a state against non-state actors in 'population-centric nontraditional warfare' (Buffaloe 2006, 17; Berman et al. 2020). Such interpretations were conceived during counterinsurgency campaigns of the decolonisation era (Galula 1964), becoming fashionable in the early 2000s at the zenith of the US military power when US planners expected to face irregular adversaries. Instead, I use the notion of 'asymmetric conflict', referring to 'a struggle between belligerents vastly incommensurate in force strength and resources' (Deriglazova 2014, 265).

One of the advantages of the 'asymmetry' notion is that it does not require employing a vague idea of 'small states' (Crowards 2002, 143). Rather, it acknowledges that the distribution of military, economic, human, cultural, and logistic resources is uneven within any constellation of actors—be it a duo, a pentarchy, or two opposing alliances. I suggest distinguishing two types of asymmetric conflicts: *a clash* and *a dyad*. A clash is a temporally isolated event when two uneven contenders fight, with a typical result that the stronger overcomes the weaker due to resource advantage. Compare this concise interpretation with a formal game theory model explaining why during non-reiterated territorial conflicts, "strong player keeps the entire territory forever, leaving the weak player with nothing" (Resnick 2013, 31).

An asymmetric dyad differs in two respects. First, the gap in capacities notwithstanding, the stronger part cannot easily coerce, let alone eliminate, the smaller, with a corollary that they co-exist and constantly renegotiate the exact terms of living together (Womack 2006, 18). Second, members of a dyad nurture different perspectives on what is at stake during the confrontation (Womack 2003, 92). Interactions within a dyad are, therefore, reiterated (non-isolated) events, which brings a distinct pattern to life: because asymmetric dyads are historically durable, the weaker part develops coping mechanisms to contain the stronger. These mechanisms range from establishing a framework to negotiate their differences in a mutually acceptable manner to strategic innovations at war. As a result, relations in a dyad constantly vacillate between two poles: temporary settled principles of coexistence and armed negotiations to reconsider these principles. Moreover, due to more pronounced stimuli to strategic learning and nibbler adaptivity at war, big polities sometimes lose to smaller opponents, which has always fascinated scholars (Mack 1975; Paul 1994; Sullivan 20112). The conventional explanation is that strong actors tend to resort to suboptimal strategies (Arreguín-Toft 2001, 95; Cassidy 2003, 3); however, there is an alternative explanation: numerous trials through historical encounters teach the weaker how to degrade and offset the rival's advantages.

The idea of asymmetry bridges the recent trend of recognising the context-dependent nature of power gradients in international relations (Long 2017, 146) with the relative nature of status in international relations (Larson et al. 2014, 9). To the point, asymmetric relations are sometimes nested; that is, a given polity is enmeshed in a complex web of uneven resource distribution in its regional and global settings, due to which the same state can simultaneously be a small and a Big power. To illustrate: in the 1970s–1980s, Vietnam was a minor power interacting with China, but it played a Big power role *par rapport* to Cambogia, whereas China, simultaneously, was a small power while interacting with the US. It is crucial to account for the nested asymmetric relations because of the status concerns: a polity important regionally but diminished globally is likely to be both opportunistically expansionist and insecure, mindful of its prestige and insulting toward others, proud of its military tradition and successes but constantly facing setbacks and losses. Entanglement in nested asymmetry provokes an acute status threat condition, making the foreign policy more erratic and status-focused because 'once status is threatened or otherwise engaged for leaders, they become increasingly willing to expend resources to save it' (Renshon 2017, 92).

Russia epitomises a state entangled in nested asymmetric relations. Throughout history, it interacted with weaker rivals typically overwhelmed due to superior resources. A typical 'gunpowder empire', the Dutchy of Muscovy ascended to regional prominence because of the riverine system between Don and Volga, which enabled prompt transportation of siege artillery, with the effect that 'controlling the only mobile gun park in those lands, the czar could dominate them all' (McNeil 1989, 31). Later, Russia's incredible 20,000 square miles a year growth between 1613 and 1918 (Montefiore 2017, xix) was mainly achieved through asymmetric clashes with peoples in the Urals and Siberia (1581–1875), Circassia (1763–1864), and Central Asia (1868–1884). Although penetration in each direction extended for centuries, Russia mostly faced separate ethnic entities (e.g., Kabardians, Ingush, Chechens during the thrust into Circassia) who lacked cultural transmission mechanisms to elaborate effective opposition. This, combined with ruthless colonial policies (Pierce 1960; Allworth 1967; Bassin 1999; Richmond 2013), sealed the fate of the local populations and contributed to the image of the unstoppable Russian steamroller.

Russia's move westward and southward was much less spectacular, even though Russia enjoyed the benefits of asymmetry while vying for preponderance with its principal rivals. When comparing Russia's population with its adversaries (see Table 2.1), it becomes evident that Russia had more human resources than Sweden throughout their competition for the Baltic shore (1600–1700). One could argue that given the scale of discrepancy, Russian vic-

TABLE 2.1 Population (in millions)

Year	Russia	Sweden	Ottoman state	France	UK
1600	12	0.9	7.2	20.9	n/a
1700	14	1.3	8.4	21.5	n/a
1750	18.2	1.78	n/a	24.5	n/a
1800	35.5	2.3	26	29.1	8.8
1850	74	3.4	37[a]	36.3	17.9
1900	125	5.1	17	41.4	32.5

a Data comes from 1831 census, being the closest available

SOURCES: **FOR RUSSIA:** PARKER (1969, 254) FOR YEARS 1600, КАБУЗАН (1990, 84–86) FOR 1700 AND 1750, *СТАТИСТИЧЕСКИЕ ТАБЛИЦЫ РОССИЙСКОЙ ИМПЕРИИ* (1863, 320–328) TRIANGULATED WITH MICHELL (1872, 349) FOR 1800 AND 1850, AND THE 1897 CENSUS RESULTS FOR 1900. **FOR FRANCE,** DUPÂQUIER (1979, 10) FOR YEARS 1600, LOUIS AND BLAYO (1975, 95–99) FOR THE YEARS 1700 AND 1750, FINE AND SANGOÏ (1991, 17) FOR THE YEARS 1800, 1850, AND 1900. **FOR SWEDEN:** PALM (2016) FOR YEARS 1600 AND 1700, ALL THE SUBSEQUENT FROM *OFFICIAL STATISTICS OF SWEDEN* (2024). **FOR THE OTTOMAN STATE,** THE DATA COME FROM KURU (2019, 215), SHAW (1978, 326 AND 337), AND KARPAT (1978, 274). ALL DATA **FOR THE UK** COME FROM MICHELL (1975, 24).

tory should have come earlier than the 1721 Nystad treaty. The Ottoman state was another regional opponent Russia failed to eclipse throughout 1600–1750 (see, for instance, the terms of the 1711 treaty of Pruth and 1739 treaty of Nis favourable to the Ottomans) when the power gradient was relatively small. Even when the gap in human resources grew, Russia failed to achieve its ultimate goal of controlling the Straits because the Sublime Porta developed a way to block Russia's advancement into its lands by cooperating with the United Kingdom and France. The fact that both Sweden and the Porta survived four centuries of conflicts with Russia suggests that, unlike peoples in the Caucasus or Siberia, each managed to establish an asymmetric dyad with the Russian giant and effectively contain it.

Russia's military fortunes were even less impressive when it faced France, the United Kingdom, and Germany, which, after the unification by the Kingdom of Prussia, constituted 'a catalyst for radical changes in tactics, operations, military organisation, and state policy' (Showalter 2001, 92). Although smaller than Russia by the demographic yardstick (at least since 1800), France, Germany, and the UK overcame this disadvantage through a combination of four factors: technological modernisation, financial solvency, effective logistics, and skilful diplomacy. A net result was that St. Petersburg acted as a smaller actor, especially when facing Paris and London, e.g., congresses at Aix-la-Chapelle (1748), Paris (1856), or Berlin (1878). Russia was especially vulnerable—and its pretensions to Great power status shattered most pronouncedly—when it dared to act unilaterally. In fact, between 1700 and 1914, the two most consequential moves by St. Petersburg, the Holy Alliance, which followed the Napoleonic wars, and the Entente Cordiale, exemplified respect for and instrumentalisation of international norms and alliances. Thus, Russia perfectly illustrated nested asymmetry in the larger circle of European politics: a considerable power when dealing with weaker opponents and a tumbling giant facing first-tier powers.

The situation changed after 1945 when the three European powers declined and had to balance the Soviets by relying on Washington. Moscow applied its weight to coerce small states like Bulgaria or Czechoslovakia into its sphere of influence. It resorted to overwhelming force in 1953, 1956, and 1968, sending a clear signal to small nations of Central Europe that resource asymmetry was impossible to ignore. Moscow was also poised for a 'massive strike [that would] create the conditions for winning the war by the classic method of seizing enemy territory [in Western Europe]' (Luňák 2000). It also capitalised on the nuclear arms possession, reinforcing its GPI by announcing loud and clear it was, as Khrushchev boasted, 'the great powers [that] have arms with a global reach of such strength and in such quantity that the entire world could be destroyed quickly and easily'.[2] Still, its imposing hard power notwithstanding, Moscow trailed behind Washington in many respects due to US advances in technology, finances, global logistics, and diplomacy (the US broad network of containing alliances in Europe, the Pacific, and the Middle East). As a result, nested asymmetry reemerged: the USSR was powerful against regional players but weak against the global contender.

Soviet foreign policy throughout the Cold War illustrates the GPI to its fullest: Moscow lacked soft power capabilities to pursue its interests in interstate relations, and most Soviet moves outside the immediate European neighbourhood were unsatisfactory. However, the Kremlin used all means—from success in the space race to Soviet literacy level to a perceived missile gap—to promote an impression that it could overcome the USA. Sergei Khrushchev provides a remarkable anecdote about dimensions of status signaling used in Moscow: during the 1959 visit to the US, the airstrip of the Washington airport was too short to accommodate the giant TU-114, so 'We concluded unequivocally: here again, we are better, they must be jealous' (Хрущев 2010, 358). Curiously, the English translation of the memoirs was performed on an abridged version and

2 *Memoirs of Nikita Khrushchev, Vol. 2: Reformer, 1945–1964* (University Park: The Pennsylvania State University Press, 2006): 521.

does not relate Khrushchev's visit with the same words. Instead, there are other episodes relative to the concerns about the technological prestige of the USSR (Memoirs of Nikita Khrushchev 2006, 98–99, 102). Efforts to be perceived as a superpower prompted the Soviets to venture into Cuba and the Middle East in the 1960s and tropical Africa in the 1970s. They exemplified opportunistic behaviour and eagerness to support whoever and for whatever costs to recruit another unreliable client (Jackson 2023). This brought little tangible gains, provoking an overstretch and faulty decision-making instead. In particular, the humiliation of Egypt's 1973 *renversement des alliances* impacted Soviet planning to the point that the 1979 invasion of Afghanistan was decided upon to preclude 'doing a Sadat on us' (Westad 1996, 130). The incursion in Afghanistan was a pure act out of status concerns with little practical value and devastating outcomes, for it turned out to be the beginning of the Soviet implosion (Carrère d'Encausse 2019, 90).

Russia's policies since 1991 reflect the GPI performed within the nested asymmetric relations. Even weak and disoriented because of the double transition, Moscow resorted to fierce posturing toward weaker neighbours (Moldova, Georgia, Kazakhstan, and Ukraine) and sporadic moves against the West—as during the 1999 incident at Pristina airport. After Putin consolidated his rule, the GPI had been further elaborated, streamlined, and consistently applied in domestic and foreign policy. As recent analysis suggests, 'the policies pursued by Putin over his nearly two decades in office is not about any particular political genius, but identification of a clear, broad goal—make Russia into a recognised great power' (Weber 2020, 252).

In particular, Putin's policy focused on four strategies: (1) fighting the global terrorist threat, (2) vigorous pursuit of membership in international organisations (WTO, BRICS, G-20); (3) enhancing Europe's reliance upon Russia's oil and gas, and (4) efforts to convert military power into international status. The 2008 military reform and massive investment into military-grade R&D were designed to revitalise Russian military might and close the technological gap with other armies. Simultaneously, Putin elaborated a set of policies ranging from commemorative to youth policy to create an image of Russia as a powerful military state and used the 'manifestations of military power as [...] evidence of its status as Great power' (De Haas 2010, 108). A shift in public opinion reflects the continuous bolstering of the GPI military component: the percentage of respondents convinced that military supremacy and nuclear arsenal are the prime sources of international respect toward Russia increased from 13 to 38 per cent between 1997 and 2016 (Левада-Центр 2016).

The 2022 invasion of Ukraine might be at least partially attributed to Russia's use of the Great power ideology. In particular, driven by a persistent quest to be accepted as a major global power, Moscow grew increasingly frustrated in the late 2000s because it witnessed the ineffectiveness of its strategies to reify its GPI. Namely, the cooperation with the West within the framework of the US-led crusade against Islamic terrorism did not significantly improve Russia's status. The reason is the nature of international prestige: status is positional goods that cannot be increased for all participants simultaneously; since many states joined the 'War on Terror', Russia's status gains diminished. For similar reasons, the second strategy, membership in international organisations, alone did not suffice to raise Russia's status, especially given the competition with other 'emergent countries' like Indonesia or Brazil. At the same time, Russia failed to convert revenues from energy supplies into durable modernisation, remaining a mere provider of raw materials to the West, which did not improve its status.

As a result, Moscow opted for the conversion of military power into prestige. Two additional factors contributed to the preference for this strategy: (1) Putin's worldview, compounded by the Russian strategic culture, which deemed the Western influence in the neighbourhood pernicious (Eitelhuber 2009, 11–12), and (2) nested asymmetry that historically helped Russia to bolster its image as an important actor in the eyes of the first-tier powers by crushing weaker regional opponents. In 2022, to convince the domestic public and international actors of Russia's status, Putin resorted to a classical Russian strategy perfected since the 17th century during ferocious campaigns against feeble foes. To put it bluntly, Russia decided to invade Ukraine in order to convince Washington and its European allies that Russia still counted in international politics and extort further concessions.

In its initial conception, the 2022 invasion of Ukraine reflected a typical Russian approach to asymmetric conflict: first, overpower the opponent with massive force deployment; next, install a friendly regime; and, finally, achieve long-lasting political and economic benefits by exploiting the newly acquired resources for material and symbolic gains. In the 20th century alone, Moscow applied this strategy in the 1940 annexation of the Baltic states, in 1956 and 1968 pacifications of Hungary and Czechoslovakia, in the 2000 instauration of Akhmad Kadyrov's regime in Chechnya, in the 2008 detachment of South Ossetia, and in the 2014 annexation of Crimea. However, by Spring 2022, it became apparent that Moscow woefully miscalculated the resolve of Ukrainians to fight back and their ability to acquire resources from the West. A lightning special force operation turned into years-long positional warfare.

Conclusions and Implications

Given the theoretical framework introduced in this paper, one feature of the Russo-Ukrainian war should be highlighted: it represents not an asymmetric clash but the armed deliberation regarding the terms of the coexistence between the members of an asymmetric dyad. To elaborate: despite Russia's constant efforts to present Ukraine as a run-away province, historically, Ukraine existed as a separate polity or a part of another (not Russian) state longer than it did as a Russian colony. To bring the fact into proper chronological perspective, Moscow ruled over most of Ukraine for only 216 years (parts of Ukraine under Russia's control lost autonomy in 1775), while other parts (e.g., Galicia) were incorporated into the Soviet Union as late as 1945. Bringing another chronological perspective, by 2025, Ukraine exists as a separate polity for almost as long as entire Ukraine belonged to the Soviet Union. The point is that throughout reiterated historical encounters with Russia, Ukraine invented coping strategies to contain Russia's assimilative drive, elude Russia's control, and maintain its cultural and political autonomy. Numerous surprises of the Russo-Ukrainian war, ranging from the effective eviction of the Russian Black Sea Fleet from Crimea through innovative marine unmanned vessels to civilian resilience to relative political stability despite the grinding conflict, suggest Ukraine can confront Russia.

Another essential insight is the preliminary outcome of Russia's gamble against Ukraine. If the present analysis is correct, Putin aimed to enhance Russia's Great power status by projecting power in the neighbouring state. Indeed, the long-term consequences of the Russo-Ukrainian war must be analysed after it is concluded. However, as of 2025, there are two conflicting interpretations of whether Putin succeeded in his prestige-bolstering strategy. According to the first one, 'The idea of a poor, weak, and resentful post-Soviet regional power [...] was replaced by an idea of Russia that competed on equal terms with the great power United States and the new great power China' (Røren 2023, 38). The alternative interpretation suggests that the war against Ukraine 'had profoundly damaging consequences for Russia's status as a great power by diminishing its global role, isolating it diplomatically, cutting it from presence in elite clubs, and exposing the flaws of its military' (Šćepanović 2024, 90). This paper concurs with the second interpretation: Russia's dismal military performance in Ukraine since 2022 has significantly undermined the cornerstone of Russia's great power image, namely the military might. This, in turn, prompted the question: what kind of superpower Russa is if it fails to overcome a weaker neighbour? The idea transgressed the boundaries of academia and penetrated the general readership (Kuzio 2022; Soric 2022; Von Drehle 2022). Given that the war rendered the Kremlin's energy-leveraging strategy vis-à-vis Europe ineffective (Krugman 2023), left its foreign policy in shambles (Stent 2023), and provoked colossal demographic losses (Kolesnikov 2024), one should likely expect further diminution of Russia's status. As an additional illustration, refer to the *2024 Asia Power Index*. There, Russia (31.1 points) is sandwiched between Australia and South Korea in the 'Middle Power' category, trailing well behind genuine superpowers, the US (81.7) and China (72.7) (see Lowy Institute 2025). Admittedly, since the 18th century, Russia has preferred to be considered a European, not an Asian, power. However, the drop in regional status is a drop, nonetheless.

Russia's nested asymmetry and the ensuing status threat analysis suggest several theoretical innovations and policy-making implications.

To begin with, it overlaps with Gel'man's suggestion (Chapter 4, this volume) to bring historical sociology, especially the bellicist perspective, to Russian and Central-European studies. The Tillyan account, long accepted to explain the state-making in Western Europe, has been applied to other regions (Centeno 1997; Tin-bor Hui 2005; Kurtz 2013; Kasperse and Strandsbjerg 2017), but several notable exclusions apart (Ganev 2005; Maleševic 2012), it remains outside of the scholars' toolbox to analyse Russia and Central Europe. However, the military-technological edge over the steppe people contributed to the Muscovy state's initial consolidation. When wars against more recalcitrant opponents like Sweden and the Sublime Porta exposed the inadequacy of the traditional Russian recruitment system, the government established new-formation infantry, cavalry, and side-armed dragoon regiments, resorting to the expertise of Western officers (Ágoston 2011, 283). Since 1600, Russia has been introducing, imitating, and doubling down on European warfare principles, from the sail navy in the 18th century to the militarisation of railroads and concentrated firepower in the early 20th. From a larger perspective, Russia's trajectory resembles that of the Ottoman Empire, first displaying military might rooted in the early introduction of siege artillery but eventually falling behind and spending centuries desperately trying to catch up with the West by mimicking its military technology and vacillating between adopting and rejecting its political institutions (Ralston 1990). The pattern is undeniable even for post-1991 Russia with its limited 2008 Serdyukov military reform and several aborted "Westernisation" initiatives under Yeltsin and Putin. Besides, the "war-making—state-making" nexus is a no less important subject for scholarship on Ukraine, where the 2014 and 2022 invasions provoked a plethora of state-enhancing steps, but their outcomes are to be systematically studied.

Moreover, the present study suggests it is necessary to bring Russia's GPI to the forefront of scholarly endeavours since it opens two novel venues:

(1) for IR specialists, especially those adhering to neoclassical realist and constructivist schools, GPI offers an additional probe into the Russian elite's worldview and policy decisions.
(2) although ideology cannot substitute institutional or empirical analysis, it is equally inconceivable to ignore the role of ideology in politics and policy. Soviet foreign policy, for instance, although irreducible to Communism, would be incomprehensible without its ideological trappings (Radchenko 2024). Likewise, Russia's foreign policy bears distinct handprints of the GPI. And just as not everything Communist leaders uttered was to be accepted at face value, GPI-produced proclamations do not necessarily represent reality.

This brings the last, admittedly most contentious issue, which policy-makers should be most sensitive to. Contrary to centuries-long pretensions by Russian elites, Russia is not a Great but a regional power. Its role, impact, and status should be compared not to Germany, France, or the UK but to Turkey. The Turkish case traces the best scenario to normalise Russia's relations with the West. Sprung from an erstwhile threatening but eventually eclipsed empire, itself plagued with cycles of West-oriented emulation and traditionalist reaction against it, Turkey became a functional modern state with its cultural peculiarities and distinct agendas that can be accommodated within the stable regional and world order. The crucial distinction between Ankara (at least prior to Erdogan's neo-Osmanism policy) and Moscow is the absence of entrenched GPI, which remains a central piece of the Russian myth of uniqueness and messianism. Therefore, this paper's historical and ideological analysis corroborates the conclusion by Aliaksei Kazharski (Chapter 7, this volume) that 'Russia needs to be recognised for what it is: one of many peripheral European cases that should not be entitled to any special treatment'. Unfortunately, the special treatment and 'Russia first policy' plagued six successive US presidential administrations and was detrimental to prospects of peaceful Europe (Vindman 2025), so it is time to confront Russia's GPI with a reality check, which, if consistently applied, will dissipate the pernicious illusion.

In other words, it is advisable to call Russia's bluffs, contain its political influence, and stop tolerating constant and nefarious interloping into neighbouring sovereign states' affairs. Admittedly, in the short run, this policy is bound to provoke more confrontations and even conflicts with Russia precisely because the strict policy toward the Kremlin will signal Russia's further status deterioration. However, since 2000, the US and the EU have been too mindful of Russia's prestige, which was raising the Kremlin's expectations ever so high, effectively widening the gap between the perceptions of status Russia ostensible is worthy of and its actual capacities. The GPI helped consume concessions as granted and inflated further expectations. It is time to break the vicious circle. When doing so, policy-makers should be aware that 'within the status threat condition, low power increases the tendency to escalate [... while] high power emerges as a key element of high-quality decision making' (Renshon 2017, 163). Thus, it will be a task for Washington and Brussels, London and Paris, Berlin and Warsaw to manage the outburst of Russia's anger when its GPI complex is dismantled. Whether the Trump administration is eager and able to do it or whether Europe must accomplish the fit alone is a separate matter. Yet Russia, deluded by the GPI, remains a regional threat.

Bibliography

Adomeit, Hannes. 1995. "Russia as a 'great power' in world affairs: images and reality." *International Affairs* 71(1): 35–68.

Ágoston, Gábor. 2011. "Military Transformation in the Ottoman Empire and Russia, 1500–1800." *Kritika: Explorations in Russian and Eurasian History* 12(2): 281–319.

Allworth, Edward, ed. 1967. *Central Asia: A Century of Russian Rule*. Columbia University Press.

Althusser, Louis. 1971. "Ideology and Ideological State Apparatuses." In *Lenin and Philosophy and Other Essays*. Monthly Review Press.

Anno, Tadashi. 2019. *National Identity and Great-Power Status in Russia and Japan: Non-Western Challengers to the Liberal International Order*. Routledge.

Aretin, Karl Otmar, Freiherr von. 1986. "Russia as a Guarantor Power of the Imperial Constitution under Catherine II." *The Journal of Modern History* 58: 141–160.

Arreguín-Toft, Ivan. 2001. "How the Weak Win Wars: A Theory of Asymmetric Conflict." *International Security* 26(1): 93–128.

Auslin, Michael. 2005. "Japanese Strategy, Geopolitics and The Origins of The War, 1792–1895." In *The Russo-Japanese War in Global Perspective*, Vol. 1, edited by John Steinberg, Bruce Menning, David Schimmelpenninck van der Oye, David Wolff, and Shinji Yokote. Brill.

Baev, Pavel. 2010. "The Russian Federation: striving for multipolarity but missing the consequences." In *Great Powers and Strategic Stability in the 21st Century Competing Visions of World Order*, edited by Graeme P. Herd. Routledge.

Bassin, Michael. 2004. *Imperial Visions: Nationalist Imagination and Geographical Expansion in the Russian Far East, 1840–1865*. Cambridge University Press.

Baugh, Daniel. 2011. *The Global Seven Years War, 1754–1763: Britain and France in a Great Power Contest*. Routledge.

Berman Eli, Joseph Felter, Jacob Shapiro, Vestal McIntyre, and John McLain. 2020. *Small Wars, Big Data: The Information Revolution in Modern Conflict*. Princeton University Press.

Beyrau, Dietrich. 1987. "Russische Interessenzonen und europäisches Gleichgewicht 1860–1870." In *Europa vor dem Krieg von 1870: Mächtekonstellation—Konfliktfelder—Kriegsausbruch*, edited by Eberhard Kolb. Oldenbourg Wissenschaftsverlag.

Bock, Andreas, Ingo Henneberg, and Friedrich Plank. 2015. "If you compress the spring, it will snap back hard: The Ukrainian crisis and the balance of threat theory." *International Journal* 70(1): 101–109.

Buffaloe, David. 2006. "Defining Asymmetric Warfare." *The Land Warfare Papers* 58: 1–41.

Carrère d'Encausse, Hélène. 2019. "1979. Commencement de la fin de l'URSS?" *Le Débat* 5(207): 88–93.

Cassidy, Robert. 2003. *Russia in Afghanistan and Chechnya: Military Strategic Culture and the Paradoxes of Asymmetric Conflict*. US Army War College Press.

Centeno, Miguel Angel. 1997. "Blood and Debt: War and Taxation in Nineteenth-Century Latin America." *American Journal of Sociology* 102(6): 1565–1605

Charafeddine, Rawan, Hugo Mercier, Fabrice Clément, et al. 2015. "How Preschoolers Use Cues of Dominance to Make Sense of Their Social Environment." *Journal of Cognition and Development* 16(4): 587–607.

Clunan, Anne. 2009. *The Social Construction of Russia's Resurgence: Aspirations, Identity, and Security Interests*. John Hopkins University Press.

Crowards, Tim. 2002. "Defining the Category of 'Small' States." *Journal of International Development* 14(2): 143–179.

CRS Report. 2024. "Great Power Competition: Implications for Defense—Issues for Congress." *Congressional Research Service* 43. Effective April 24, 2025. https://sgp.fas.org/crs/natsec/R43838.pdf.

Curanović, Alicja. 2021. *The Sense of Mission in Russian Foreign Policy*. Routledge.

De Haas, Marcel. 2010. *Russia's Foreign Security Policy in the 21st Century: Putin, Medvedev and Beyond*. Routledge.

de Talleyrand, Charles Maurice. 1881. *The Correspondence of Prince Talleyrand and King Louis XVIII: During the Congress of Vienna*. Richard Bentley and Son.

Deriglazova, Larisa. 2014. *Great Powers, Small Wars: Asymmetric Conflict Since 1945*. John Hopkins University Press.

Duffy, Christopher. 1981. *Russia's Military Way to the West: Origins and Nature of Russian Military Power, 1700–1800*. Routledge.

Dukes, Paul, Graeme Herd, and Jarmo Kotilain. 2009. *Stuarts and Romanovs: The Rise and Fall of a Special Relationship*. Dundee University Press.

Dupâquier, Jacques. 1979. *La population française aux XVIIe et XVIIIe siècles*. Presses universitaires de France.

Eitelhuber, Nikita. 2009. "The Russian Bear: Russian Strategic Culture and What it Implies for the West." *Connections* 9(1): 1–28.

Ellman, Michael. 2023. "Russia as a great power: from 1815 to the present day, Part II." *Journal of Institutional Economics* 19: 159–174.

Fine, Agnès and Jean-Claude Sangoï. 1991. *La population française au XIXe siècle*. Presses Universitaires de France.

Frank, Andre Gunder. 1994. "Soviet and East European 'socialism': A review of the international political economy on what went wrong." *Review of International Political Economy* 1(2): 317–334.

Galula, David. 1964. *Counterinsurgency Warfare: Theory and Practice*. Praeger.

Ganev, Venelin. 2005. "Post-communism as an episode of state building: A reversed Tillyan perspective." *Communist and Post-Communist Studies* 38(4): 425–445.

Gazprom Annual Report. 2019. "Наша работа—общее будущее." Effective April 24, 2025. https://sustainability.gazpromreport.ru/2019/5-social/5-5-spiritual-values/.

Gilpin, Robert. 1981. *War and Change in World Politics*. Cambridge University Press

Gioe D.—Smith M. (eds.), *Great Power Cyber Competition Competing and Winning in the Information Environment* (New York: 2024).

Goldman, Marshall. 2008. *Petrostate: Putin, Power and the New Russia*. Oxford University Press.

Götz, Elly. 2016. "Neorealism and Russia's Ukraine Policy, 1991-present." *Contemporary Politics* 22(3): 301–323.

Hancock, Kathleen. 2007. "Russia: Great Power Image versus Economic Reality." *Asian Perspective* 31(4): 71–98.

Hedenskog, Jacob, Vilhelm Konnander, Bertil Nygren, Ingmar Oldberg, and Christer Pursiainen, eds. 2005. *Russia as a Great Power Dimensions of Security Under Putin*. Routledge.

Helm, Brendan, Dina Smeltz, and Arik Burakovsky. 2020. "US Experts Anticipate Future Decline for Russia Among the Great Powers They Forecast Diminishing Russian Influence and Stagnating Economic and Military Strength." *Chicago Council on Global Affairs*. Effective April 24, 2025. https://www.jstor.org/stable/resrep26631.

Hill, Alexander. 2023. "Russia in Africa: Resurgent Great Power or Bellicose Pretender?" *The Journal of Slavic Military Studies* 36(3): 319–322.

Hunter, Robert. 2024. "On 75th Anniversary, NATO is at a serious crossroads", *Responsible Statecraft*. Effective April 24, 2025. https://responsiblestatecraft.org/75th-anniversary-of-nato/.

Jackson G., *A Lost Peace: Great Power Politics and the Arab-Israeli Dispute, 1967–1979* (New York: 2023).

Kagan, Frederick. 2002. "Russia's Small Wars, 1805–1861." In *The Military History of Tsarist Russia*, edited by Frederick Kagan and Robin Higham. Palgrave.

Kanet, Roger. 2007. *Russia: Re-Emerging Great Power*. Palgrave Macmillan.

Karpat, Kemal. 1978. "Ottoman Population Records and the Census of 1881/82–1893." *International Journal of Middle East Studies* 9: 237–274.

Kasperse, Lars Bo and Jeppe Strandsbjerg, eds. 2017. *Does War Make States?* Cambridge University Press.

Kokoshin, Andrei. 2002. "What is Russia: a Superpower, a Great Power or a Regional Power." *International Affairs: A Russian Journal* 6. Effective April 24, 2025. https://ciaotest.cc.columbia.edu/olj/iarj/iarj_02_06a.html#note28.

Kolb, Eberhardt. 1990. *Der Weg aus dem Krieg: Bismarcks Politik im Krieg und die Friedensanbahnung 1870–1871*. Oldenbourg Wissenschaftsverlag.

Kolesnikov Andrei. 2024. "Russia Is Burning Up Its Future." *Foreign Affairs*, March 7. Effective April 24, 2025. https://www.foreignaffairs.com/russian-federation/russia-burning-its-future.

Krugman, Paul. 2023. "Putin's Energy Offensive Has Failed", *The New York Times*, April 6. Effective April 24, 2025. https://www.nytimes.com/2023/04/06/opinion/putins-energy-natural-gas.html.

Kurtz, Markus. 2013. *Latin American State Building in Comparative Perspective: Social Foundations of Institutional Order*. Cambridge University Press.

Kuru, Ahmet. 2019. *Islam, authoritarianism, and underdevelopment: a global and historical comparison*. Cambridge University Press.

Kuzio, Taras. 2022. "Putin's failing Ukraine invasion proves Russia is no superpower," *Atlantic Council*, November 1. Effective April 24, 2025. https://www.atlanticcouncil.org/blogs/ukrainealert/putins-failing-ukraine-invasion-proves-russia-is-no-superpower/.

Larson, Deborah Welch, T.V. Paul, and William Wohlforth. 2014. "Status and World Order." In *Status in World Politics*, edited by Larson, Deborah Welch, T.V. Paul, and William Wohlforth. Cambridge University Press

Larson, Deborah Welch and Alexei Shevchenko. 2003. "Shortcut to Greatness: The New Thinking and the Revolution in Soviet Foreign Policy." *International Organization* 57(1): 77–109.

LeDonne, John. 2004. *The Grand Strategy of the Russian Empire, 1650–1831*. Oxford University Press.

Lieven, Dominic. 1983. *Russia and the Origins of the First World War*. Palgrave Macmillan.

Lieven, Dominic. 1992. *Nicholas II: Twilight of the Empire*. St. Martin Press.

Lodge, Richard. 1926. "An Episode in Anglo-Russian Relations during the War of the Austrian Succession." *Transactions of the Royal Historical Society* 9: 63–83.

Long, Tim. 2017. "It's not the size, it's the relationship: from 'small states' to asymmetry." *International Politics* 54: 144–160.

Louis, Louis and Yves Blayo. 1975. "La population de la France de 1740 à 1860." *Population* 30(1): 71–122.

Lowy Institute. 2025. *Asia Power Index 2024 Edition*, Effective April 24, 2025. https://power.lowyinstitute.org/.

Luňák, Petr. 2000. "The Warsaw Pact War Plan of 1964." *Parallel History Project*. Effective April 24, 2025. https://phpisn.ethz.ch/lory1.ethz.ch/collections/coll_warplan/intro_lunak48fe.html?navinfo=15365.

Mack, Andrew. 1975. "Why Big Nations Lose Small Wars: The Politics of Asymmetric Conflict." *World Politics* 27(2): 175–200.

Maleševic, Siniša. 2012. "Did Wars Make Nation-States in the Balkans? Nationalisms, Wars and States in the 19th and early 20th Century South East Europe." *Journal of Historical Sociology* 25(3): 299–330.

Malozemoff, Andrew. 1958. *Russian Far Eastern Policy, 1881–1904*. University of California Press.

Marston, Daniel. 2001. *The Seven Years' War*. Osprey Publishing.

Marx, Karl and Friedrich Engels. 1969. *Werke, Band 3: Die deutsche Ideologie*. Dietz Verlag.

Mastanduno, Michael. 2009. "System Maker and Privilege Taker: US Power and the International Political Economy." *World Politics* 61(1): 121–154.

McCain, John. 2016. "Speech", *Congressional Record, Proceedings and Debates of the 114th Congress* 162. S6870–6872. Effective April 24, 2025. https://www.congress.gov/114/crec/2016/12/08/CREC-2016-12-08.pdf.

McKay, Derek and H.M. Scott. 2012. *The Rise of the Great Powers 1648–1815*. Routledge.

McNeil, William. 1989. *Age of Gunpowder Empires, 1450–1800*. American Historical Association.

Mearsheimer, John. 2023. "Israel-Hamas, Ukraine-Russia and China: John Mearsheimer on why the US is in serious trouble", *Centre for Independent Studies* (2023). Effective April 24, 2025. https://www.youtube.com/watch?v=62FCVJycwSA.

Medetsky, Anatoly. 2016. "Russia Becomes a Grain Superpower as Wheat Exports Explode." *Bloomberg*, October 7. Effective April 24, 2025. https://www.bloomberg.com/news/articles/2016-10-06/russia-upends-world-wheat-market-with-record-harvest-exports.

Memoirs of Nikita Khrushchev, Vol. 2: Reformer, 1945–1964, edited by Sergei Khrushchev. 2006. Pennsylvania State University Press.

Memoirs of Nikita Khrushchev, Vol. 3: Statesman, 1953–1964, edited by Sergei Khrushchev. 2006. Pennsylvania State University Press.

Michell, Robert. 1872. "Summary of Statistics of the Russian Empire", *Journal of the Statistical Society of London* 35(3): 341–372.

Michell, Brian. 1975. *European Historical Statistics, 1750–1975* (Macmillan: 1975).

Miyoshi Jager, Sheila. 2023. *The Other Great Game: the Opening of Korea and the Birth of Modern East Asia*. Harvard University Press.

Montefiore, Simon Sebag. 2017. *The Romanovs: 1613–1918*. Vintage.

NATO 2022 Strategic Concept. 2022. Effective April 24, 2025. https://www.nato.int/nato_static_fl2014/assets/pdf/2022/6/pdf/290622-strategic-concept.pdf.

Naumov, Victor. 2015. "Empress Elizabeth I, 1741–1762." In *The Emperors and Empresses of Russia: Rediscovering the Romanovs*, edited by Donald Raleigh. Routledge.

Nester, William. 2014. *The French and Indian War and the conquest of New France*. University of Oklahoma Press.

Neumann, Iver. 2008. "Russia as a great power, 1815–2007." *Journal of International Relations and Development* 11: 128–151.

Nicolson, Harold. 1964. *The Congress of Vienna: a Study in Allied Unity, 1812–1822*. The Viking Press.

Nish, Ian. 1966. *The Anglo-Japanese Alliance: the Diplomacy of Two Island Empires, 1894–1907*. Athlone Press.

Obama, Barack. 2015. "Press Briefing by Press Secretary Josh Earnest." *The American Presidency Project*. Effective April 24, 2025. https://www.presidency.ucsb.edu/node/312480.

Official Statistics of Sweden. 2024. "Population and Population Changes 1749–2023." Effective April 24, 2025. https://www.scb.se/en/finding-statistics/statistics-by-subject-area/population/population-composition/population-statistics/pong/tables-and-graphs/population-statistics---summary/population-and-population-changes.

Palm, Lennart Palm. 2016. "Sweden's 17th century—a period of expansion or stagnation"? *Institutionen för historiska studier*: 1–39.

Parker, William Henry. 1969. *Historical Geography of Russia*. Aldine Publishing Co.

Paul, T.V. 1994. *Asymmetric Conflicts: War Initiation by Weaker Powers*. Cambridge University Press.

Petersson, Bo. 2014. "Still Embodying the Myth? Russia's Recognition as a Great Power and the Sochi Winter Games." *Problems of Post-Communism* 61(1): 30–40.

Pierce, Richard. 1960. *Russian Central Asia 1867–1917: A Study in Colonial Rule*. University of California Press.

Posen, Barry and Stephen Van Evera. 2023. "Lessons from Russia's invasion of Ukraine." *Defense Priorities* (2023). Effective April 24, 2025. https://cis.mit.edu/publications/analysis-opinion/2023/lessons-russias-invasion-ukraine.

Radchenko, Sergei. 2024. *To Run the World: The Kremlin's Cold War Bid for Global Power*. Cambridge University Press.

Radvanyi, Jean and Marlene Laruelle. 2023. *Russia: Great Power, Weakened State*. Rowman & Littlefield.

Ralston, David. 1990. *Importing the European Army: The Introduction of European Military Techniques and Institutions to the Extra-European World, 1600–1914*. University of Chicago Press.

Rambaud, Alfred. 1895. *Russes et Prussiens: guerre de sept ans*. Berger Levrault et Cie.

Rangsimaporn, Paradorn. 2009. *Russia as an Aspiring Great Power in East Asia: Perceptions and Policies from Yeltsin to Putin*. Palgrave Macmillan.

Raquel Freire, Maria. 2011. "USSR/Russian Federation's Major Power Status Inconsistencies." In *Major Powers and the Quest for Status in International Politics*, edited by Thomas Volgy, Renato Corbetta, Keith Grant, and Ryan Baird. Palgrave MacMillan.

Renshon, Jonathan. 2017. *Fighting for Status: Hierarchy and Conflict in World Politics*. Princeton University Press.

Resnick, Uri. 2013. *Dynamics of Asymmetric Territorial Conflict: the Evolution of Patience.* Palgrave MacMillan.

Reuters. 2025. "Gazprom swings to net loss of $12.9 billion under Russian standards in 2024, IFX reports." Effective April 24, 2025. https://www.reuters.com/business/energy/gazprom-swings-net-loss-129-billion-under-russian-standards-2024-ifx-reports-2025-03-17/

Rich, David Alan. 1998. *The Tsar's Colonels: Professionalism, Strategy, and Subversion in Late Imperial Russia.* Harvard University Press.

Rich, Paul. 2009. "Russia as a great power." *Small Wars and Insurgencies* 20(2): 276–299.

Richmond, Walter. 2013. *The Circassian Genocide.* Rutgers University Press.

Rogger, Hans. 1983. *Russia in the Age of Modernisation and Revolution, 1881–1917.* Routledge.

Røren, Pål. 2023. "The Belligerent Bear: Russia, Status Orders, and War." *International Security* 47(4): 7–49.

Rosefielde, Steven. 2004. *Russia in the 21st Century: The Prodigal Superpower.* Cambridge University Press

Rumer, Eugene and Celeste Wallander. 2003. "Russia: Power in Weakness?" *The Washington Quarterly* 27(1): 57–73.

Šćepanović, Janko. 2023. "Still a great power? Russia's status dilemmas post-Ukraine war." *Journal of Contemporary European Studies* 32(1): 80–95.

Schumann, Matt and Karl Schweizer. 2012. *The Seven Years War: A Transatlantic History.* Routledge.

Shaw, Standford. 1978. "The Ottoman Census System and Population, 1831–1914." *International Journal of Middle East Studies* 9: 325–338.

Showalter, Dennis. 1996. *Frederick the Great: A Military History.* Longman.

Showalter, Dennis. 2001. "The Prusso-German RMA, 1840–1871." In *The Dynamics of Military Revolution, 1300–2050*, edited by MacGregor Knox and Williamson Murray. Cambridge University Press.

Smith, Daniel, Marcus Corbin, and Christopher Hellman. 2001. *Reforging the Sword: Forces for a 21st Century Security Strategy.* Center for Defense Information.

Smith, Hanna. 2014. "Russia as a Great Power: Status Inconsistency and the Two Chechen Wars." *Communist and Post-Communist Studies* 47(3–4): 355–363.

Soric, Miodrag. 2022. "Russia's army: An overestimated power", *Deutsche Welle*, September 29, Effective April 24, 2025. https://www.dw.com/en/russias-army-an-overestimated-power-in-the-war-against-ukraine/a-63264441.

Stent, Angela. 2023. "Vladimir Putin's Big Backfire", *Politico*, February 26. Effective April 24, 2025. https://www.politico.com/news/magazine/2023/02/26/putin-great-power-shrunk-russias-world-00084374.

Stone, David. 2006. *A Military History of Russia: from Ivan the Terrible to the War in Chechnya.* Praeger Publishers.

Strakhovsky, Leonid Ivan. 1947. *Alexander I of Russia: The Man Who Defeated Napoleon.* Greenwood Press.

Sullivan, Patricia. 2012. *Who Wins? Predicting Strategic Success and Failure in Armed Conflict.* Routledge.

Talyor, Alan John Percivale. 1954. *The Struggle for Mastery in Europe, 1848–1918.* Oxford History of Modern Europe.

Taras, Raymond, ed. 2024. *Exploring Russia's Exceptionalism in International Politics.* Routledge.

The Insider. 2023. "Не в ресурсе. Как Россия потеряла статус энергетической сверхдержавы." Effective April 24, 2025. https://theins.ru/ekonomika/259228.

The Moscow Times. 2024. "Первую группу российских детей отправят на отдых в Северную Корею в июле." Effective April 24, 2025. https://www.moscowtimes.ru/2024/05/26/pervuyu-gruppu-rossiiskih-detei-otpravyat-naotdih-vsevernuyu-koreyu-viyule-a131866.

Tin-bor Hui, Victoria. 2005. *War and State Formation in China and Early Modern Europe.* Cambridge University Press.

Vindman, Alexander. 2025. *The Folly of Realism: How the West Deceived Itself About Russia and Betrayed Ukraine*. Public Affairs.

Volgy, Thomas, Renato Corbetta, Keith Grant, and Ryan Baird. 2011. "Major Power Status in International Politics." In *Major Powers and the Quest for Status in International Politics*, edited by Thomas Volgy, Renato Corbetta, Keith Grant, and Ryan Baird. Palgrave MacMillan.

Von Drehle, David. 2022. "War proves that Russia is no longer a superpower", *The Washington Post*, March 15. Effective April 24, 2025. https://www.washingtonpost.com/opinions/2022/03/15/ukraine-war-proves-russia-no-longer-a-superpower/.

Wawro, Geoffrey. 2003. *The Franco-Prussian War.* Cambridge University Press.

Weber, Yuval. 2020. "Measuring Hierarchy in the European Union and Eastern Partnership Countries." In *Resilience and the EU's Eastern Neighbourhood Countries: From Theoretical Concepts to a Normative Agenda*, edited by Gilles Rouet and Gabriela Carmen Pascariu. Palgrave Macmillan.

Weigley, Russell Frank. 1991. *The Age of Battles: The Quest for Decisive Warfare from Breitenfeld to Waterloo.* Indiana University Press.

Westad, Odd Arne. 1996. "Concerning the situation in 'A': New Russian Evidence on the Soviet Intervention in Afghanistan." *Cold War International History Project Bulletin* 8/9: 128–184.

Wetzel, David. 2012. *The Duel of Nations: Germany, France, and the Diplomacy of the War of 1870–1871.* University of Wisconsin Press.

White House. 2017. *National Security Strategy of the United States of America* (2017). Effective April 24, 2025. https://trumpwhitehouse.archives.gov/wp-content/uploads/2017/12/NSS-Final-12-18-2017-0905.pdf.

White House. 2022. *National Security Strategy*. Effective April 24, 2025. https://www.whitehouse.gov/wp-content/uploads/2022/10/Biden-Harris-Administrations-National-Security-Strategy-10.2022.pdf.

Wolf, Larry. 1994. *Inventing Eastern Europe: The Map of Civilization on the Mind of the Enlightenment*. Stanford University Press.

Womack, Brantly. 2003. "Asymmetry and Systemic Misperception: The Cases of China, Vietnam and Cambodia during the 1970s." *Journal of Strategic Studies* 26(2): 91–118.

Womack Brantly. 2006. *China and Vietnam: the Politics of Asymmetry*. Cambridge University Press.

Zubok, Vladislav. 2007. *A Failed Empire: the Soviet Union in the Cold War from Stalin to Gorbachev*. University of North Carolina Press.

National Foreign Assessment Center. 1981. "USSR: The Cost of Aid to Communist States." Effective April 24, 2025. https://www.cia.gov/readingroom/docs/CIA-RDP83M00914R001900220025-5.pdf.

Анисимов, Максим. 2020. *Российская дипломатия и Семилетняя война*. Академический проект.

Гривач, Алексей and Андрей Денисов. "Успехи и неудачи энергетической сверхдержавы." *Россия в глобальной политике* 2. Effective April 24, 2025. https://globalaffairs.ru/articles/uspehi-i-neudachi-energeticheskoj-sverhderzhavy/.

Кабузан, Владимир.1990. *Народы России в XVIII веке. Численность и этнический состав*. Наука.

Левада-Центр. 2016. "Образ России на международной арене." Effective April 24, 2025. https://www.levada.ru/2016/11/14/obraz-rossii-na-mezhdunarodnoj-arene/.

Левада-Центр. 2020. "Великая держава." Effective April 24, 2025. https://www.levada.ru/2020/01/28/velikaya-derzhava/.

Марцинкевич, Борис. 2020. "Росатом—элемент концепции России как энергетической сверхдержавы." *Геоэнергетика.ru*. Effective April 24, 2025. https://www.geoenergetics.ru/2020/03/25/rosatom-element-koncepcii-rossii-kak-energeticheskoj-sverxderzhavy/.

Миних, Эрнст. 1891. *Россія и русскій дворъ въ первой половинѣ XVIII вѣка*. Русская старина.

Министерство иностранных дел Российской Федерации. 2005. "О заседании Совета Безопасности России с участием Президента России В.В.Путина." Effective April 24, 2025. https://mid.ru/ru/foreign_policy/news/1733401/.

Путин, Владимир. 1999. "Россия на рубеже тысячелетий." *Независимая газета*. Effective April 24, 2025. https://www.ng.ru/politics/1999-12-30/4_millenium.html.

Статистические таблицы Российской империи, издаваемые по распоряжению министра внутренних дел. Вып. 2. Наличное население империи за 1858 год. 1863.

Стейнберг, Джон. 2004. “Причины поражения русской армии в русско-японской войне: оперативная точка зрения.” In *Русско-японская война 1904–1905. Взгляд через столетие*, edited by Олег Айрапетов. Алгоритм.

Троицкий, Николай. 1997. “Внешняя политика 1879–1894 гг.: Россия и Союз 3-х императоров.” In *Россия в XIX веке: Курс лекций*. Высшая школа.

Уралов, Сергей. 2021. “Россия—гелиевая сверхдержава.” *SONAR 2025*. Effective April 24, 2025. https://www.sonar2050.org/publications/rossiya--gelievaya-sverhderjava/.

Хасанова, Марина. 2023. “Глобальное лидерство сверхдержавы: плюсы и минусы.” *Лаборатория аналитики ИАМП Дипакадемии МИД РФ*. Effective April 24, 2025. https://russiancouncil.ru/blogs/laiamp/globalnoe-liderstvo-sverkhderzhav-plyusy-i-minusy/.

Хрущев, Сергей. 2010. *Трилогия об отце, Т.2: Рождение сверхдержавы*. Время.

Чечулин, Николай. 1896. *Внѣшняя политика Россiи въ началѣ царствованiя Екатерины II, 1762–1774*. Типографiя Главнаго управленiя удѣлов.

Яровая, Татьяна and Данила Сидяков. 2020. “Россия и Китай—сверхдержавы 21-го века.” In *Россия и Китай: история и перспективы сотрудничества: материалы X международной научно-практической конференции*. Благовещенск—Хэйхэ.

CHAPTER 3

A Strong Weak State, or the Central Paradox of Russian State-Building

Dinissa Duvanova

1 Introduction

The Russian war in Ukraine gave rise to many questions about the economic and institutional foundation of autocratic aggression and the unexpected resilience of Russian autocracy under international and domestic pressures. When Russia invaded Ukraine in February 2022 and escalated threats to NATO members, many experts were desperately searching for signs of Russia's economic decline, regime breakdown, and the downfall of Vladimir Putin. Despite the hopes that economic sanctions, popular protests, and elite subversion will destabilize Putin's regime, two years into the war the Russian economy appears unscaled. The government effectively crushed the anti-war protests. The political regime withstood a coup attempt by a recalcitrant billionaire marching toward Moscow with his private army. In March of 2024, 88 percent of Russian voters cast their ballots in support of President Putin. It is unnerving to see the Russian dictatorship maintain a strong grip on civil society, the economy, and the armed forces. It becomes clear that Putin's regime's ability to mobilize resources and effectively pursue its domestic and foreign objectives will determine the future of the war and the region.

Such an unexpected resilience of Putin's regime constitutes the central paradox of Russia's "weak strong man" (Frye 2021). Francoise Dauce in this volume argues that academic narratives have long framed Russia as an inherently fragile state, underestimating its institutional adaptability and elite coordination mechanisms, contributing to misconceptions about Russia's expected wartime resilience. How did the allegedly dysfunctional institutions of kleptocratic governance mount a brutal military invasion and effective resistance to international sanctions and internal opposition to the war? Since the fall of the USSR research on the Russian economy and politics, has been documenting evidence of Russia's economic struggles, derailed institutional reforms, and bad governance. This evidence led to conclusions about the autocratic regime's weakness. Russia's resilience to external and internal pressures presents a serious challenge to the influential narratives that emerge from decades of scholarship.

 | DOI:10.1163/9789004747364_004

Such narratives were based on analyses of oligarchic capture, regulatory inefficiencies, electoral manipulations, and weak civil society. They portrayed Russia as a weak capacity state plagued by institutional failures and dependence on popular support and approval. Russian studies now face the formidable task of reconciling the demonstrated strength, effectiveness, and oppressiveness of the Russian state with the narratives of economic dysfunction, institutional weakness, and the existence of internal threats to autocratic stability.

The chapter reviews the major conclusions of the decades of research on the Russian political economy to understand why the war and the Russian state's resilience to its economic, social, and political effects have caught researchers off guard. I review the scholarship on Russian economic reforms, state-building, and sources of the regime's legitimacy to identify how the notions of Russia's economic and political weakness missed the key developments in its authoritarian state-building. I identify three dominating narratives that make it harder to grapple with the realities of Russia's wartime economy and politics. The first is the narrative of an inefficient, and rent-ridden economy. The second is the belief in the institutional weakness of the Russian state. The third is the story that roots Russian authoritarianism in genuine popular support. I argue that although these narratives advanced our understanding of the Russian political economy, they also prevented us from seeing more clearly the growing state capacity, institutional efficiency in achieving the economic goals of the political regime, and an elaborate repressive apparatus to undermine and suppress political dissent. The events that unravelled since February 2022 call for shifting our analytical perspectives towards the formal institutions of state action and state power and re-focusing our conceptual apparatus on the sources of effectiveness, stability, and flexibility of modern autocracies. The chapter concludes by speculating how the full recognition of Russian institutionalization and resilience may further advance the study of authoritarianism, economic statism, and mechanisms of political regimes' stability and survival.

This chapter contributes to the multifaceted perspective on Russia's institutional resilience, governance strategies, and ideological narratives discussed in this volume. My critique of the conventional wisdom that Russia is a weak state is collaborated by others arguing that Russian authoritarian governance model is strategically resilient and institutionally adaptive. Vladimir Gel'man's chapter, for instance, reinforces my arguments by demonstrating how elite cohesion and bureaucratic endurance sustain Russian autocracy, even in wartime conditions. Kirill Shamiev's chapter adds depth to this perspective by highlighting the role of the military-security complex in reinforcing authoritarian stability, while also acknowledging personal loyalties and corruption that undermine Russia's strategic decision-making. Ivan Kłyszcz and Katalin

Miklóssy, however, challenge the idea of a coherent, effective Russian state by revealing how Russian IR and Western academic discourse often mythologize the view of Russia's Great Power status, rather than acknowledge its weakness. Yuliia Kurnyshova and Ivan Gomza further complicate this chapter's argument by demonstrating that Russia's governance model is not only about repression and institutionalization but also deeply rooted in civilizational and Great Power narratives that sustain domestic legitimacy and justify military aggression. These ideological constructs, as Kłyszcz and Gomza argue, mask the regime's economic and military vulnerabilities, presenting an alternative view that suggests Russia's resilience is built more on perception management than on actual institutional strength. These latter contributions implicitly challenge my assertion that Russia is not a weak state, they also reveal internal contradictions, ideological distortions, and external misperceptions that nuance the understanding of Russian state power and authoritarian survival.

2 Statism and Economic Inefficiency

The first dominant narrative in scholarly accounts of Putin's Russia centers around the inefficient and underperforming Russian economy. It dates to the early years of the post-communist transition and scholarly identification of the major obstacles to Russia's prosperity and growth. From the Soviet Union, Russia inherited an economy run by the bureaucratic command-and-control method and a public bureaucracy that many considered too extensive, politicized, and unprofessional (Bessinger, 1988; Aslund, 1995, Rutland 2009). The collapse of the USSR ended the system of state economic management, in which the communist party intervened in enterprise-level decisions. Russia dismantled the central planning process, liberalized trade and capital flows, and privatized state property. The economy contracted, inflation soared, industrial production collapsed, and the living standards fell. During the 1990s, Russia experienced a devastating economic recession and found itself in a regulatory vacuum rapidly filled with anti-competitive and outright criminal business activities. The "wild" 1990s were notorious for their criminal racketeering, shadow economy, endless bureaucratic inspections, and dysfunctional courts burying business disputes in mountains of paperwork (Johnson, at al. 1998; Frye and Zhuravskaya 2000).

The state-markets dichotomy that underpins the neoliberal economic paradigm dominated the analysis of the Russian political economy of the day. It became central in shaping the economic failures narrative. From the neoliberal perspective, the Russian economy faltered because market reforms' trajectory was derailed by the elite's interest peddling and public demands for social spending (Shleifer and Treisman 2000). Large-scale privatization and inconsistent liberalization created multiple opportunities for fast enrichment for enterprise managers and government officials, leading to the rise of notorious "oligarchs" who amassed enormous ill-gotten riches and meddled in politics (Guriev and Rachinsky 2005; Ericson 2012; Schimpfössl 2017; Novokmet et al. 2018). The "hard" structural reforms were stalled by the early "winners" of the economic transition (Hellman 1998). Partial reforms distorted the markets. Continuing government involvement created wasteful rents and corruption. It also facilitated the fusion of political and economic powers (Hellman et al. 2000; Holmes 2006). The lack of the rule of law undermined property rights and prevented efficient resource allocation (Frye 2004). Insecure property rights led to private capital flight, asset stripping, and underinvestment.

The partial reform equilibrium was disrupted by the 1998 Asian financial crisis that brought Russia to default. The currency lost three-fourths of its pre-crisis value, cutting imports, jobs, and disposable incomes. Inadvertently, the default paved the road to economic recovery and highlighted the urgency of structural reforms to improve the business climate. When Prime Minister Putin appointed by President Yeltsin in 1999 won national elections and became the President in 2000, Russia spearheaded much-needed economic reforms (Rutland 2008). The reforms aimed at improving economic governance, modernizing the financial system, and attracting investment. By the mid-2010s, the Russian economy was well integrated into world markets and finance. The country embraced regulatory mechanisms to deal with private enterprise and its government adhered to conservative fiscal and monetary policies. Russia developed a modern banking and financial system, effective tax collection, and revamped corporate governance. In 2012 it joined the WTO. Yet, scholars continued to debate whether Russia could modernize (Ledeneva 2013).

While international development agencies recorded improvements in annual indicators of Russia's regulatory climate (World Bank 2020a; Shwab and Zahidi 2020), enterprise-level data and subnational analyses showed considerable inertia and continuing hurdles for business (Frye et al. 2009; Pyle 2011; Karas et al. 2015; Yakovlev 2006; Yakovlev and Zhuravskaya 2013). Although "hard" reforms moved forward, the overall assessment of Russian business climate and economic efficiency remained poor. Reflective of the rampant corruption, continuing harassment of the private sector by the political authorities, and self-serving economic policies of central and regional governments the scholars came to describe the Russian economy as "patrimonial capitalism" (Robinson 2011), "piranha capitalism" (Markus 2016), or crony capitalism (Aslund 2019). Under the increasingly authoritarian rule of President Putin the state capture of the 1990s gave rise to business capture by state officials. *Reiderstvo*, or hostile takeovers, with the assistance of state security agencies, replaced the racketeering of the 1990s (Rochlitz 2014; Markus 2016). The courts were now in cahoots with the corrupt state officials. Corruption-ridden and patronage-promoting bureaucracy served as the key mechanism of rent appropriation by the economic elites (Burkhardt and Libman 2018, Aslund 2019).

Yet, as the reforms of the 2000s were taking place, economic growth resumed, GDP nearly doubled, foreign direct investment (FDI) flowed in, and real disposable incomes in 2000–2008 grew 10% annually (Moscow Times 2019). From the neo-liberal governance perspective, the Russian economy was still underperforming. The government continued to over-regulate. Property rights remained insecure; rent opportunities undermined efficient resource allocation. Following the energy price boom of 2000–2008, Russia accumulated large currency reserves, which were used to create state holdings and acquire private companies (Goldman 2010). Institutional reforms, financial discipline, and growing economic outputs did not square well with the expanding state ownership, unfair business practices, and state predation on private business interests. It looked like Russian economic growth was defying the neoliberal ideas about the foundational principles of a healthy market economy.

One way to explain the economic growth Russia experienced after the turn of the century is the country's extensive reliance on natural resource exports (Goldman 2010, Rutland 2014, Rothacher 2021). This reliance on energy exports reinforced the notion of Russia's economic vulnerability—its dependence on global energy prices and existing energy markets. The second analytical tool in explaining the discrepancy between the luck of liberal markets and strong growth trajectory is institutional subversion and institutional adaptation. To give one example of how institutional subversion works to reconcile poor practices with good indicators, consider the research by Sharafutdinova and Dawisha (2017). These authors show that much of Russia's FDIs had been the returned domestic capital outflows whose owners resorted to institutional arbitrage to secure property rights abroad because of the lack of secure property rights at home.[1] Accounts of institutional subversion and oil dependence essentially placed the sources of Russia's economic performance in the global economic order. A logical conclusion was that Russia could not do well in isolation from the world economy.

1 Consider the implications of this finding for the effectiveness of Western financial sanctions. If a large portion of Russia's FDI had been of domestic origin, cutting the flow of foreign investment would not have a desired chilling effect on the Russian economy.

An opportunity to evaluate Russia's dependence on the global economic system came with the 2014 annexation of Crimea and the subsequent imposition of international sanctions. The sanctions coincided with a precipitous drop in global oil prices. Between 2014 and 2016 the price of crude dropped 70 percent. In 2015, the Russian economy contracted by two percent and the real income dropped by 7.1 percent (OECD 2020). The inflation-adjusted incomes declined for six years and in 2019 stood 11 percent below the 2013 level (Moscow Times 2020).[2]

External economic pressures turned the Russian leadership towards nationalism, import substitution strategy, and boosting agricultural production (Rutland 2023). As a result of these policy changes, the state's role in the economy grew stronger, and Russia's reliance on Western trade and investment weakened (Connolly 2018). Putin's regime's response to the international economic pressures in 2014–2016 reduced the economy's reliance on foreign investors and suppliers, ultimately contributing to Russia's economic resilience to external shocks. The neo-liberal critiques of Russia's statism, lack of competitiveness, and efficiency had missed the fact that state control of financial, industrial, and social service sectors had been a response to economic crises and was aimed at making the Russian economy more resilient to external pressures.

Moreover, thinking about Russian political-economic governance in terms of state-market dichotomies does not help us understand why in 2022 among the first government responses to the newly imposed sanctions was a package of measures for lifting regulatory oversight and tax burden on the small and medium enterprise (SME) sector. In March 2022, SMEs were granted interest payment deferrals, automatic business license extensions, and a moratorium on all planned inspections (Government Decrees #337, #353, and #336). Between March 1 and September 1, 2022, Russian tax authorities also suspended regulatory compliance enforcement, such as freezing bank accounts, inspections of foreign currency transactions, and insolvency liquidations. Further SME support measures included freezing the property tax rates, simplifying commercial transport licensing, and suspending previously planned changes in product labelling.

If the economic weakness narrative is correct, such measures as cutting the country off the international payment system, freezing foreign assets, discounting oil exports, and interrupting trade in high-technology goods should have

2 Gurvich and Prilepskii (2015), estimated that sanctions accounted for only one-third of the initial drop in Russia's GDP.

sent Russia into a spiral of financial meltdown, macroeconomic chaos, and collapse of production. The solid performance of Russia's wartime economy[3] has shown the value of dissenting arguments recognizing that strengthening markets can coexist with growing state authority (Wengle 2015). Such unorthodox accounts had seen the Russian economic policy as well-designed to serve the state-building goals (Miller 2018). Through economic reforms and political manoeuvres, Russian autocracy amassed an obedient clientele of state officials, law enforcers, and business leaders. It fine-tuned the amount of resource plundering by state officials to balance the political loyalties of corrupt bureaucrats and economic elites on the one hand and investments in economic growth and military might on the other (Duvanova 2023). Russian political-economic order has supported a much more agile autocracy than domestic democratic forces and the international community might have hoped for.

3 Informal Institutions and State Weakness

The second narrative that dominates our understanding of post-communist Russia is the narrative of state weakness and "bad governance" (Way 2015; Gel'man 2017 2022). Corruption, spotty regulatory enforcement, official predation, and informal power relations are conserved as the clear symptoms of state weakness. Driven by the desire to understand the institutional foundations of Russia's illiberal capitalism that combined international openness and deregulatory initiatives with the state's increasing economic footprint and widespread official predation, political scientists delved into the nature of political rule in Russia. Some suggested that Russia became yet another personalistic autocracy (Frye 2021) or neo-patrimonial state (Robinson 2011), like many other autocratic regimes worldwide. Personalistic autocrats conflate the personal economic interests of a narrow group of political elites with the national economic objectives and often run their countries as individual fiefdoms. Other scholars concentrated on the unique features of the Russian state and economy. They described Russia as being ruled by "patronal" politics (Hale 2014), a criminal kleptocratic cabal (Dawisha 2014), crony (Aslund 2019), or "piranha" capitalism (Markus 2016). Such accounts highlight peculiarly intertwined formal institutions of the modern state and informal clientele of economic and political

3 For the analysis of micro-economic and industrial policy mechanisms that helped sustain the Russian war-time economy see Yakovlev (2023a) and Yakovlev et al. (2023).

support. They tend to treat formal institutions of the Russian state as weak or compromised by the logic of autocratic greed and informal power exchanges.

As Aliaksei Kazharski highlights in his contribution to this volume, the tendency to label such informal institutions as uniquely Russian overlooks the fact that similar informal networks have played critical roles in the governance structures of many semi-peripheral states. From the patronage networks of post-communist Eastern Europe to the neo-patrimonial systems in parts of Southern Europe, informal governance mechanisms have been a common tool for maintaining elite control in hybrid regimes. In this light, Russia's reliance on informal power relations is not indicative of a weak state but rather reflects a well-adapted model of authoritarian governance that has historically allowed semi-peripheral states to navigate political and economic volatility.

The earlier works on post-communist transitions had cast their theories in terms of institutional failures and subversions (Bunce 1999; Treisman 2001; Hellman 1998). At the end of the 1990s, a review article by Cynthia Roberts and Thomas Sherlock concluded that state capacity lies at the root of Russia's "derailed" political-economic transition and called for "bringing the state back in" (Roberts and Sherlock 1999). In following this advice, the scholars of Putin's regime made major advancements in the study of informal institutional arrangements that undermined the logic and function of economic governance, state administration, and political contestation (Ledeneva 2013; Taylor 2018; Sharipova 2018). Corruption, predation, and clientelism have become regarded as the core mechanisms of Russia's institutional order. The literature scrutinized informal power relations that fostered interpersonal loyalties, collusive arrangements, and self-serving dealings. It's been argued that informal arrangement subverted the formal institutional makeup of legislative processes, public administration, and law enforcement. Corruption-ridden and patronage-promoting bureaucracy had been "an indispensable part of the mechanism of neo-patrimonial governance within the framework of the power vertical" (Gel'man 2016, p. 461). Scholars have documented bribe extortion schemes, patronage and nepotism in legislative and government positions, and high-ranking state officials' connections in private business (Radaev 2000, Ledeneva 2006, Dawisha 2014, Szakonyi 2019). These findings are often used to reinforce the notion of the poor quality of Russian governance and the low capacity of Russian state institutions.

In a larger comparative study of governance, state capacity is linked to better policy outcomes. Professional, rule-bound, and resourceful institutions are recognized as the necessary tools for effective economic policy. As scholars continued scrutinizing party-building (Reuter 2017), instrumental use of elections (Reuter et al 2016, Szakonyi 2020), legal protection of property rights

(Gans-Morse 2017a), tax regimes (Easter 2012; Berenson 2018), state security apparatus (Taylor 2011), and foreign policy (Busygina 2017) they necessarily engaged the issues of institutional effectiveness and state capacity. When Russian institution-building projects failed to live up to the expectations informed by the ideal type of liberal democracy, it was often seen as evidence of institutional weakness, rather than institution-building of a different sort. The narrative of the low-capacity institutions became so prominent, that decisions as mundane as reporting a crime to law enforcement constituted an empirical puzzle (Gans-Morse 2017b). While scholarly focus on the personalistic nature of power relations and weak institutions remains valuable, it should be balanced with recognition of significant state capacity-building efforts in certain sectors.

To be fair, it is hard to ignore the prominent role corruption plays in Russia. Political corruption features prominently in the academic and journalistic accounts of Russian politics. Insinuations, speculations, direct accusations, and prosecution of corruption have become so ubiquitous, that it is hard to find a prominent Russian politician immune to this curse. The list of high-ranking officials fleeing corruption investigations or serving sentences starts with Mayors Sobchak and Luzhkov; regional governors A. Khoroskavin (Sakhalin), V. Gayzer (Komi), N. Belykh (Kirov), V. Nakhlupin (Crimea); ministers A. Ulyukaev (Ministry of Economic Development), M. Abyzov (minister without portfolio), V. Ishaev (Minister for the Far East Development and ex-governor of Khabaovsk); legislators V. Belousov and R. Arashukov; not to mention a plethora of private sector operatives. Many of these cases have been politically motivated. The high-profile cases of Mikhail Khodorkovskii, and Sergey Furgal show that authorities frequently used criminal investigations of fraud, tax evasion, embezzlement, and even murder to silence those who challenge their monopoly on power (Sakwa 2014). The Transparency International corruption perception index continues to place Russia at the bottom of the world's anti-corruption ranking. In 2023 the country was ranked 141st out of 180. Yet, through Putin's era, the government continued to engage in anti-corruption initiatives and public administration reforms.

One of the most fundamental efforts toward improving state governance was Aleksei Kudrin's civil service reform. The 2004 Law on the State Civil Service and accompanying legislation codified formal educational and professional requirements, competitive merit-based hiring, clear hierarchically arranged and functionally defined positions, and institutional safeguards against undue political interferences in public service. The reform transformed the civil service in terms of personnel qualification, operational clarity, social prestige, and the desirability of public employment. Compared to the pre-reform period,

Russian bureaucrats have become more educated, better paid, disciplined, and more professional (Kalinin 2019; Fortescue 2020). Strict formal educational requirements were set forward for hiring. Promotion and career advancement were linked to formal performance indicators (Gimpelson 2009). The finding that Russian officials often manipulate their performance reports (Kalgin 2016) suggests that official performance indicators became consequential for rewards and promotion. Analysing the biographical data on high-level ministerial officials in the post-reform period, Fortescue (2020) found no evidence of patrimonial networks that would likely produce cascading effects during personnel reshuffling or promotions. A 100-fold increase in fines for bribery and personal asset disclosure requirements for public officials put in place during Medvedev's administration, made civil service more transparent and accountable (Klimenko 2014). Financial disclosure requirements resulted in fewer incumbents seeking re-election (Szakonyi 2023). These accounts directly questioned the narrative of dysfunctional institutions and the weak state.

The war in Ukraine presented yet another challenge to the state's institutional weakness narrative. The supposedly dysfunctional and failing institutions of the Russian state and inefficient Russian economy were able to sustain two years of costly unpopular military aggression in Ukraine. Even though the notion of the dysfunctional state squares well with the contradicting record of economic growth and corrupt governance, it does not help us understand the sources of the Russian state's resilience. A focus on personal power, nepotism, or corruption leads to erroneous expectations of Russia's rapid military failure, economic collapse, and institutional implosion. If the Russian dictatorship is held together by informal patronage or cabals of security apparatus and criminals, why the mechanisms of predation, lawlessness, and asset stripping did not implode with the imposition of sanctions and isolation from the West? How, despite the growing scarcity of resource rents, does the Russian state continue to effectively direct economic, administrative, and coercive resources? We cannot answer these questions without recognizing the impressive growth and strengthening of formal institutions of state governance and advances in building state capacity.

There is one caveat though. In the past decade capacity and effectiveness have replaced democracy and freedom as internationally recognized developmental objectives. They came to be regarded as normative goods, universally desirable improvements to the human condition. Russia's effective warmaking against Ukraine, however, clearly demonstrates that institutional capacity should not be treated as a normative good. The stronger state institutions in Russia might decrease the amount of unpredictability, promote order, and at times, enhance the public goods provision. Still, the primary objective for

Russia's capacity-enhancing institutional developments had been to serve the autocrat who effectively and efficiently suppresses freedoms, plunders national resources, prosecutes the regime's opponents, and wages war.

4 Regime's Dependence on Popular Support

The third narrative that came to dominate studies of Russian politics predicates regime survival on the continuing support and approval on the part of the Russian public. It sees Russian citizens as a force that has the potential to challenge and ultimately overthrow the government if the latter does not cater to public demands. This narrative builds on the belief that maintaining legitimacy is important because a) repression of dissent is costly and b) elites may use popular mobilization to challenge regimes' monopoly.

The precarious position of Russia's ruling elites who rose to power through corruption and theft of public resources undercuts the moral source of the regime's political legitimacy. Generally, economic inequalities tend to narrow the social bases for political regimes. The billions of dollars made by oligarchs in the rigged mid-1990s privatization and the astronomical salaries and personal assets of managers of state companies made people question the legitimacy of private wealth (Frye 2006) and disparage the kleptocratic regime. Yet the continuing social differentiation of the Putin era did not increase poverty and coincided with growing incomes (Remington 2011).

Studies of public opinion, voting, and protest behaviour have directly and indirectly shaped the prevailing notion that Russian autocracy depends on and is sustained by widespread popular support for Putin's regime. Alongside other essential mechanisms of autocratic regime building, such as electoral manipulations (Myagkov et al. 2009; Higashijima 2022), autocratic party building (Reuter 2017), and the use of media to shape collective consciousness (Sharafutdinova 2021), genuine popularity of Putin and support for the government had been recognized as the central features of regime stability.

An agreement emerged around the notion that the negative experiences of the 1990s recession undermined public support for democracy and aided the rise of Russian authoritarianism. In the 2000s, people experienced the opposite. During Putin's first two presidential terms, high energy prices and conservative macroeconomic policy produced impressive economic improvements. Good economic performance has been a strong predictor of Putin's popularity (Treisman 2011, Greene and Robertson 2017). Although much of the economic prosperity was due to the surge in global oil prices, this did not prevent the people from attributing good lives to autocratic rule. Opinion surveys and electoral results supported the conclusion that support for the government in the early 2000s was largely due to the good performance of the Russian economy.

The 2008 financial crisis and declining oil prices had a cooling effect. Putin's popularity declined slightly (Treisman 2014). But such an effect was short-lived. To explain continuing support for the incumbent government in times of slower economic growth, economic setbacks, and international sanctions against the Russian annexation of Crimea the literature focused on the rise of nationalist rhetoric, propaganda, and rallying around the flag effect (Hale 2018; Frye 2019, Greene and Robertson 2019). As the autocratic state was losing its ability to buy off popular approval with generous federal spending, it seemed to shift to the new legitimation strategy. By fostering a sense of nationalism, and vilifying internal and external critics, autocracy justified further rollback of political rights and civil liberties. The pro-regime propaganda had been proven effective in maintaining high approvals. After the 2022 sanctions devalued the currency, restricted consumer choice, ability to travel internationally, and put millions in danger of military conscription, Putin's approval remains strong (Frye 2023).[4] Even the 2023 coup attempt did not affect the levels of regime support (Zakharov 2024).

Analyses of the flip side of government approval—protest behaviour—had also contributed to the narrative of regime dependence on social support. In the post-Soviet period, Russian citizens often protested to express their economic and political grievances and to place demands on the authorities (Robertson 1010, Lankina 2016). Researchers explored the connection between protest behaviour, elite strategies, and electoral politics and suggested that protests may have enhanced the legitimacy of autocratic government (Robertson 1010, Smyth et al 2013; Frye and Borisova 2019, Owen 2020). By allowing protests to proceed, authorities maintain the semblance of participatory politics and constitutional order. An in-depth analysis of Russian labour relations by Crowley (2021) concluded that the fear of labour protests profoundly affected regimes' policies and institutional choices, reinforcing the notion that autocratic survival depends on popular support. Yet, the recent analysis of Russian protests found that they generally fail to bolster support for the opposition (Tertyshnaya 2020), calling into question the underlying notion that Russian autocracy is vulnerable to popular domestic challenges.

4 One can legitimately doubt the validity of survey research methods in autocracies that repress the free expression of political opinion and criminalize antiregime opinions. There is also evidence that perceptions of popularity that might be deliberately shaped by the state-controlled media may affect Putin's approval rating (Buckley et al. 2024).

The 2012 mass arrest in Bolotnaya Square in Moscow, and foreign intervention in Georgia, Syria, and Ukraine in 2014 evidenced the growing capacity of Russia's coercive power and political resolve to use it to achieve domestic and foreign policy objectives. Russia's response to the antiregime protests in 2011 and 2021 used a mixture of police brutality, intimidation of the opposition, and undercutting the organizational bases of civil society. The research found that pre-emptive repression in Russia effectively undermined protest mobilization (Tertytchnaya 2023). This should caution us to see the use of oppression as "merely a sign that other tools for keeping Putin in power are failing" (Frye 2021).

The latest wave of antiwar protests and their suppression have shown that the Russian state had amassed sufficient repressive power to maintain Putin's regime in the face of declining economic rents and a disillusioned populace.[5] After the start of the 2022 war, a wave of antiwar protests has swept across the country. Observers saw this as an indication of Putin's weakness and imminent decline (Laruelle 2022). Facing growing opposition to the war, the government criminalized antiwar speech and detained 18,910 activists in 2022 (Ovid-info 2024). As a result of increasing repression, protest activity nearly disappeared in 2023. When in February 2024 the authorities announced Aleksey Navalny had died in an arctic penal colony, Russian authorities arrested 387 people who attempted to publicly mourn the opposition leader's death (Amnesty International 2024).[6] While historically protests in Russia spiked during electoral periods, the 2024 presidential election saw no uptake in the protest activity. Despite the strong belief that Putin's regime depends on the voluntary compliance and support of the Russian people, repression has been the key to its survival in 2022–2024.[7] Interpreting the use of repressive methods as a sign of regimes' weakness prevented us from recognizing the rise of repressive tactics as essential mechanisms of autocratic rule in Russia.

---

5 Chapters by Andrey Makarychev and Kirill Shamiev in this volume further delve into the increasingly repressive nature of Russian state. Considering Makarychev's argument, Russia's autocratic survival is neither accidental nor merely institutional—it is deeply embedded in the state's ability to dictate the boundaries between citizens' life and death, normalcy and emergency. Shamiev's chapter details the essential role of the state's reliance on the security services to enforce stability and regime survival, even in the face of military failures.

6 The news about Navalny's detention in January 2021, in contrast, galvanized over 350 separate protest events around the country bringing out tens of thousands of protesters. Over 3000 people were imprisoned then (Human Rights Watch 2021).

7 It should be recognized that the declining public support for Putin's regime for the most part did not result in open dissent but instead manifested in social atomization and depoliticiz-

Conclusion

Russia's political-economic order underpins the regime's power to engage in military aggression, withstand Western sanctions, and suppress popular discontent at home. Two years of the war that caused the most devastating destruction of human life, physical infrastructure, and natural resources in Europe since WWII, have demonstrated that despite the decades of predatory and wasteful economic governance, cronyism, corruption, and institutional subversion, Russian autocracy had emerged as a high-capacity state resilient to internal and external challenges. The chapter by Vladimir Gel'man in this volume tells us that the war has exposed the adaptability of Russian autocracy, where institutional adjustments and elite coordination have effectively compensated for external pressures. As Kirill Shamiev's chapter in this volume highlights, Russia's resilience is not just a matter of institutional durability but also of militarized governance that paradoxically uses personal loyalty to reinforce central civilian control over security forces. Sharing these author's recognition of somewhat unexpected Russia's institutional resilience, this chapter surveyed some central themes emerging from influential academic accounts of the political economy of Putin's Russia. Since the turn of the century, while scholars debated the status of its market reforms and political change, Russia continued to build a robust and adaptable form of authoritarianism that effectively withstood windfalls and shortages of economic rents and balanced elements of liberal and statist economic orders. In this chapter I identified the elements of scholarly discourses that directly or indirectly contributed to the notion that the Russian state, under the surface of its oil wealth and newly discovered nationalist pride, is a corruption-ridden dysfunctional institutional shell that enables self-serving elites to plunder natural wealth with impunity. By omitting many insightful details and valuable contributions I tried to stay focused on the overlooked contradictions of Russian state-building. My task was to understand why Russia's ability to fight a protracted war came as a surprise to many. I structured my review of Russia's supposed weaknesses into three interrelated and mutually reinforcing narratives of economic dysfunctionality, institutional weakness, and popular bases of regime support. I made three arguments.

First, the dichotomy between efficient markets and wasteful states is not particularly conducive to understanding the Russian political economy. Recent

---

ation, limiting the potential for collective resistance. This further highlights the fact that the recent intensification of repression is not driven primarily by the intensification of protest, but rather by the changes in the core mechanisms of political regime survival.

history has challenged the idea that a statist economy is necessarily incompatible with stimulating private entrepreneurship. The dichotomy between market and state raises more questions than it solves in the analysis of modern autocratic economic policy. To better understand the political economies that combine features and outcomes usually seen as fundamentally incompatible, we need to revamp our conceptual apparatus and theoretical understanding of modern capitalism.

Second, formal institutional mechanisms that underpin informal power relations require more attention. The first step is to recognize that Putin's plutocratic rule and oppression had been comfortably coinciding with drastic improvements in the state's capacity and formal institution-building. Formal institutions (federal bureaucracy, provincial and local governments, security agencies, military, and public education) can and should be analysed through the lenses of their explicitly defined functions, resources, and powers. This does not mean that the informal power relations should be disregarded. Yet, we cannot ignore the formal mechanisms of governance simply because we know that officials may cut corners or violate norms and institutional policies.

Third, if we want to understand the role of citizens in autocratic regimes, we need to pay closer attention to the relationship between collective attitudes and behaviour. Brutal repressions and state monopoly on information in wartime Russia invalidated theoretical shortcuts connecting individual attitudes and actions to collective outcomes. In the times when free expression is criminalized and behaviours are self-censored, we desperately need innovative approaches to study mechanisms of repression and the psychology of adaptation, compliance, and moral compromise (Yaffa 2020). When the state actively shapes individual attitudes and political choices, we must reconsider the political status of public opinion, protest politics, and electoral behaviour. The first step is to recognize the effectiveness of the state repressive apparatus that operates not only through specialized state security agencies, police, courts, and other "power" agencies, but also through the media, local authorities, educational institutions, and economic enterprises.

Ultimately, we need to go beyond seeing Russia's illiberalism, neo-patrimonialism, and repression as deficiencies. Regrettably, all good things do not necessarily come together. We must recognize that oppressive dictatorships can be agile, adaptable, and effective in achieving their goals. This recognition may bring about a two-fold advantage. In the short run, it may lead to a better understanding of the underlying causes of autocratic resilience. Russia's neo-patrimonialism and unruly bureaucracy had been the core components of the long-term survival of its strong and resilient authoritarian state. There is also a good chance autocratic rule will outlast Putin's tenure in office in the same way authoritarian regimes outlasted the personal rule of Nursultan Nazarbayev in Kazakhstan, Heydar Aliyev in Azerbaijan, and Saparmurat Niyazov in Turkmenistan. In the long run, we may discover various ways in which political regimes constrain policy and institutional choices of autocratic rulers. This is essential for understanding any possible limitations of Russia's autocratic rule.

Bibliography

Amnesty International. 2024. "Russia: Authorities Brutally Suppress Mourners of Aleksey Navalny." February 19. https://www.amnesty.org/en/latest/news/2024/02/russia-authorities-brutally-suppress-mourners-of-aleksei-navalny/.

Aslund, Anders. 1995. *How Russia Became a Market Economy*. Brookings Institution Press, Washington, DC.

Aslund, Anders. 2019. *Russia's Crony Capitalism: The Path from Market Economy to Kleptocracy*. Yale University Press, New Haven, CT.

Berenson, Marc P. 2018. *Taxes and Trust: From Coercion to Compliance in Poland, Russia, and Ukraine*. Cambridge University Press, Cambridge, UK.

Buckley, N., Marquardt, K.L., Reuter, O.J., and Tertytchnaya, K. 2024. Endogenous Popularity: How Perceptions of Support Affect the Popularity of Authoritarian Regimes. *American Political Science Review* 118 (2): 1046–1052.

Bunce, Valerie. 1999. *Subversive Institutions: The Design and the Destruction of Socialism and the State*. Cambridge University Press, Cambridge, UK.

Busygina, Irina. 2017. "State-building in Russia and the Choice for Coercion in External Relations." In *Russia–EU Relations and the Common Neighborhood*, edited by Irina Busygina. Routledge, Abingdon, UK.

Connolly, Richard. 2018. *Russia's Response to Sanctions: How Western Economic Statecraft Is Reshaping Political Economy in Russia*. Cambridge University Press, Cambridge, UK.

Crowley, Stephen. 2021. *Putin Labor Dilemma: Russian Politics between Stability and Stagnation*. Cornell University Press, Ithaca, NY.

Dawisha, Karen. 2014. *Putin Kleptocracy: Who Owns Russia?* Simon and Shuster, New York, NY.

Duvanova, Dinissa. 2023. *Thieves, Opportunists, and Autocrats: Building Regulatory States in Russia and Kazakhstan*. Oxford University Press.

Easter, Gerald. 2012. *Capital, Coercion, and Postcommunist States*. Cornell University Press, Ithaca, NY.

Ericson, Claes. 2012. *The Oligarchs: Money and Power in Capitalist Russia*. Stockholm Text; English Edition, Stockholm, SE.

Fortescue, Stephen. 2020. "Russia's Civil Service: Professional or Patrimonial? Executive-Level Officials in Five Federal Ministries." *Post-Soviet Affairs* 36 (2): 1–24.

Freinkman, Lev and Yakovlev, Andrei. 2015. "Institutional Frameworks to Support Regulatory Reform in Middle-Income Economies: Lessons from Russia's Recent Experience." *Post-Communist Economies* 27 (3): 354–369.

Frye, Timothy. 2002. "Private Protection in Russia and Poland." *American Journal of Political Science* 46 (3): 572–584.

Frye, Timothy. 2004. "Credible Commitment and Property Rights: Evidence from Russia." *American Political Science Review* 98 (3): 453–466.

Frye, Timothy. 2006. "Original Sin, Good Works, and Property Rights in Russia." *World Politics* 58 (July): 479–504.

Frye, Timothy. 2017. *Property Rights and Property Wrongs*. Cambridge University Press, Cambridge, UK.

Frye, Timothy. 2019. "Economic Sanctions and Public Opinion: Survey Experiments from Russia." *Comparative Political Studies* 52 (7): 967–994.

Frye, Timothy. 2021. *Weak Strongman: The Limits of Power in Putin's Russia*. Princeton University Press, Princeton, NJ.

Frye, Timothy and Borisova, Ekaterina. 2019. "Elections, Protest, and Trust in Government: A Natural Experiment from Russia." *The Journal of Politics* 81 (3): 820–832.

Frye, Timothy, S. Gehlbach, K. Marquardt, OJ. Reuter. 2023. "Is Putin's Popularity (Still) Real? A Cautionary Note on Using List Experiments to Measure Popularity in Authoritarian Regimes." *Post-Soviet Affairs* 39 (3): 213–222.

Frye, Timothy and Zhuravskaya, Ekaterina. 2000. "Rackets, Regulation, and the Rule of Law." *The Journal of Law, Economics, and Organization* 16 (2): 478–502.

Frye, T., Yakovlev, A., and Yasin, Y. 2009. "The 'Other' Russian Economy: How Everyday Firms View the Rules of the Game in Russia." *Social Research* 76 (1): 29–54.

Gans-Morse, Jordan. 2017a. "Demand for Law and The Security of Property Rights: The Case of Post-Soviet Russia." *American Political Science Review* 11 (1): 338–359.

Gans-Morse, Jordan. 2017b. *Property Rights in Post-Soviet Russia: Violence, Corruption, and the Demand for Law*. Cambridge University Press, Cambridge, UK.

Gel'man, Vladimir. 2016. "The Vicious Circle of Post-Soviet Neopatrimonialism in Russia." *Post-Soviet Affairs* 32 (5): 455–473.

Gel'man, Vladimir. 2017. "Political Foundations of Bad Governance in Post-Soviet Eurasia: Towards a Research Agenda." *East European Politics* 33 (4): 496–516.

Gel'man, Vladimir. 2022. *The Politics of Bad Governance in Contemporary Russia*. University of Michigan Press.

Gimpelson, Vladimir, Magun, Vladimir, and Brym, Robert. 2009. "Hiring and Promoting Young Civil Servants: Weberian Ideals Versus Russian Reality." In *Russian Bureaucracy and the State: Officialdom from Alexander III to Vladimir Putin*, edited by D.K. Rowney and E. Huskey. Palgrave Macmillan.

Greene, S., and Robertson, G. 2017. "Agreeable Authoritarians: Personality and Politics in Contemporary Russia." *Comparative Political Studies* 50 (13): 1802–1834.

Goldman, Marshall I. 2010. *Petrostate: Putin, Power, and the New Russia*, Oxford University Press.

Guriev, Sergei and Rachinsky, Andrei. 2005. "The role of Oligarchs in Russian Capitalism." *Journal of Economic Perspectives* 19 (1): 131–150.

Gurvich, Evsei and Prilepskii, Il'ia. 2015. "The Impact of Financial Sanctions on the Russian Economy." *Russian Journal of Economics* 1 (4): 359–385.

Hale, Henry. 2014. *Patronal Politics: Eurasian Regime Dynamics in Comparative Perspective*. Cambridge University Press, Cambridge, UK.

Hale, Henry. 2018. "How Crimea Pays: Media, Rallying 'Round the Flag, and Authoritarian Support." *Comparative Politics* 50 (3): 369–380.

Hellman, Joel S. 1998. "Winners Take All: The Politics of Partial Reform in Postcommunist Transitions." *World Politics* 50 (2): 203–234.

Hellman, Joel S., Jones, Geraint, and Kaufmann, Daniel. 2000. "Seize the State, Seize the Day: State Capture, Corruption, and Influence in Transition." Research working paper WPS 2444. World Bank Group. http://documents.worldbank.org/curated/en/537461468766474836.

Holmes, Leslie. 2006. *Rotten States: Corruption, Post-Communism, and Neoliberalism*. Duke University Press, Durham, NC.

Human Rights Watch. 2021. "Russia: Police Detain Thousands in Pro-Navalny Protests." January 25. https://www.hrw.org/news/2021/01/25/russia-police-detain-thousands-pro-navalny-protests.

Johnson, Simon, Kaufmann, Daniel, and Shleifer, Andrei. 1998. "The Unofficial Economy in Transition." *Brookings Papers on Economic Activity*, 2: 159–239.

Kalgin, Alexander. 2016. "Implementation of Performance Management in Regional Government in Russia: Evidence of Data Manipulation." *Public Management Review* 18 (1): 110–138.

Kalinin, Aleksey. 2019. "On the Reform of the Executive Power and the Structures of the Executive Bodies in the Russian Federation." *Obshchestvo i ekonomika* 5: 78–86.

Lankina, Tomila and Skovoroda, Rodion. 2016. "Regional Protest and Electoral Fraud: Evidence from Analysis of New Data on Russian Protest." *East European Politics* 33 (2): 253–274.

Laruelle, Marlene. 2022. "Putin is in trouble." *New York Times*. September 22.

Ledeneva, Alena. 2006. *How Russia Really Works: The Informal Practices that Shaped Post-Soviet Politics and Business*. Cornell University Press, Ithaca, NY.

Ledeneva, Alena. 2013. *Can Russia Modernise? Sistema, Power Networks and Informal Governance*. Cambridge University Press, Cambridge, UK.

Markus, Stanislav. 2016. *Property, Predation, and Protection: Piranha Capitalism in Russia and Ukraine*. Cambridge University Press, Cambridge, UK.

Miller, Chris. 2018. *Putinomics: Power and Money in Resurgent Russia*. University of North Carolina Press.

Moscow Times. 2019. "Long Read: 20 Years of Russia's Economy Under Putin, in Numbers. Ben Aris and Ivan Tkachev explore how Russia has been transformed under Putin's leadership by going through the numbers." August 19.

Owen, Catherine. 2020. "Participatory Authoritarianism: From Bureaucratic Transformation to Civic Participation in Russia and China." *Review of International Studies* 46 (4): 415–434.

Radaev, Vadim. 2000. "Corruption and Violence in Russian Business in the Late 1990s." In *Economic Crime in Russia, edited by* A. Ledeneva and M. Kurkchiyan. Kluwer Law International, Alphen ann den Rjin, NL.

Remington, Thomas. 2011. *The Politics of Inequality in Russia*. Cambridge University Press.

Reuter, Ora John. 2017. *The Origins of Dominant Parties: Building Authoritarian Institutions in Post-Soviet Russia*. Cambridge University Press, New York, NY.

Reuter, O.J., Buckley, N., Shubenkova, A., and Garifullina, G. 2016. "Local Elections in Authoritarian Regimes: An Elite-Based Theory with Evidence from Russian Mayoral Elections." *Comparative Political Studies* 49 (5): 682–697.

Roberts, Cynthia and Sherlock, Thomas. 1999. "Bringing the Russian State Back in: Explanations of the Derailed Transition to Market Democracy." *Comparative Politics* 31 (4): 477–498.

Robertson, Graeme B. 2010. *The Politics of Protest in Hybrid Regimes: Managing Dissent in Post-Communist Russia*. Cambridge University Press, New York.

Robinson, Neil. 2011. "Russian Patrimonial Capitalism and the International Financial Crisis." *Journal of Communist Studies and Transition Politics* 27 (3–4): 434–455.

Robinson, Neil. 2013. "Russia's Response to Crisis: The Paradox of Success." *Europe-Asia Studies* 65 (3): 450–472.

Rochlitz, Michael. 2014. "Corporate Raiding and the Role of the State in Russia." *Post-Soviet Affairs* 30 (2–3): 89–114.

Rothacher, Albrecht. 2021. *Putinomics: How the Kremlin Damages the Russian Economy*. Springer, Cham.

Rutland, Peter. 2008. "Putin's Economic Record: Is the Oil Boom Sustainable?" *Europe-Asia Studies* 60 (6): 1051–1072.

Rutland, Peter. 2009. *The Politics of Economic Stagnation in the Soviet Union: The Role of Local Party Organs in Economic Management* (2nd ed.). Cambridge Russian, Soviet, and Post-Soviet Studies Series. Cambridge University Press, Cambridge, UK.

Rutland, Peter. 2014. "Petronation? Oil, Gas, and National Identity in Russia." *Post-Soviet Affairs* 31 (1): 66–89.

Rutland, Peter. 2023. "The Contradictions in Putin's Economic Nationalism: From Western Partner to Fortress Russia." *Russian Politics* 8 (1): 24–47.

Sakwa, Richard. 2014. *Putin and the Oligarch: The Khodorkovsky—Yukos Affair.* I.B. Tauris, London, UK.

Sharafutdinova, Gulnaz and Dawisha, Karen. 2017. "The Escape from Institution-Building in A Globalized World: Lessons from Russia." *Perspectives on Politics* 15 (2): 361–378.

Sharafutdinova, Gulnaz. 2021. *The Red Mirror: Putin's Leadership and Russia's Insecure Identity.* Oxford University Press, New York.

Shwab, Klaus and Zahidi, Saadia. 2020. *Global Competitiveness Report 2020: How Countries Are Performing on the Road to Recovery.* World Economic Forum, Geneva, CH.

Smyth, Regina, Sobolev, Anton, and Soboleva, Irina. 2013. "A Well-Organized Play." *Problems of Post-Communism* 60 (2): 24–39.

Szakonyi, David. 2019. "Princelings in the Private Sector: The Value of Nepotism." *Quarterly Journal of Political Science* 14 (4): 349–381.

Szakonyi, David. 2020. *Politics for Profit: Business, Elections, and Policymaking in Russia.* Cambridge University Press, Cambridge, UK.

Szakonyi, D. 2023. "Indecent Disclosures: Anticorruption Reforms and Political Selection." *American Journal of Political Science* 67 (3): 503–519.

Taylor, Brian D. 2011. *State Building in Putin's Russia.* Cambridge University Press, New York, NY.

Taylor, Brian D. 2018. *The Code of Putinism.* Oxford University Press, New York, NY.

Tertytchnaya, K. 2020. "Protests and Voter Defections in Electoral Autocracies: Evidence from Russia." *Comparative Political Studies* 53 (12): 1926–1956.

Tertytchnaya, K. 2023. "'This Rally is not Authorized': Preventive Repression and Public Opinion in Electoral Autocracies." *World Politics* 75 (3): 482–522.

Treisman, Daniel. 2001. *After a Deluge: Regional Crises and Political Consolidation in Russia.* The University of Michigan Press, Ann Arbor, MI.

Treisman, Daniel. 2011. "Presidential Popularity in a Hybrid Regime." *American Journal of Political Science* 55 (3): 590–609.

Treisman, D. 2014. "Putin's Popularity Since 2010: Why did Support for The Kremlin Plunge, Then Stabilize?" *Post-Soviet Affairs* 30 (5): 370–388.

Way, Lucan. 2015. *Pluralism by Default: Weak Autocrats and the Rise of Competitive Politics.* John Hopkins University Press, Baltimore, MD.

Wengle, Susanne. 2015. *Post-Soviet Power: State-led Development and Russia's Marketization.* Cambridge: Cambridge University Press.

World Bank. 2020. *Doing Business 2020: Comparing Business Regulation in 190 Economies.* World Bank, Washington, DC.

Yaffa, Joshua. 2020. *Between Two Fires: Truth, Emotion and Compromise in Putin's Russia.* Tim Duggan Books, New York.

Yakovlev, Andrey. 2006. "The Evolution of Business–State Interaction in Russia: From State Capture to Business Capture?" *Europe-Asia Studies* 58 (7): 1033–1056.

Yakovlev, Andrey. 2023. "Strategii biznesa kak factor stabilizatsii Rossiiskoi ekonomiki." *Sotsiodigger* 4 (25): 3–11.

Yakovlev, Andrey, Freinkman, L., Yershova, N., and Agalian A. 2023. "Otsenka effektivnosti instrumentov promyshlennoi politiki v Rossii: novye podkhody." *Eko* 2: 60–83.

Yakovlev, Evgeny and Zhuravskaya, Ekaterina. 2013. "The Unequal Enforcement of Liberalization: Evidence from Russia's Reform of Business Regulation." *Journal of the European Economic Association* 11 (4): 808–838.

Zakharov, Alexei, Ora John Reuter, Vladimir Shuklin, and Denis Volkov. 2024. "Effects of a Coup Attempt on Public Attitudes Under Autocracy: Quasi-Experimental Evidence from Russia." *Post-Soviet Affairs* 40 (6): 1–9.

CHAPTER 4

Against the Grain

Research on Russian Domestic Politics after February 2022

Vladimir Gel'man

The Russian military assault on Ukraine in February 2022 caused many enormously negative consequences for both domestic and international developments in Russia, Ukraine, and Europe, ranging from extraordinary human losses and war-driven mass migration to dramatic changes in various aspects of security and the major rise of the military-industrial complex. Among many other issues, this assault greatly affected global academia in many ways, putting both Russia and Ukraine in the spotlight and generating many discussions among scholars, experts and policy makers, to say nothing of politicians, pundits, and activists. At the same time, the very environment of research on Russia underwent major restructuring after February 2022, although some tendencies had emerged even before this turning point. The impact of the assault was similar to what I labeled an "exogenous shock" in one of my previous publications (Gel'man, 2023). While some elements of this restructuring were directly caused by war-related political changes, others reflected broader trends in academia, which severely affected Russian studies and accelerated amid the ongoing war. Some contentious issues in Russian studies were already being discussed among scholars during the period after February 2022 (Bogdanova et al., 2023; Conversations with the Field, 2023; How to Study Russia, 2023), although the context, content and style of these discussions varied greatly in different academic communities.[1] This article is focused on further developments in the field from the perspective of analysis of scholarly responses to recent political, institutional, and scholarly challenges to Russian studies, which have come from various corners. These challenges include questions about the validity not only of certain information sources (such as official economic statistics and survey research), but also of theoretical frameworks, which eventually become outdated and/or require many serious revisions and modifications. Post-2022 Russian studies have faced a major decline in Russian academia (related to

---

1 On French and Finnish discussions, respectively, see chapters by Francoise Dauce and Katalin Miklossy in this volume.

 | DOI:10.1163/9789004747364_005

institutional crisis, breaking ties between Western and Russian scholarship, and displacement of Russian scholars), increasing problems of data gathering and analysis under conditions of war, the increasing disjuncture between supply and demand in policy-related research on contemporary Russia, and numerous other issues. The problem is not only how scholars should respond to these and other challenges, but also to what extent such responses may help to answer many important research questions regarding Russian politics in the future. This article aims to shed new light on these and related questions.

Two important disclaimers are necessary. First, in this article I deal exclusively with research on Russian politics and do not address research on politics in other states of post-Soviet Eurasia and Eastern Europe.[2] In fact, the consequences of the ongoing war include not only political but also geographical fragmentation of the region, and the very label "post-Soviet" has become less and less useful as an umbrella term for the various former Soviet republics. Indeed, paths of scholarly research on, say, Russian, Armenian and Kazakhstani (not to mention Ukrainian) politics have now greatly diverged theoretically, empirically, and methodologically. Second, in this article I mostly concentrate on research on Russian domestic politics, which is my primary field of interest, and do not devote much attention to research on dynamics of Russian foreign policy, which would require a separate analysis (for an overview, see Polianskii, 2024).[3]

The structure of this article is the following. After a brief historical excursion into the past and present of Russian studies, I will concentrate on analysis of recent discussions about the domestic sources of the Russian military assault on Ukraine and its causes and effects, as well as on the scholarly search for the roots and remedies of the ongoing war. Then, I will briefly cover some new methodological and empirical developments in research on Russian domestic politics and discuss the need for new frameworks for analysis of Russian politics in a broader theoretical and comparative perspective.[4] Some suggestions and recommendations for further research on Russian domestic politics will be outlined in the conclusion of the article.

2 On Polish and Hungarian studies, see the chapter by Katalin Miklossy in this volume.
3 See also chapters by Ivan Gomza and Ivan U. Kłyszcz in this volume.
4 On these issues, see also the chapter by Dinissa Duvanova in this volume.

1 Then and Now: The Changing Trajectory of Russian Studies

Historically, Russian studies emerged as a separate interdisciplinary scholarly field in the United States, and later on in Western Europe, during the period of the Cold War due to the policy-driven demand of "know your enemy" (Engerman, 2009). Over time, this field established itself under the rubric of Sovietology. Western researchers produced quite a number of very important scholarly pieces despite limited access to data and relative isolation from mainstream political science as well as huge concentration on security-related issues (Breslauer, 1992). However, Sovietologists not only failed to expect and explain major changes in the Soviet Union during Gorbachev's perestroika but were also not ready to understand the logic behind the Soviet collapse and many other associated processes. It is no wonder that in the 1990s Sovietology was heavily discredited and soon disappeared (Rutland, 1993). The new field of post-Soviet Russian studies emerged amidst several contradictory trends. On the one hand, as Russia was no longer considered an "enemy" in the West, policy-driven interest in the field declined, and several research centers and major projects shrunk if not ceased to exist. In addition, the global refocusing of political science research from area studies to large-N comparative analyses in the 1990s (Bates, 1997) left little room for Russian studies against the background of increasing interest in other regions, ranging from China to the Middle East. On the other hand, the new generation of scholars which now dominates the field of research on post-Soviet Russian politics, and which was trained in comparative political studies (rather than in area studies), was able to effectively use advanced research methods and techniques of data analysis. They successfully seized the emerging opportunities of political openness and data availability in Russia for new scholarly projects, which have resulted in many notable publications (La Lova, 2023). Russia-based scholars as well as certain Russia-born representatives of a new global academic diaspora also successfully contributed to the field, and, despite numerous problems (Gel'man, 2015a), there was a clearly visible rise in political research in Russian academia in the 2000s and 2010s. These promising changes gave birth to numerous collaborative projects. Among them, one should mention the Laboratory of Comparative Social Research,[5] a part of the global World Values Survey network (previously led by the late Ronald F. Inglehart), based in the Higher School of Economics (HSE), and the International Center for the Study of Institutions and Development (initially, a joint venture between the HSE and the Harriman Institute

5 See https://lcsr.hse.ru/en/ (accessed 28 April 2025).

at Columbia University).[6] Both projects became highly visible and productive scholarly enterprises. This is why Timothy Frye's (2017) bold statement that Russian studies were thriving rather than dying, and that the scholarly outputs of this field demonstrated great academic success vis-à-vis other area studies, was not an exaggeration.

The Russian military assault on Ukraine in February 2022 seriously undermined if not completely overturned these tendencies. First, it ended the age of political openness and data availability in Russia and put many research projects on Russian politics into question. For example, the Survey of Russian Elites, the major longitudinal study of attitudes of Russian foreign policy elites conducted by American scholars in collaboration with Russian pollsters from 1993 to 2020 (Rivera, 2020), was forced to stop as there was no longer any way to interview elite representatives in Russia, who became unavailable for conversations during the war. Second, Russian scholarship in social sciences in general and in political science in particular faced major institutional and personnel losses during the war—some research centers were either closed[7] or changed their profiles, and many leading scholars left Russia for indefinitely long (Al'bitskii, 2024). Still, some talented Russia-based scholars continued to produce first-class research on Russian domestic politics under these worsening conditions (Golosov, 2024; Kynev, 2024); however, the overall major decline of Russian political science (as well as of other social science disciplines) is most probably irreversible (Zavadskaya, Gerber, 2023). Amid the increasing complexity of academic cooperation between Western scholars and their Russian partners (Kangas et al, 2023), and a very problematic lack of opportunities to conduct fieldwork in Russia, the use of intermediaries in Kazakhstan or Armenia to maintain international contacts and the exchange of ideas seems like an imperfect substitution at best. Due to these changes, it is unsurprising not only that some Russia-based scholars and experts have left the field but also that there has been a visible decrease in interest in research on Russia among many Western scholars—some of them even chose to shift their scholarly focus to other areas, such as Central Asia.

Aside from these problems, many of which appear irresolvable at least as of yet, there is a major need for substantive changes in Russian studies. Many scholarly concepts and approaches which dominated the scene of research on Russian politics over recent decades have become outdated both in factual and

---

6 See https://iims.hse.ru/en/csid/ (accessed 28 April 2025).

7 For example, the Center for Comparative History and Political Studies at Perm State University was closed in April 2022, and the Department of Political Science at the European University at St. Petersburg was closed in June 2024.

in substantive terms. There are virtually no scholars who continue to consider Russia to be a "normal country" (Shleifer, Treisman, 2004, 2014) after February 2022. Even the recent argument that Russia's political regime has turned from the standard form of authoritarianism ("fear dictatorship") to "spin dictatorship" almost exclusively based on extensive use of propaganda (Guriev, Treisman, 2022) lost its appeal immediately after the appearance of this framework for analysis. At the same time, these new challenges may be seen as a call for major changes in Russian studies—there are several scholarly tasks which need to be resolved simultaneously:

(1) Revision of old theories, aimed at their adjustment to ongoing political, economic, and international dynamics, and development of new ones;

(2) Search for new scholarly methodologies and upgrades to research methods appropriate for the study of the ongoing war and other related phenomena;

and

(3) Finding new empirically convincing evidence, which would help make scholarly analyses more empirically solid and rigorous and avoid unsubstantiated claims.

To what extent can the current research on Russian politics cope with these difficult tasks? What has been done already and what lies ahead?

2 Old Questions under New Conditions

After February 2022 Russian studies appears to have undergone a major tide of securitization: even topics not directly connected with the war (ranging from the environment to public health) are now considered through this lens of analysis, while research on topics, primarily related to military and security issues,[8] received a major boost. Under conditions of war, this shift was unavoidable; however, its impact on research on Russian politics is ambivalent, and whether or not this impact will be productive for scholarship will depend upon many factors. The securitization of Russian studies has definitely affected the scholarly search for answers to two eternal questions of Russian politics which first emerged in the nineteenth century—"who is to be blamed?" (in this case, for the Russian authoritarian drift and for the military assault on Ukraine)

8 On these issues, see also the chapter by Kirill Shamiev in this volume.

and "what is to be done?" (in this case, in order to stop numerous war-related disasters and counter numerous pernicious trends). As of yet, scholars' answers to both questions are rather imperfect.

Scholarly assessments of Russian politics often fall between excessively shallow and excessively deep explanations, and the analysis of current affairs is no exception. One pole is excessively shallow, as it revolves almost exclusively around the personality of Putin and his past trajectory in the Soviet security apparatus, his experience at the St. Petersburg city hall during the 1990s, and the like (Åslund, 2019). Analyses of this type consider the many pernicious developments in Russia to be induced only by Putin and his close entourage, and the Russian assault on Ukraine is perceived mostly as a projection of the preferences, misperceptions, fears, and grievances of one man in the Kremlin. Such a view on Russia, in a way corresponds with a statement made by Russia's top official, chair of the State Duma Vyacheslav Volodin, who once proclaimed: "if there is Putin, there is Russia; if there is no Putin, there is no Russia".[9] Such a view, however, is certainly insufficient for full-fledged explanations, as it is overly focused on a particular figure. Scholars rarely equate the Bolshevik Revolution of 1917 with just Lenin and Trotsky, or the Soviet collapse in 1991 with just Gorbachev and Yeltsin. In a similar way, consideration of contemporary Russia only in terms of "Putinology" (Frye, 2021) is rather unproductive, as recent developments in Russia were caused not only by Putin but also by major segments of the Russian political class. The Survey of Russian Elites quite convincingly illustrates these dynamics in the preferences of Russian elites between 1993 and 2020 (Rivera, 2020), as both political support for authoritarianism and aggressive militarism were shared by a significant proportion of the Russian elites well before the February 2022 attack on Ukraine. Without denying Putin's pivotal role in the conversion of these ideas into authoritarian drift and military actions, one must admit that Putin was not alone. The ideas of the possibility and desirability of durable authoritarianism and of the assault on Ukraine, in a way, reflected the visions, preferences and desires of the Russian political class as a whole (on the analysis of evolution of Russian elites, see Yakovlev, 2021).

Quite the opposite of the excessively shallow view is the excessively deep view on Russian politics. It tends to comprehensively condemn Russia—both the state and society—as eternally, inherently, and irreversibly imperialistic, militaristic, and anti-Western. Thus, the military assault on Ukraine is per-

---

9 See 'Volodin: rossiyane vosprinimayut ataki na prezidenta RF kak ataki na svoyu stranu', *TASS*, 22 October 2014 https://tass.ru/politika/1525655 (accessed 28 April 2025).

ceived as a side effect of Russia's past, of its imperial and/or colonial legacies, of real and/or imagined features of Russian culture, of its geo-political and geo-economic determinism, of its eternal drive to anti-modernisation, and the like. However, this approach, deeply embedded in many historical (Pipes, 1974) and contemporary (Hedlund, 2005) writings, is misleading. It not only ignores or downplays many successful episodes of Russian liberalisation in the past—both pre-1917 and in the late twentieth century—but also assumes that Russia as a country cannot be improved by any means and is worthy only of total elimination in one way or another. The perception of Russia as a quintessential and eternal global evil is understandable today due to the fury of many observers after Russia's violent attacks on Ukraine. However, this approach should not be considered a reliable analytical tool because it aims at accusations (Gudkov, 2022; Snyder, 2022) instead of explanations, and impartial scholarly analysis should not be replaced by emotional punditry. Speaking in a more comparative-historical way, no serious analysts today tend to condemn Germany for Nazism or believe that Germany cannot be improved after Nazism in any possible way, even though such a view was rather popular during the period of World War II and immediately after it.

Overall, the dichotomy of excessively shallow and excessively deep explanations for Russia's post-Communist trajectory to a certain degree reflects the fundamental gap that has developed in recent decades between structure-induced and agency-driven approaches to the analysis of Russian politics (Gel'man, 2015b; Orttung, 2023). However, the fruitful integration of these approaches (which should be complementary rather than mutually exclusive) remains a major weakest link in the research agenda of Russian studies, which is still focusing on blame—either Putin is guilty, or Russia is guilty, or both. Instead, evidence-based explanations addressed to mid-range theories would be more productive for a scholarly understanding of Russian politics.

As to the question "what is to be done?", the number of policy-oriented products by scholars of Russia, addressed both to policy makers and to the mass public, skyrocketed after 2022. Policy memos and briefs, analytic reports, and projected scenarios, as well as op-eds, public debates, pamphlets and open letters, often produced in very emotional styles, flooded the field and at least for a while dominated the intellectual landscape of Russian studies. This trend demonstrated a sharp contrast with the period of the Cold War, when many Sovietologists (with some notable exceptions, such as Zbigniew Brzeziński or Richard Pipes) did not particularly attempt to influence policy making towards the Soviet Union and rather pragmatically used the demand of policy relevance to support their own academic ventures (Engerman, 2009). It is hard to judge to what extent academia had and will have an impact on post-2022 policy towards

Russia in the US and/or in Europe (most probably, not much), but the increasing gap between the normative stances of many scholars and experts and the pragmatic interests of decision-makers is very observable and highly likely to continue in the future.

Against this background, several directions of the search for the roots of Russia's trajectory and remedies against the war deserve further exploration. First, discussion of the post-Communist transition in Russia during the 1990s suddenly restarted in 2024: after Alexei Navalny's death, his followers released a series of videos which explored the argument that Russia's reforms in the 1990s were the point of departure for authoritarianism and the revival of imperial militarism in Russia (Navalny, 2023).[10] These videos provoked a heated public response, mostly politically motivated, and this discussion was perceived not only as shifting the blame for the drift into authoritarian militarism in Russia to the reformers of the 1990s but also as a major revision of existing narratives of post-Communism by the younger generation of the Russian opposition (Zharkov, 2024). Despite these political tensions, the role of the legacy of the 1990s in Russia's political trajectory in the twenty-first century remains under-explored and under-theorised. The liberal mainstream considers the "roaring" 1990s in Russia to have been a generally justified, even if imperfect form of transition (Shleifer, Treisman, 2004), which was severely distorted and later on failed mostly due to the poor choice of Putin as Yeltsin's successor (Åslund, 2019). In turn, their leftist opponents argue that the market reforms of the 1990s paved the way to the rise of "neo-liberal autocracy" (Kagarlitsky, 2002) and that Russia's trends in the Putin era were a logical continuation of such a neo-liberal approach (Matveev, 2024). Both of these narratives, however, are rather imperfect. They tend to ignore the comparative dimension of transition—in fact, market reforms, with their subsequent decline in living standards, rise of inequality and ubiquitous corruption, were somewhat similar in all post-Communist countries, whereas Russia's post-transition trajectory was qualitatively different in many ways. In addition, participants in these discussions tend to overemphasize the role of structures or agencies in Russia's drift to authoritarianism and militarism without attempts to find a healthy scholarly balance between them (Orttung, 2023). Still, the turn to more in-depth accounts of Russia's contemporary history as a source of recent political developments should be welcomed, as there is a need to discuss these matters in a more systematic way (Gel'man, 2025).

---

10 See, in particular, *Istorya bol'shogo zagovora: kto i kak zakhvatil Rossiyu. Predateli, film 1*, 16 April 2024 https://www.youtube.com/watch?v=-_wM7LpOnPQ (accessed 28 April 2025).

Second, "decolonization" became a new catchword of Russian studies, incredibly popular in the post-2022 environment. It resonates with the "progressive" leftist narrative, which tends to place all the blame for the poor state of affairs in various corners of the globe on the shoulders of former European colonial powers as well as of the United States. In this respect, decolonization has contributed to accusations that Russia is pursuing a neo-imperialist approach towards the countries of the former Soviet Union. The call for decolonization, which initially concentrated on broadening the cultural focus of research instead of overconcentration on the Kremlin and the Russian elites, and on taking into account many developments in other parts of Eurasia, soon became an element of politics within academia. It led to loud demands to divert resources from the study of Russia to the study of other parts of Eurasia, ranging from Ukraine to Central Asia (Shaipov, Shaipova, 2023). Such demands, despite all arguments in their favor, appear problematic in the light of the war-driven "know your enemy" approach and implied policy relevance of research. Some scholars and experts tend to propose the punishment and humiliation of Russia under the guise of "decolonization" (Budraitskis, 2023), which seems more like a political statement than a research program. Furthermore, some critically minded authors have even considered the "decolonization" of Russia through the lenses of high expectations of its territorial disintegration in the future (Etkind, 2023), although reputable specialists have noted that these expectations are groundless and not based on any reasonable evidence (Golosov, 2024; Kynev, 2024).

Third, the analysis of sources and mechanisms of Russia's political and economic resilience despite exogenous shocks and war-driven domestic and international challenges has become a very important direction of research on Russian politics. Indeed, neither the special military operation's unfulfilled promises of demilitarization and denazification of Ukraine, nor major economic sanctions and the loss of revenues from oil and gas export, nor any other war-related challenges and crises have undermined the Russian political regime's ability to pursue its goals in the domestic and international arenas. At least at the time of writing of this article, the Kremlin appears nearly invincible despite increasing problems on the economic, technological, demographic and other fronts. While some scholars tend to consider the major role of the nation-wide state bureaucracy and business actors in Russia's resilience (Yakovlev, 2024), others pay particular attention to the role of elites in Russia's regions and cities and their impact on the persistence of the political, economic and societal status quo (Busygina, 2024; Klimovich, 2024; Shkel, 2025). There is no doubt that evidence-based empirical analyses of this type, based on a positive rather than normative scholarly perspective, are not only policy relevant but tell us more about the political economy of the war and the resilience of authoritarianism in Russia and beyond than numerous endless public discussions of the potential sources of Russia's defeat or victory.

3 Challenges and Traps

The veil of secrecy, which covers many details of post-2022 Russian domestic politics, alongside many problems of interpretation of available yet imperfect and incomplete data, has become a major hindrance for empirical research. In a way, such problems are not so dissimilar to those of Sovietology during the Cold War, when a shortage of information regarding the Soviet Union contributed to numerous misinterpretations among scholars and experts (Engerman, 2009). It is no wonder that some scholars have expressed major concern that causal identification research design, which now prevails in various social science disciplines, will be no longer available in post-2022 Russian studies (Libman, 2023). However, in the twenty-first century the nature of these problems and their solutions has changed a great deal.

To some extent, the rise of digital research on Russia in various formats (Gritsenko et al., 2021) and the further development of computational social science techniques in Russian studies (Bessudnov, 2024) compensated for numerous shortages and gaps in information vital for research, especially under post-2022 conditions. Scholars and experts were able to decipher various digital data made by state actors, businesses and ordinary people, and effectively analyze them through the use of diverse open data, ranging from social media networks to official statistics (such as excess mortality records, first during the COVID-19 pandemic and then during the war). The Russian assault on Ukraine became a catalyst for these processes and further contributed to the rise of online area studies in general and Russian studies in particular, with their own research agendas. Moreover, post-2022 Russian studies tend to employ new open source intelligence methods (also known as OSINT analytics), which were previously used mostly by hackers, journalists, and activists.[11] These methods have high potential for a fruitful integration into political research on Russia and beyond (Limonier, 2023). Although the new sub-field has faced many challenges and raised some concerns (Kalsaas, 2023), there is no doubt that digital research

11 On these issues, see also the chapter by Anselm Schmidt and Sanshiro Hosaka in this volume.

will develop further, and Russian studies may become a testing ground, if not a new launching pad, for these methodological innovations.

Research on the political attitudes of Russian people became yet another contentious issue in the field. Immediately after February 2022, observers were puzzled by the fact that mass surveys demonstrated a high degree of public support for the Russian assault on Ukraine among ordinary Russians, in the manner of "rally around the flag," amid a decline in the response rate in these surveys (Reisinger et al., 2023). Scholars of survey research proposed some new methodological solutions to these problems (Rosenfeld, 2023), continued their collaboration with Russia-based pollsters (such as the Levada Center), and conducted new studies of Russians' attitudes (Frye et al, 2024). Moreover, they successfully utilized some opportunities for the use of quasi-experimental survey design in survey research under new conditions (Zakharov et al, 2024).[12] However, proponents of interpretive research questioned the validity of survey data in Russia (and of quantitative social science in general) and called for reliance upon qualitative interviews and ethnographic observations as the most relevant research methods for the new situation in Russia (Morris, 2023). Indeed, scholars from the Public Sociology Laboratory, who extensively used these methods for in-depth analysis of changing attitudes of ordinary Russians towards the war, gathered outstandingly valuable field materials (Erpyleva, Kappinen, 2024). However, by all academic standards, various research strategies are usually considered complementary rather than mutually exclusive, and this is why different efforts to gather evidence on present-day Russian politics should be welcomed.

Although this discussion is important in methodological terms, its implications went far beyond approaches and techniques of data gathering and analysis of attitudes of the Russian mass public. Irrespective of the positivist or interpretive orientations of scholars, one may ask a more fundamental question: to what extent do ordinary people matter in politics in general and under conditions of repressive authoritarian militarism in particular? Or, rather, may the mass public in Russia and beyond be considered just an instrument in the hands of powerful elites (or counter-elites), while its political agency is so negligible that the masses may not be seriously taken into account in the study of the Russian political regime (Gel'man, 2015b)? Such a question has not yet

12 The regular monthly survey by the Levada Center was conducted during the exact days of Yevgeny Prigozhin's mutiny in June 2023, and scholars received a unique opportunity to compare survey data on public support for persons and institutions just before and after the mutiny. The data analysis demonstrated no major changes, however.

become part of the research agenda of Russian studies, although related matters are definitely worth more systematic in-depth analysis, both theoretically and empirically.

While scholars of Russian studies attempted to cope with methodological and empirical challenges under post-2022 conditions, their efforts to conceptualize developments in Russia (not only after February 2022, but also in a broader scholarly perspective) have faced two major traps. First is the trap of making big and sweeping generalisations of current trends and ongoing changes amid numerous uncertainties and incomplete and imperfect information. Due to these constraints, many generalisations of this kind, especially when they consider current affairs against the background of scholarly interpretations of the entirety of Russian history (Etkind, 2023), risk being highly premature and insufficiently empirically grounded. Such tendencies are mostly driven not by academia as such but rather by demands made by the media of scholars, who tend to be perceived as omniscient and all-knowing gurus, if not prophets. Some scholars, in turn, are eager to play these roles, which are often incompatible with academic research. Such demands are very tempting, and some academic celebrities achieve major recognition due to their loud public claims, despite (or even thanks to) the questionable academic validity and numerous biases of these claims: in this respect, post-2022 Russian studies are not an exception (Snyder, 2022; Chomsky, 2023).

Second, there is an ongoing spread of the disease of scholarly forecasting of the future of Russia, Ukraine, Europe, and the entire world amid the ongoing war. Such an enterprise (driven by the same media demands) is understandable but hardly productive for scholarship, and forecasting is always at risk of producing numerous wrong expectations. After all, there is no magic crystal ball in the scholar's arsenal, and even the best and brightest research centers hardly resemble the Hogwarts School of Witchcraft and Wizardry. Although the most experienced and clever scholars hedge their bets and offer several mutually exclusive scenarios in their forecasts (Kotkin, 2024), the overall diverting of scarce intellectual and material resources to making predictions about Russia is not always useful (for an overview of previous experiences of forecasting Russia's future, see Bacon, 2012). Furthermore, scholars are often not much better in their forecasting than, say, fiction writers.[13] Judging by the experience of

13 In his posthumously published novel *The Command Authority* (2013), the bestselling American fiction author Tom Clancy correctly predicted the Russian annexation of Crimea and the military conflict between Russia and Ukraine over Donbas. No scholars had made any predictions of this kind, as the major changes came unexpectedly for many observers.

Sovietology, one might argue that predictions of the Soviet collapse, occasionally made well before Gorbachev's perestroika, also faced the same trap. While Andrei Amalrik (1970) was factually incorrect in expecting the Soviet collapse by 1984, triggered by Soviet-Chinese military rivalry, he offered a very perceptive substantive interpretation of the dividing lines within the Communist system and its major vulnerabilities, and many of them proved to be relevant. In contrast to him, Hélène Carrère d'Encausse (1979) almost correctly predicted the Soviet collapse by 1990 in factual terms, but her expectations were based on rather misleading interpretations of Soviet domestic politics—she predicted a rise in militant Islamism and violent separatism in Soviet republics of Central Asia, driven by demographic changes.[14] As one can see, both predictions were wrong, albeit for very different reasons. One must admit that for good or for ill, any scholarly predictions about Russia's future under current conditions may become incorrect both in factual and substantive terms.

In Lieu of a Conclusion: Avenues for Further Research

To summarize, developments in Russian studies after February 2022 came rather contrary to the extraordinary nature of the ongoing war, in spite of its own academic trajectory, and did not meet either the best or the worst expectations under changing domestic and international political conditions. Research on Russian domestic politics has faced numerous problems and still lacks many reliable instruments for their analysis, but scholars continue their efforts to gather, analyze and interpret new data on various aspects of Russian politics and society, and have laid some grounds for further advancements. One should not expect an immediate academic breakthrough in the field given all the political, informational and institutional constraints stated above, but based on information about ongoing research projects, one can be moderately optimistic about the future of Russian studies. In this respect, some modest suggestions may be useful for scholarly discussions and for developing avenues for further research.

The exogenous shock of February 2022 led to a major shift in many discussions on Russian politics towards a normative focus—condemning the Kremlin became nearly universal among international scholars and experts. It gave

14 After the Soviet collapse in 1991, Carrere d'Encausse received outstanding academic recognition and became a member and later on a permanent secretary of the French Academy, despite (or perhaps because of) the fact that the academic value of her forecast was dubious at best.

birth to numerous reactions and responses to the war through which specialists in the field demonstrated their stances towards it. However, such an approach, virtuous as it may be, cannot substitute for a scholarly research agenda. Meanwhile, the normative perspective is not always productive for research as we have to analyze not how Russian politics should (or, rather, should not) develop but how and why Russian politics actually functions in its current form. Regardless of our normative views, we are forced to answer the question "why?", whether we want it or not. We have to address questions of how the Kremlin manages the war, the economy, and society at large, why it is able to cope with so many simultaneous domestic and international challenges, what the strengths and vulnerabilities of Russian political mechanics are, and the like. In other words, refocusing the research agenda from a normative to a positive perspective of scholarly research on Russian politics is a primary task for specialists.

Furthermore, contrary to the tendency to perceive political developments in contemporary Russia as a unique and exceptional case of failed democratisation and building of a new aggressive neo-imperial state, there is a serious need to put this case into a broader theoretical and comparative-historical framework of analysis in a systematic way. Examples of this approach in the study of post-Soviet Russian politics before 2022 have been relatively rare (Derluguian, 2005; Hanson, 2010; Gaidar, 2012), but the potential for this kind of research is very high. For example, the ongoing war against Ukraine could probably be best conceptualized in Tillyan historical sociological perspective as an episode of state-making and unmaking. Tilly's maxim of "war made the state, and the state made war" (Tilly, 1975: 42) is very relevant in the current context in regard to both Russian and Ukrainian post-imperial state building under conditions of entrenched and protracted bad governance in these two countries (Gel'man, 2022). The problem is that such an agenda of in-depth comparative-historical research would require addressing key questions of post-imperial nationalism, of international integration (or, in case of Russia, of detachment), of breakdowns of modernisation processes, and the like. Studies of this kind are quite time-consuming, and they cannot bring significant fruit in the short run, here and now, especially given all the pressures in academia and beyond. However, this agenda is worth pursuing as a potential source of major academic advancements in the study not only of contemporary Russian politics, but of politics and societies across the globe.

Finally, despite the very gloomy current state of affairs in Russian politics, and the many problems and concerns involved in researching it, there should be no grounds for falling into the deadly sin of sloth, which is always counter-productive, including in academia. Scholars of Russian studies may consider the ongoing crisis of the war and its related pernicious effects not only as a sign of the hopelessness and senselessness of Russia. Instead, one might consider this crisis as an opportunity to use the multi-dimensional challenge of our times to raise new scholarly questions and search for new answers for further research in Russian studies and beyond. If this approach to research on Russian politics prevails in the foreseeable future, then political science research in general and Russian studies in particular will be on the right track.

References

Al'bitskii, Egor, 2024, "Obshchestvo ne prinimaet. S nachala voiny iz Rossii uekhalo svyshe 150 ekonomistov, politologov i sotsiologov", *Novaya gazeta—Evropa*, 4 July
https://novayagazeta.eu/articles/2024/07/04/obshchestvo-ne-prinimaet (accessed 28 April 2025).

Amalrik, Andrei, 1970, *Will the Soviet Union Survive until 1984?* Harper and Row.

Bacon, Edwin, 2012, "Writing Russia's Future: Paradigms, Drivers, and Scenarios", *Europe-Asia Studies*, 64 (7): 1165–1189 https://doi.org/10.1080/09668136.2012.698046

Bates, Robert H., 1997, "Area Studies and the Discipline: A Useful Controversy?" *PS: Political Science and Politics*, 30 (2): 166–169 https://doi.org/10.2307/420485

Bessudnov, Alexey, 2024, "Using Digital Trace Data in Russian Studies", *SocArXiv Papers* https://osf.io/preprints/socarxiv/825cu_v1 (accessed 28 April 2025)

Bogdanova, Elena, Dzenovska, Dace, Morris, Jeremy, Muravyeva, Marianna, Pallot, Judith, 2023, "Post-Soviet Studies: Crisis of Concepts, Conventions, and Compromises", *Laboratorium*, 15 (2): 93–114 https://doi.org/10.25285/2078-1938-2023-15-2-93-114

Breslauer, George, 1992, "In Defense of Sovietology", *Post-Soviet Affairs*, 8 (3): 197–238 https://doi.org/10.1080/1060586X.1992.10641352

Budraitskis, Ilya, 2023, "On the Debates Regarding the "Decolonization" of Russia", *Italian Institute for International Political Studies*, 3 April https://www.ispionline.it/en/publication/on-the-debates-regarding-the-decolonization-of-russia-123484 (accessed 28 April 2025).

Busygina, Irina, 2024, "The Sources of Territorial Resilience in Putin's Russia", *Post-Soviet Affairs*, 40 (4): 250–261 https://doi.org/10.1080/1060586X.2024.2351781

Carrère d'Encausse, Helene, 1979, *Decline of an Empire: The Soviet Socialist Republics in Revolt*, Newsweek Books.

Chomsky, Noam, 2023, *Noam Chomsky on Russia–Ukraine War*, 9 May
https://www.youtube.com/watch?v=tD_JdwimnjU (accessed 28 April 2025).

Conversations with the Field (2023), "Conversations within the Field: Russia's War against Ukraine and the Future of Russian Studies", *Post-Soviet Affairs*, 39 (1–2), special issue https://www.tandfonline.com/toc/rpsa20/39/1-2

Derluguian, Georgi, 2005, *Bourdieu's Secret Admirer in the Caucasus: A World-System Biography*, University of Chicago Press.

Engerman, David C., 2009, *Know Your Enemy: The Rise and Fall of America's Soviet Experts*. Oxford University Press.

Erpyleva, Svetlana, Kappinen, Sasha, eds., 2024, ""We Need to Carry On": Ethnography of Russian Regions During Wartime", *The Russia Program, George Washington University, Academic Policy Paper Series*, no. 5 https://therussiaprogram.org/ps_lab_ethnography (accessed 28 April 2025).

Etkind, Alexander, 2023, *Russia against Modernity*, Polity Press.

Frye, Timothy, 2017, "Russian Studies Is Thriving, Not Dying", *The National Interest*, 3 October https://nationalinterest.org/feature/russian-studies-thriving-not-dying-22547 (accessed 28 April 2025).

Frye, Timothy, 2021, *Weak Strongman: The Limits of Power in Putin's Russia*, Princeton University Press.

Frye, Timothy, Hale, Henry, Reuter, Ora John, Rosenfeld, Bryn, 2024, "Putin's Hidden Weakness: New Evidence Shows Many Russians Support Him—but Not His Policies", *Foreign Affairs*, 25 March https://www.foreignaffairs.com/russian-federation/putins-hidden-weakness (accessed 28 April 2025).

Gaidar, Yegor, 2012, *Russia: A Long View*, MIT Press.

Gel'man, Vladimir, 2015a, "Political Science in Russia: Scholarship without Research?", *European Political Science*, 14 (1): 28–36 https://doi.org/10.1057/eps.2014.33

Gel'man, Vladimir, 2015b, *Authoritarian Russia: Analyzing Post-Soviet Regime Changes*, University of Pittsburgh Press.

Gel'man, Vladimir, 2022, *The Politics of Bad Governance in Contemporary Russia*, University of Michigan Press.

Gel'man, Vladimir, 2023, "Exogenous Shock and Russian Studies", *Post-Soviet Affairs*, 39 (1–2): 1–9 https://doi.org/10.1080/1060586X.2022.2148814

Gel'man, Vladimir, 2025, "In Search of the Roots of Failure: Reassessing Russian Politics in the 1990s", *Russian Politics*, 10 (1): 51–70 https://doi.org/10.30965/24518921-01001003

Golosov, Grigorii V., 2024, *Politicheskie rezhimy i transformatsii: Rossiya v sravnitel'noi perspective*, Ruthenia.

Gritsenko, Daria, Wijermars, Mariëlle, Kopotev, Mikhail, eds., 2021, *The Palgrave Handbook of Digital Russia Studies*, Palgrave Macmillan https://doi.org/10.1007/978-3-030-42855-6

Gudkov, Lev, 2022, *Vozvratnyi totalitarism*, Novoe literaturnoe obozrenie.

Guriev, Sergei, Treisman, Daniel, 2022, *Spin Dictators: The Changing Face of Tyranny in the 21st Century*, Princeton University Press.

Hanson, Stephen E., 2010, *Post-Imperial Democracies: Ideology and Party Formation in Third Republic France, Weimar Germany, and Post-Soviet Russia*, Cambridge University Press.

Hedlund, Stefan, 2005, *Russian Path Dependence: A People with a Troubled History*, Routledge.

How to Study Russia (2023), "How to Study Russia", *Russian Analytical Digest*, 293, 3 March (special issue) https://css.ethz.ch/content/dam/ethz/special-interest/gess/cis/center-for-securities-studies/pdfs/RAD293.pdf (accessed 28 April 2025)

Kagarlitsky, Boris, 2002, *Russia under Yeltsin and Putin: Neo-Liberal Autocracy*, Verso.

Kalsaas, Johanne, 2023, "Area Studies Online? Opportunities and Challenges When Researching "Digital Russia" during the War on Ukraine", *Russian Analytical Digest*, 293, 3 March, 6–10 https://css.ethz.ch/content/dam/ethz/special-interest/gess/cis/center-for-securities-studies/pdfs/RAD293.pdf (accessed 28 April 2025)

Kangas, Anni, Mäkinen, Sirke, Dubrovskiy Dmitri, Pallot, Judith, Shenderova, Svetlana, Yarovoy, Gleb, Zabolotna, Oksana, 2023, "Debating Academic Boycotts and Cooperation in the Context of Russia's War Against Ukraine", *New Perspectives*, 31 (3): 250–264
https://doi.org/10.1177/2336825X231187331

Klimovich, Stanislav, 2024, "Federalism at war: Putin's blame game, regional governors, and the invasion of Ukraine", *Post-Soviet Affairs*, 40 (4): 262–277
https://doi.org/10.1080/1060586X.2024.2352978

Kotkin, Stephen, 2024, "The Five Futures of Russia: How America Can Prepare for Whatever Come Next", *Foreign Affairs*, 103 (3): 64–83

Kynev, Alexander, 2024, *Kto i kak upravlyaet regionami Rossii*, Ruthenia.

La Lova, Lanabi, 2023, "Methods in Russian Studies: Overview of Top Political Science, Economics, and Area Studies Journals", *Post-Soviet Affairs*, 39 (1–2): 27–37
https://doi.org/10.1080/1060586X.2022.2162293

Libman, Alexander, 2023, "Credibility Revolution and the Future of Russian Studies", *Post-Soviet Affairs*, 39 (1–2): 60–69 https://doi.org/10.1080/1060586X.2022.2148446

Limonier, Kevin, 2023, "What Open Source Investigations (OSINT) Can Bring to Russian Studies", *Russian Analytical Digest*, 293, 3 March, 4–5 https://css.ethz.ch/content/dam/ethz/special-interest/gess/cis/center-for-securities-studies/pdfs/RAD293.pdf (accessed 28 April 2025).

Matveev, Ilya, 2024, "From the Chicago Boys to Hjalmar Schacht: The Trajectory of the (Neo) liberal Economic Expertise in Russia", *Problems of Post-Communism*, 71 (6): 508–517 https://doi.org/10.1080/10758216.2023.2291359

Morris, Jeremy, 2023, "Political Ethnography and Russian Studies in a Time of Conflict", *Post-Soviet Affairs*, 39 (1–2): 92–100 https://doi.org/10.1080/1060586X.2022.2151275

Navalny, Alexey, 2023, "Moi strakh i nenavist", *Navalny.com*, 11 August https://navalny.com/p/6651/ (accessed 25 February 2025).

Orttung, Robert W., 2023, "Balancing Structure and Agency in Studying Russia's Future",

Russian Analytical Digest, 293, 3 March, 11–12 https://css.ethz.ch/content/dam/ethz/special-interest/gess/cis/center-for-securities-studies/pdfs/RAD293.pdf (accessed 28 April 2025).

Pipes, Richard, 1974, *Russia under the Old Regime*, Scribner.

Polianskii, Mikhail, 2024, "Russian Foreign Policy Research and War in Ukraine: Old Answers to New Questions?" *Communist and Post-Communist Studies*, 57 (2): 156–172 https://doi.org/10.1525/cpcs.2024.2112378

Reisinger, William M., Zaloznaya, Marina, Woo, Buyng-Deuk, 2023, "Fear of Punishment as a Driver of Survey Misreporting and Item Non-Response in Russia and Its Neighbors", *Post-Soviet Affairs*, 39 (1–2): 49–59 https://doi.org/10.1080/1060586X.2022.2150490

Rivera Sharon Werning, ed., 2020, *Survey of Russian Elites 2020: New Perspectives of Domestic and Foreign Policy*, Hamilton College https://hamilton.edu/documents/SRE2020ReportFINAL.pdf (accessed 28 April 2025).

Rosenfeld, Bryn, 2023, "Survey Research in Russia: In the Shadow of War", *Post-Soviet Affairs*, 39 (1–2): 39–48 https://doi.org/10.1080/1060586X.2022.2151767

Rutland, Peter, 1993, "Sovietology: Notes for a Post-Mortem", *The National Interest*, 31, 109–122 https://www.jstor.org/stable/42894863

Shaipov, Artem, Shaipova, Yuliia, 2023, "It's High Time to Decolonize Western Russia Studies", *Foreign Policy*, 11 February https://foreignpolicy.com/2023/02/11/russia-studies-war-ukraine-decolonize-imperialism-western-academics-soviet-empire-eurasia-eastern-europe-university/ (accessed 28 April 2025).

Shkel, Stanislav, 2025, *Elections, the Ethnic Factor and Patronal Politics in the Russian Regions: Anatomy of Loyalty*, Routledge.

Shleifer, Andrei, Treisman, Daniel, 2004, "A Normal Country", *Foreign Affairs*, 83 (2): 20–38.

Shleifer, Andrei, Treisman, Daniel, 2014, "Normal Countries: The East 25 Years After Communism", *Foreign Affairs*, 93 (6): 92–103.

Snyder Timothy, 2022, "We Should Say It: Russia Is Fascist", *New York Times*, 19 May https://www.nytimes.com/2022/05/19/opinion/russia-fascism-ukraine-putin.html (accessed 28 April 2025).

Tilly, Charles, 1975, "Reflections on the History of European State-making", in Charles Tilly, ed., *The Formation of National States in Western Europe*, Princeton University Press, 3–87.

Yakovlev Andrei, 2021, "Composition of the Ruling Elite, Incentives for Productive Use of Rents, and Prospects for Russia's Limited Access Order", *Post-Soviet Affairs*, 37 (5): 417–434 https://doi.org/10.1080/1060586X.2021.1966988

Yakovlev, Andrei, 2024, "Is Russia Shifting Toward Mobilization Economy or Forming New, Loyal Business Elite?" *Russia Matters*, 30 May https://www.russiamatters.org/analysis/russia-shifting-toward-mobilization-economy-or-forming-new-loyal-business-elite (accessed 28 April 2025).

Zakharov, Alexei, Reuter, Ora John, Shuklin, Vladimir, Volkov, Denis, 2024, "Effects of a Coup Attempt on Public Attitudes under Autocracy: Quasi-experimental Evidence from Russia", *Post-Soviet Affairs*, 40 (6), 472–480 https://doi.org/10.1080/1060586X.2024.2359311

Zavadskaya, Margarita, Gerber, Theodore, 2023, "Rise and Fall: Social Science in Russia before and after the War", *Post-Soviet Affairs*, 39 (1–2): 108–120 https://doi.org/10.1080/1060586X.2022.2164450

Zharkov, Vasilii, 2024, "Dvadtsatyi s'ezd v emigratsii", *Moscow Times*, 14 June https://www.moscowtimes.ru/2024/06/14/dvadtsatii-sezd-v-emigratsii-a134023 (accessed 28 April 2025).

Åslund, Anders, 2019, *Russia's Crony Capitalism: The Path from Market Economy to Kleptocracy*, Yale University Press.

CHAPTER 5

A New Corporeal Order

Russia Between Biopower and Necropolitics

Andrey Makarychev

1 Introduction

The concept of biopolitics usually refers to different modalities of governing (managing, controlling, monitoring, regulating, categorising, but also care taking and incentivizing) human bodies. In a broader sense, biopolitics as a name for submission of individual bodies to state regulations is meant to integrate life (and thus the realm of corporeality) into political calculations and agendas. Biopolitics as a multifaceted phenomenon might signify a logic of power, a paradigm of governance and / or a cultural phenomenon. In all its dimensions, biopolitics is a helpful lens to analyse dynamics of power since it presumes that all major political constructs operate primarily as bodily / corporeal categories, be it democracy, populism, liberalism, or totalitarianism. This approach contributed to political analysis through inscribing human bodies and lives into analysis of ideologies, institutions, and elite groups.

However, so far very little is known about how biopolitical concepts can be used for analysing wars waged by undemocratic regimes. The questions that were driving me in my research were: how does biopolitics work in Russia in times of war? What does biopolitical concept elucidate in Russian domestic and foreign policies? Which version / modality of biopower better explains Russian politics? Biopolitical expertise is one of the missing elements in the debate about the origins of Russia's invasion in Ukraine and its long-term effects for international security. In this chapter my main intention is to partly fill this gap through extending biopolitical theorisings from its Western liberal hotbed to the field of Russian studies for a better understanding of the nexus between illiberal sovereignty and wars. The operationalisation of the idea of biopolitics for studying Russia anew is part of a broader engagement with different critical theories helpful for deeper insights into the functioning of power relations in non-democratic countries.

In this chapter I make three main claims that correspond to three sections below. First, I show that biopolitical concepts, being genealogically European, can be effectively applied to non-Western regimes where they demonstrate

 | DOI:10.1163/9789004747364_006

their explanatory potential for unpacking illiberal practices of subjugation of human beings to sovereign power that exploits their vulnerabilities. Biopolitical arguments grounded in Western philosophical traditions are by and large transferable to Russian studies, yet in some cases the specificity of Putin's regime requires approaches and conceptualisations stretching beyond the conventional biopolitical theorising. For this I propose the concept of Corporeal Order which I contextualise, and apply to characterise a core component of Putin's regime whose evolution has to be analysed not only as a shift from authoritarianism to military dictatorship, but also from the viewpoint of using biopower as a substitution to legal norms and public institutions. This Corporeal Order might be discussed in terms of biopolitical totalitarianism and is grounded in an approach to the population as a biomaterial and biomass, or—in the vocabulary of Giorgio Agamben—*homines sacri* and bare lives.

My second point concerns the concept of sovereignty which Giorgio Agamben interprets as a power to create multiple zones of indistinction—between inclusion and exclusion, norm and exception, the inside and the outside, the rule of law and chaos, and ultimately between bio- and necropolitics. My argument is different—I interpret illiberal sovereignty as monopoly on deciding on a) what belongs to the sphere of biopolitics and what is attributed to the domain of necropolitics, and b) how bio- and necropolitics overlap to the point of a merger.

Thirdly, the bulk of the academic literature discusses how sovereign power produces—through discourses and policy practices—hierarchies and creates, according to Agamben, *homines sacri*. What I add to this theorising is analysis of sovereign power as an object of deconstructive discourses that challenge its immanence and expose the self-destructing—and therefore necropolitical—nature of illiberal sovereignty.

My theoretical frame is a combination of biopolitics and critical discourse analysis. What lies at the intersection of these two intellectual traditions is the idea of *biopolitical translation* which could be understood in at least two different ways. One is transformation of geopolitical categories into biopolitical ones: discourses on spaces and bodies might be viewed as *intertextual* in the sense that they performatively intertwine, reinforce, and legitimize each other (Schlosser 2007). In this regard Russia's invasion in Ukraine, being a geopolitical event, at the same time might be narrated from bio- and necropolitical perspectives (Troncota 2017, 243–271).

Yet translation can also be understood as modification and transformation of biopolitical categories themselves, so that new meanings become different from the originals. Thus, the concept of naked / bare life, coined by Agamben as a metaphor for production of docile bodies by the sovereign power, was re-signified by the performative artist Piotr Pavlensky as a symbol and a sign of protest against Putin's dictatorship (Journeyman Pictures 2017). The "naked party" event held in Moscow in fall 2023—to be discussed below—adds a different meaning of bareness as a sign of private lifestyle that is defenseless against the sovereign power and its Corporeal Order.

2 Biopolitics in Times of War: Russia in the Limelight

In this section I critically engage with the extant literature on biopolitics seen from the vantage point of conflicts and crises, and then show why and how Russia's military intervention in Ukraine ought to be approached from a biopolitical viewpoint.

3 The West as a Biopolitical Culprit

As seen from the political science perspective, biopolitics offers a helpful lens for a nuanced—and alternative to the traditional focus on institutions, ideologies and elites—understanding of nationalism as community-(re)making, the rules of belonging and disciplining, mechanisms of inclusion and exclusion, the internal- versus-external dynamics of norms acceptance and rejection. Being focused on post-ideological protection and caretaking, biopolitics is a peculiar mode of making collective identities in large social groups cemented by shared or common attitudes to and understanding of biopolitical norms. Biopolitics produces national, ethnic or religious identities grounded in corporeal practices of control over human bodies and their physical existence. By using techniques of "normalisation", biopolitics differentiates between the outside and the inside, and thus (re)draws boundaries of the collective body of a nation or an empire. The concept can help us better grasp reasons for actions aimed at consolidating and stabilising relations of power through regulations based on norm-based rationality, as well as challenging it. As a normative phenomenon, biopolitics seeks to find out what lives are worth of protection / support in "our" biopolitical community, and what lives count as biopolitically meaningful.

In the existent academic literature, the application of biopolitical concepts for discussing wars is embedded in the broader Foucault's legacy of tackling security as part of the liberal governmentality paradigm, with its knowledge-based rationalisation of uncertainties (Wichum 2013). Two points are important in this regard. First, in the Foucauldian tradition the very concept of security—and insecurity as well—might be analytically operational only within liberal regimes of power (McLoughlin 2012, 702). Second, Foucault's followers do not think that wars ought to be approached as "states of exception"; on the contrary, they normalise wars suggesting "that it is in fact the logic of war that best describes how power works in the social relations of modern liberal government" (Kiersey 2009).

Giorgio Agamben articulated a different vision of biopolitics as deeply inscribed into—rather than divorced from—sovereign power. As seen from his perspective, Western democracies tend to react to crises and extraordinary situations similarly to non-democratic regimes, which ultimately blurs distinctions between Western and non-Western, as well as liberal and non-liberal governments.

An alternative conceptualisation is rooted in Achille Mbembe's works: he deploys war in necropolitical rather than biopolitical register of concepts (Mbembe 2003), and relates them with the post-colonial condition. Yet again, the variety of necropolitical discourses has Western genealogies that are politically accusatory of the (neo)liberal capitalism for racialisation and subjugation of its subalterns (Stumer 2018).

Despite terminological and conceptual distinctions, all three theories (Foucault's, Agamben's and Mbembe's) stem from a presumption that the liberal cosmopolitan West is a major source of wars and violence, which is reflected in the concept of "murderous biopolitics" (Valverde 2010, 139). In other words, liberal biopower is portrayed as a fertile ground for wars and violence (Reid 2004). This logic reached its apex in Agamben's direct support of Russia's aggression against Ukraine (Agamben 2024).

Apparently, long before Russia's invasion of Ukraine this standpoint was criticized for overlooking "the most exemplary place of modern biopolitics, the politics of the great totalitarian states. (Due to that—Author) contemporary political thought has failed to situate the totalitarian phenomenon in the horizon of biopolitics … Agamben's provocative claim is that until this is done Nazism and fascism will remain with us" (Oksala 2010, 34). Russia's war on Ukraine as part of a broader confrontation of Putin's regime with the West has added new arguments to this criticism: nowadays the entire structure of the Euro-Atlantic security order is under attack by a country that openly dissociated itself from the global liberalism and challenged it as a radical outsider. In the Kremlin's logic, the war "purified" the collective body of the nation from "traitors" and "the fifth column" as a way to "strengthen" the country, which attests to the domination of organicist understanding of the nation as an extended version of family, with the hegemonic masculinity as a key operational concept.

Moreover, against the backdrop of Russia's aggression against Ukraine some core arguments of post-Foucauldian biopolitical scholarship tend to merge with the logic of Westsplaining and its realist underpinnings. In particular, this concerns the idea of liberal peace sometimes portrayed as allegedly being "predicated on the pacification/elimination of all forms of political difference", which—as some authors assume—reveals "the martial face of liberal power" and leads to "liberal terror" (Evans 2010, 421, 428). In the aftermath of Russia's aggression against Ukraine a similar argument was used by pro-Kremlin authors to misinterpret sanctions as a biopolitical warfare launched by the West against Russian population, including vulnerable groups whose health depends on Western medical equipment and pharmaceutics (Alexeeva-Karnevaly 2022).

These accusations against liberalism look disbalanced and misguided, particularly considering a series of Russia's military interventions in Georgia, Syria, and Ukraine. Russia's invasion in Ukraine has questioned the validity of pointing to the Euro-Atlantic West as a major source of violence, militarisation and warmongering in the word. Contra Mbembe, it is not Western powers but Russian colonialism that systematically and intentionally takes lives of the subjugated peoples. Contra Agamben, distinctions between democracies and non-democracies not only persisted as a major factor in international security politics, but became even more politically meaningful. The attribution of the politics of death, or necropolitics, to Western political tradition, as Agamben does pointing to the culmination of "radical death" in Auschwitz (Deranty 2008, 175), looks increasingly problematic. Mass atrocities committed by Russian occupying troops in Ukraine questioned the veracity of Agamben's argument and demonstrated that it is exactly the necropolitical characteristics ascribed by many scholars to the West that nowadays are immanent components of the neoimperial policy pursued by Russia, a country that overtly decoupled itself from the Euro-Atlantic West (Terry 2024). By the same token, the portrayal of the liberal West as a generator of security threats is disavowed by the strategy of Ukraine to join NATO and the EU as the only feasible way to protect itself from Russia's attacks and safeguard its territorial integrity.

The war in Ukraine has fundamentally challenged the relevance of academic voices who for decades have been persistently building logical connections between necropolitical violence and neoliberal capitalism. Representative of this mindset is the deployment of death into the logic of free market, capitalist economy and "global militarised industries" (Haskaj 2018, 1154). Russia's aggression against Ukraine revealed the falsity of such a reasoning in two interrelated aspects. First, it was not the abstract global neoliberalism, but a transgressive illiberal regime led by the Russian president that unleashed the war with all its necropolitical effects. Second, it is countries of neoliberal democracy, with their military resources, that became major helpers to Ukraine as the victim of the unprovoked attack.

At the same time, the war corroborated Agamben's gloomy predictions of a forthcoming "catastrophe of biopolitics" erupting from the transformation of humanity into animality (Whyte 2013, 4). What is even more relevant for the case of Ukraine is that Agamben related the expectation of a forthcoming global crisis with the depoliticization of societies and the predominantly "domestic focus on the sustenance of life" (Ibid.). Yet what Agamben attributed to the global level turned out to be an effect of Russian politics, and it is Russian regime that showed its "inability to function without transforming itself into a lethal machine" (Short 2005, 23).

Against this background, the Foucauldian thesis of inevitability of wars as an underlying component of Western modernity appears to politically justify great power management, spheres of influence and other components of the Cold War—like types of world order. The concomitant argument of "generativity of war"—"war came to function as a schema for social and political order/ordering" (Spieker 2011, 187)—might be politically accepted by major world powers, but looks much more controversial from the viewpoint of victims of aggressions. Moreover, the characterisation of wars as normalising events and preconditions for national survival and legitimation of sovereignty-as-acquisition (Spieker 2011, 190) seems to be structurally similar to Putin's discourse. Thus, interpreting Foucault, Julian Reid argued that "the participation of populations in war is therefore reconceived not as the product of a right of seizure, but as a positive, life-affirming act … (B)iopolitical techniques provide them with new-found powers to wage war inter-socially (which makes war—Author) the source of the forces that comprise populations biopolitically … (W)ar becomes conceived as a condition of possibility for life" (Reid 2008, 76).

Yet how did it happen that a regime that invested lots of efforts and resources in biopolitical practices, ended up with a murderous war against its neighbor? In the next section I discuss the genealogy of the nexus between illiberal sovereignty and biopower under Putin's presidency.

4 Sovereignty and Biopolitics in Russia: A Short Genealogical Story

In the extant biopolitical literature sovereignty is discussed as a monopoly on violence within national borders, and as political will-based power to take decisions on exceptions concerning the definition of friends and foes. By the same token, sovereignty is seen as production of the permanent state of exception ('the camp' as Agamben's metaphor) and the ensuing 'bare life'. From Foucault's perspective, sovereignty is a law-based competence of power to be distinguished from governmentality.

The 1990s in Russia were marked by the almost complete absence of the state from the spheres of biopolitics and corporeality, including alcohol and drug consumption, sexual behavior, family relations and education which were relegated to the domain of private lives. Under Putin's presidency the situation began to gradually change, and the state reclaimed its right to intervene in and regulate personal life, intimacy, and sexuality. The ban on adoption of Russian children by foreign families in 2012 was quite indicative in this respect: this was the first obvious case of the government's direct and blunt appropriation of children's lives not as objects of the Foucauldian care, but as a resource for the state. Therefore, Putin's regime might be characterized by a transformation from a permissive understanding of the body to its opposite, in which biopolitics was ultimately instrumentalized and politicized by sovereign power. Family and sexuality became an object of state regulation, along with educational practices of indoctrinated 'patriotism' and sportive mega-events which legitimized athletes' bodies as a resource for what Agamben would have called the state's 'Glory' (Agamben 2007). The flip side of this biopolitics were practices of incarceration and imprisonment, enforced psychiatric treatment inherited from the Soviet times, differentiation of people with an intentionally established unequal status in society—"foreign agents", members of "undesirable organisations", associates / partners of "unfriendly states". Moreover, in some regions of Russia (in particular, in Chechnya) the idea of the biopolitically interventionist state was extended to practices of de-humanisation and animalisation of alternative lifestyles (Grozny 2022).

A similar transformation occurred with the idea of pastoral power which in the Foucauldian version was not intended to have an authoritarian or illiberal meaning, and was rather related to practices of governmentality. Yet in Russia, the Orthodoxy demonstrated its anti-democratic potential, which merged with the desire of the state to biopolitically subjugate people's bodies. Moreover, on behalf of the Russian Orthodox Church the patriarch Kirill formulated its mission in necropolitical categories by calling the flock to support the "holy war" against Russia's Slavic neighbor.

The same story happened to the idea of the "Russian world", which originally back in the 1990s was a name of a policy towards compatriots living abroad, in principle comparable with other relations between the state and diaspora in many countries across the world. However, less than two decades after its inception the "Russian world" turned into a justification for invasion in Ukraine with its mass-scale atrocities committed by the Russian occupying troops.

Given these transformations, some authors characterize Putin's rule as "bio-nationalist", meaning by that reliance on different techniques of controlling human bodies for the sake of national consolidation (Naterstad 2023). However, the evolution of the ruling regime transformed this bio-nationalism into what could be more properly described as bio-imperialism which, in its turn, ends up with necro-imperialism. All this makes Russia's biopolitics qualitatively different from its liberal models, which demonstrates how far Russia is from Western traditions of the state—body relations. Arguably, a central role of the Corporeal Order in Putin's regime of power can be explained by the failure of other methods of building statehood, including ideology, economy, and institutions. It is the resource of the body and human physicality, around which the ideas of commonality, loyalty and belonging are being constructed. Categorisation and differentiation of human bodies is an indispensable component of the Russian Corporeal Order. For example, the proverbial "foreign agent" can be seen as a "foreign body" whose status in the hierarchy of different bodies in the biopolitically controlled space is marginal.

On the one hand, the state-constructed Corporeal Order shows hypersensitivity to some areas to which it should not pay attention at all, including private lives of citizens, sexual preferences, family issues, or reproductive behavior (Independent Institute of Philosophy 2023). On the other hand, the state is insensitive to mass death and even normalises it: for example, the loss of about one million people during the 2020–2021 pandemic routinised death, which created preconditions for its acceptance and later inclusion in political strategies. Consequently, the war against Ukraine was marked by a high level of symbolic glorification of what might be dubbed "patriotic death" with its overt necropolitical connotations.

5 Sovereignty and the Corporeal Order

In this section I turn to my second argument exposing the core characteristic of Russia's Corporeal Order as a key feature of sovereignty that reclaims the right to possess individual bodies and eradicates the remnants of responsibilisation, individualisation, incentivisation and care taking. In the extant literature there were already attempts to unveil the coercive underside of Russia's model of biopower: "Paradoxically, the more biopolitics penetrates the social, the more it necessitates violence" (Rigi 2007, 43). In this regard the sovereign reappropriation of human bodies seems to be incompatible with the neoliberal foundations of the Foucauldian model of biopolitics. The concept of the Corporeal Order shifts the idea of freedom from individuals to the level of the state and its central government. It drastically shrinks spaces of embodied politics beyond this order, thus marginalizing and sidelining bodily practices that don't fit in the orderly and glorious—in Agamben's sense—biopower.

I theorise the concept of the Corporeal Order as a type of dispositive which, in the Foucauldian tradition, agglomerates "discourses, institutions, regulatory decisions, laws, administrative measures, scientific enunciations, philosophical, moral, and philanthropic propositions; in short: as much the said as the unsaid, these are the elements of the dispositive" (Bussolini 2010, 91). In line with this philosophical approach, the Corporeal Order has institutional and discursive backgrounds, is built on practices established by the sovereign power, and legitimised by the multiplicity of social and cultural actors. Two points are of primordial importance for unpacking this concept.

First, as I argued in the introduction, the new Russian Corporeal Order is grounded in distinguishing between what belongs to the politics of life and what defines the politics of death. Two examples are illustrative of this argument. The first one was lucidly manifested in a high-profile mega-event hosted by Russia during the war against Ukraine and titled the Games of the Future. It addressed the younger generation and aimed at digressing attention from the insecurities of the military campaign against Ukraine to the celebration of sportive lifestyle and "happy life" mediated by the sovereign state. The Games of the Future held in Kazan in 2024 give a good example of performative biopolitics as a pivotal component of Russian Corporeal Order. The opening ceremony of the event set the stage for blending physical and cyber sports, and in the meantime created a biopolitical frame that links athletic activities, physical fitness, leisure, gaming, environmental technologies, digital lifestyle, and computer medicine. The ceremony was staged as a politically sterile and desecuritised performance in which war was completely silenced and substituted by futuristic themes, such as smart life-enhancing algorithms, ecological solutions for water supply, green urbanism, and new undersea sources of energy (Dizzar_ 2024).

A second telling example of the sovereign proclivity to define what is included in the domains of the necropolitics of war and the biopolitics of caretaking was Putin's remark addressed to the acting head of the Kursk oblast attacked by the Ukrainian Armed Forces in August 2024. When the regional governor started his report with information on the depth of the Ukrainian incursion into the Russian territory, Putin interrupted him by saying "It is up to the military authorities to find out how deep or wide is the penetration. I am expecting you to tell us about how you help people in need" (Diplomatrutube 2024).

These two examples are illustrative gestures of the propensity of the sovereign power to define certain issues as belonging to the register of life and caretaking, as distinguished from war and the corresponding politics of death. They are meant to manipulatively differentiate biopolitics from necropolitics and preconditioned by the sovereign power of drawing the boundary between them.

Secondly, the Corporal Order established by Putin's regime is grounded in the ability to intentionally blur the boundaries between the spheres of bio- and necro-. From a discursive perspective, the idea of indistinction might be seen as a conflation of diametrically opposing concepts that ultimately overlap and converge. Crises and conflicts, including wars, seem to accelerate the ability of sovereign power to create zones or areas of indistinctions. Exactly as civilisation and barbarity might co-exist and give a cumulative effect (Rosenberg 2022), bio- and necropolitics might be seen as the two sides of the same coin. My claim is that the conflation of bio- and necropolitics is an inherent characteristic of the sovereign power that appropriates bare lives (Cairo 2006, 289) in two different ways—biopolitical (manifested as a punitive reaction to expression of corporeality emancipated from the space of state-centric security) and necropolitical (as production of dead bodies).

In line with Agamben's biopolitical thinking, a 'zone of indistinction' erases the dividing line between *zoe* and *bios*, citizens and outlaws, legality and illegality, law and violence, and ultimately life and death (Downey 2009, 112). Therefore, a merger of biopolitics (a politics of life and care) and necropolitics (a politics of death) is a core characteristic of illiberal sovereignty and its Corporeal Order, a system of disciplinary and punitive rules, regulations and procedures that have no clearly identifiable location—they might emerge in a court room, a prison, or a governmental office (Morozov 2017). This ubiquitous bio-/necropolitical binomial might be seen as a reversal of Ranciere's theorizing of democratic politics of resistance as having no specific place and therefore able to pop up everywhere. Contra Ranciere, one may claim that it is practices of bio- and necropower that can emerge anywhere in society where the state appropriates human bodies (Radio Svoboda 2023a) through multiple techniques of biopolitical normalisation (Sem na sem 2024).

The phenomenon of war is particularly important for biopolitical debate since it "brings together the imperative to kill, to take life, with the imperative to care for or cultivate life: deviations or abnormalities have to be killed off in order that life develop as normal and healthy existence" (Binkley 2018, 110). This logic of conflation is opposed to attempts to delineate the two concepts—bio- and necropolitics—as mutually exclusive, and claim that "biopolitics is premised on the disavowal of death" (Murray 2016, 490).

Some biopolitical "theorists have moved decidedly in the direction of the decision or the event … the sudden, miraculous, traumatic flash of lightning that, in the terms of the situation in which it erupts, has no justification and no rationale, and can only be experienced as an exception" (Barbour 2010, 147). From the vantage point of this reasoning, war might be approached as a situation of the sovereign exception which relativizes distinctions between law and fact (Agamben 1997, 112); in other words, the new status quo—for instance, the occupation of Ukrainian territories—is legitimized by Russia through annexation and the materiality of the Russian troops on the ground. "Since it is the sovereign who decides whether this crisis exists, and since this crisis is always one that is extra-constitutional, no law can ever anticipate, that is to say limit, the sovereign in advance" (Gratton 2006, 454). The same argument was well accentuated in post-colonial literature: "sovereignty consists fundamentally in the exercise of a power outside the law" (Bignall and Svirsky 2012, 4), and "law and non-law lose their distinction" (Huang 2014, 42).

Crucial at this juncture is a linkage between sovereign exception and biopolitics in the sense that exceptionality presupposes differential treatment of human lives and bodies (Gratton 2006, 455). But this nexus might be reformulated as follows: whenever there is a sovereign exception, there is a precondition for biopolitics to transform into necropolitics which flourishes due to "a zone of indistinction and continuous transition between man and beast" (Płonowska Ziarek 2008, 99), as well as between bodies and flesh (Gercek 2020). Not incidentally, the metaphor of flesh is widely used in Russian independent and critical discourses to characterise Putin's attitude to the population as cannon fodder (Populiarnaya politika 2023a), which paves the way to fascist aesthetics as an inalienable part of Russia's Corporeal Order (Urzhanov 2024).

The sovereign power interfering in the human body and life in Russia came to its necropolitical finale in the form of the war in Ukraine, when the state does not just appropriate the bodies of the population—it determines who will live, how and where, and who will not live. It is at this juncture that biopolitics transforms into necropolitics: the victims of the war become a component of political calculations, and the cult of death is imposed on the population. The pro-Kremlin television talk shows openly glorify a 'sacred war' against Ukraine which requires sacrificing lives for the homeland. Such a high level of symbolisation of death in Russia blends geo-, bio- and necropolitics with such tropes as a single and indivisible "body of the nation" cemented by the cult of the supreme and irremovable ruler as an indispensable component of Russian Corporeal Order.

5.1 *Russia's Bare lives: The "Naked Party"*

The story of the "almost naked party"—a private event held in one of night clubs in Moscow in December 2023—and its effects are highly illustrative of a collision between the emancipatory lifestyle of Russian celebrities and the Corporeal Order sustained by the state. At the same time, this episode illustrates attempts to use biopolitics as a tool helpful for protecting "traditional Russian values" (Pavlovich and Borovikhin 2021).

Photos of semi-dressed Russian celebrities leaked through social media sparked harsh reaction from the authorities who lambasted the party organizers and participants for disrespecting Russian moral conservatism and offending Russian soldiers waging war in Ukraine. As a reaction to the public outcry, many participants (high-profile artists, singers, and performers) were forced to publicly apologize for their 'misbehavior', and some of them lost lucrative contracts.

The "naked party" became a metaphor for the disciplinary power of the state over human bodies which is central for the new Corporeal Order established by Putin's regime. As a gesture of complete loyalty to the sovereign power Nastya Ivleeva, a famous Russian media celebrity and the organizer of the party, in 2024 visited the occupied territories and gave an interview in which she reconsidered her previous views critical of the Kremlin, and expressed full support for the invasion of Ukraine (Luka Ebkov 2024). The nexus of life and death was crucial here: the necropolitical glorification of war became an indispensable flip side of the biopolitics of care about conservative corporeality and its moral foundations.

5.2 *The World Youth Festival: Asserting Life and Justifying War*

Another example of indistinction of bio- and necropolitics is the World Youth Festival held in Sochi in 2024. On the one hand, the script of this mega-event was replete with multiple references to the life-centric and life-asserting vision of the future, with reiterated references to family, home, friends, and love (mishamsk 2024). On behalf of the Rosatom corporation, the issues of waste management, overpopulation, effects of the pandemic, growing inequality, overproduction, and sustainable development were included in the Festival's agenda (Komsomol'skaya Pravda 2024a). Margarita Symonyan, the editor-in-chief of Russia Today, underscored the biopolitical component of the Festival as a manifestation of Russian soft power, calling Russia "a safe haven for normal people" (Komsomol'skaya Pravda 2024b).

Yet on the other hand, the Festival's scenario contained multiple references to the Ukraine war. Among its keynote speakers was Evgeniy Balitskiy, the Kremlin-appointed head of the Zaporizhzhie oblast who claimed that the occupied territories "is our land where Russia blood was shed ... Our land is everywhere where people speak Russian" (Komsomol'skaya Pravda 2024c), which represents a lucid example of intentional conflation of biopolitics referring to population and necropolitics as a product of sovereign power. Denis Pushilin, the Russia-appointed head of the Donetsk People's Republic, was another protagonist of the Festival's script who exemplified the same logic of indistinction (Komsomol'skaya Pravda 2024d). Even more representative in this respect was Maria L'vova-Belova (Komsomol'skaya Pravda 2024e), Presidential Commissioner for Children's Rights who is in the wanted list of the International Criminal Court—along with Putin himself—for her role in relocation of Ukrainian children from the occupied territories to Russia. Another participant of the Festival was Patriarch Kirill who called on praying for "holy Russia" and "victory" over invisible enemy that morally corrupts humanity under the falsely understood ideas of human rights (Komsomol'skaya Pravda 2024f). Both L'vova-Belova as children's ombudsman and Kirill as the head of the Orthodox 'pastoral power' incarnate the indistinguishable conjunctions of biopower over human bodies and necropower over people's physical lives, which sustains and cements Russian Corporeal Order.

6 Deconstructive Narratives of Popular Biopolitics

In this section I single out and dwell upon alternative narratives of the bio- / necropolitics nexus that challenge the sovereign power and its Corporeal Order. One narrative offers a medicalised vision of the violence pertinent for Putin's regime, while another looks at how the politics of death undermined the grounds of the sovereign rule.

6.1 *The Biocentric Deconstruction of the Imperial Glory*

As many academics deem, biology infiltrates politics as calculable and ordered mode of thinking, with the corresponding medical language (Elden 2008, 24) widely used to describe political events and phenomena. Indeed, it is not rare to hear from policy commentators that Russian society is sick and needs a therapy (TV2 2022), and that Putin is a psychopathic type of personality (Vlast' vs Vlaschenko 2023). Events like Covid-19 and the war in Ukraine make medicalized metaphors more frequent (Radio Svoboda 2024) in the discursive domain of popular biopolitics.

Irina Yakutenko is one of the voices known for her regular public talks in which she discusses the widely spread practices of dehumanisation of Others and the ensuing readiness of society to accept violence. Decisions to go to war, she avers, are not necessarily rational, and might be taken and implemented due to the blockage of logical reasoning and risk assessment (Kak teper' 2023). The biopolitical emphasis on traditional lifestyles, according to her theory, strengthens the sense of existential security as the precondition for biological survival and differentiation from others. The price of exiting an in-group and joining another group is high, which creates majority (Yakutenko 2023) prone to de-sensibilisation and adaptation to violence and the reduction of emotional engagement with ethical or moral issues.

Another public speaker in the field of popular biopolitics is Ilya Kolmanovsky who explains how human beings as particular types of animals consolidate against others. The language of animality towards the enemy is a biological phenomenon of coping with existential threats and an indicator of life instinct. It creates a state of exception as prioritization of physically defeating and disarming the source of danger while temporarily discarding concerns about ethical or moral considerations. Collective identity is a mechanism of in-group favoritism and may be conducive to violence against those who are perceived as strangers and outsiders (Bol'she vsekh nado 2022). This propensity to use brutal force and pro-active unmotivated aggression might be boosted by geographic and cultural proximity of those against whom the aggression is targeted (FREEDOM 2023).

For creating a divide between 'us' and 'them', dictatorial regimes make people accept that killing others is inherent to the mechanisms of survival. What matters for dictators is multiple justifications for enmity towards others and destruction of people's agency as an ability for rational calculation, which makes barbarity—publicly presented as militarised patriotism—possible (Radio Svoboda 2023b; see also Yuliia Kurnyshova's chapter in this volume). Many dictators, according to this reasoning, succeed when they manage to convince people that there is something above individual rights and liberties—something where each person inherently belongs (Prodolzhenie sleduet 2022). The state propaganda intentionally keeps the source of threats uncertain, and appeals to such elementary biological instincts as, for example, fear of Ukrainians as 'super soldiers' transformed by Western biological labs, and revenge for "our children" (Populiarnaya politika 2023b) who were killed or wounded at the war. The tribal structure of Russian society, in Kolmanovsky's view, facilitates this task of instigating a biological instinct among the population, since for survival it is better to be with those who possess brutal power. In this regard war is a consolidating factor that creates a vicious circle: the more Russians die in the battlefield, the more justifiable the whole decision to launch the war looks like (Kotrikadze Dziadko 2022). In Kolmanovsky's opinion, there is no anti-dote to this self-sustaining logic: there might be "only culmination and collapse". In his prediction, a hypothetical de-escalation on the side of the authorities might not necessarily decrease the level of readiness to spill blood in society (Radio Van 2023).

To sum up, these narratives assume that political behavior is biologically conditioned, and political order mirrors the biological one. Hence, to understand politics, one needs to have some knowledge in biology, since, according to this argument, political behavior is shaped by human instincts for power and domination by sexual and gender hierarchies. This logic equates politics with a biological necessity to survive, making life an ontological category that implies violence. This narrative deconstructs Putin's conception of sovereignty as a glorious power sustained by a peculiar 'regime of truth' through uncovering basic instincts of power holders and their subjugated population. It challenges the sublime heroic narrative of Russia about its metaphysical greatness, and illuminates its biologically determined instincts.

6.2 *Russia's Necropolitical Dystopia*

Another deconstructive narrative approaches the phenomenon of necropolitics through the Foucauldian metaphor of a 'beheaded King', implying a collapse of the sovereign power with the ensuing triumph of bare life. On the surface this scenario might appear to be consonant with Agamben's idea of the "global civil war" (Flohr 2014), yet under a closer scrutiny it looks anti-Agambenian in the sense that "bare lives" cease to be subalterns and substitute sovereign power through direct physicality of pure violence.

Mikhail Epshtein's conceptualization of "the culture of the end" as a performative phenomenon lays a pertinent frame for understanding the Russian version of necropolitics. The most important argument in Epshtein's idea of "the culture of the finitude" is that pro-Kremlin—and Kremlin-generated—apocalyptic narratives of a possibility of a nuclear war can't be reduced to a mere bluff; in his view, the necropolitical momentum lucidly articulated and produced by Putin's regime is deeply rooted in Russian culture and more specifically, literature (TV2 2023).

Vladimir Sorokin's novel "The Heritage" published in 2023 is a proper illustration of the necropolitical turn in Russian culture. Sorokin imagines a future Russia after a nuclear war and the following collapse of the central state followed by violent fragmentation of the country into several territories waging perpetual wars against each other and in the meantime being *de facto* colonized by China. Russian culture lost its authenticity, and Russian language transformed into a hybrid newspeak with multiple Chinese loan words. What remained of Russia immersed into archaic tribalism with the cult of physical force and obscene violence at its center. Life is not only de-humanised and devalued; in Sorokin's necropolitical imagery human corpses are used as firewood for trains and locomotives. This symbolic equalisation of human bodies with extractive resources is a highly symptomatic metaphor of predominance of necropolitical appropriation over biopolitics. In "The Heritage" violence is everywhere: military gangs kill each other and attack civilian population without being able to explain why they do so. Pure violence manifests itself in mass-scale sexual abuse and rape. Sorokin's intentional eroticisation of sex violence is meant to point to the Lacanian *jouissance* in which the idea of bare life is re-signified as anatomopolitics of individual desires unconstrained by social norms and conventions.

6.3 *The Necropolitical Activism of Party of the Dead*

Party of the Dead (PD) is an example of radical political art aimed at ironic deconstruction of the death-centric nature of Putin's regime and its reliance on necropolitical practices of appropriating lives and bodies of the population. Maxim Evstropov, the founder of the group, speaks of death as a deeply political phenomenon in the sense of "exploitation of the dead by those who are alive" (Spaika Media 2021) for the sake of cementing relations of power. Evstropov refers to "necroresource" in a broader sense to include oil as a product of dead organisms, and expands his understanding of necropolitics to destruction of sanctioned European food by Russian authorities.

The PD's visual narrative is marked by a strong sense of intentional ambiguity. On the one hand, this narrative is deconstructive of Putin's necropolitical discourse sustained by excitement about—and glorification of—war in the official media (TV2 2018). In Evstropov's words, it is for the sake of worshiping the dead that new corpses appear (Activatica 2022). There are plenty of sarcastic slogans produced by the PD as reactions to the "death drive" coming from the Kremlin: "We don't abandon ours. Only their corpses", "We have one earth", or "Those others have died, and this one will" (an allusion to Putin's mortality).

On the other hand, the PD art project is inspired by the imagination of the "good dead", as opposed to the "useful dead" on behalf of whom the regime justifies its malign necropolitics. Such slogans as "The dead don't go to wars and don't kill" (The Insider 2022) illustrate this. On a broader note, Evstropov conceptualizes the PD discourse as "necrocommunism" in the sense that the usage of the provocative metaphor of dead bodies is a way to express the disdain for nation states and their attributes, and to lambast state authorities' cultural conservatism. In this respect the PD's slogan of 'Liberty, Equality, Mortality' looks like a mockery of European liberalism, but in the meantime, it also expresses a strong sense of disdain for Putin's necropolitical regime.

6.4 *Pussy Riot: Aesthetic Resistance to Putin's Necropolitics*

The Pussy Riot art project is another apposite example of radical critique of the male-dominated Corporeal Order based on violence and physical coercion, and ultimately conducive to war. Sarcastic deconstruction of the "male state" and such its supporters as the Orthodox Church (Pussy Riot 2022c) is the core focal point of Pussy Riot's performances—either through hyper-sexualization of female bodies as objects of the male possessive gaze and desire ("Plastic") (Pussy Riot 2022a), or through visual animalisation of male bodies as symbolising rape and aggression (Pussy Riot 2021). Pussy Riot's pro-Ukrainian musical videos (Pussy Riot 2022b) are direct extensions of their gendered anti-Putin artistic utterances. As a feminist art group, they are deeply engaged with necropolitical themes, artistically performing what Epshtein called the cult of death inherent to Putin's regime producing such threats as nuclear war (Pussy Riot 2019b) or environmental disasters. Pussy Riot not only unveils the violent nature of both Soviet and Russian regimes of power (Pussy Riot 2019a), but also symbolically strikes back. Thus, the script of "Putin's Ashes" (Pussy Riot 2023) is built on Putin's imagined funeral, while in the "Knife" (Pussy Riot 2020) the young female protagonist promises to kill her offender for the sake of restoring justice. These deconstructive gestures are meant to aesthetically imagine and represent a space of symbolic resistance to the whole range of oppressive practices—from domestic violence to atrocities in Ukraine.

Conclusion

The biopolitical turn—the focusing on the centrality of the body as a key category of analysis of power relations—is apparently a global phenomenon: one may find some seeming parallels between Russia's sovereign biopower and similar trends in Western countries, particularly among right-wing populists adhering to the organicist understanding of the nation to be cemented with family values and religious indoctrination. Against this background, the question of whether Russia is a part of a much broader trend or should be tackled as a specific case is open to debate. On the one hand, given the multiplicity of forms and patters of biopower, there is always an ample space for comparing Russia with biopolitical practices in other countries building their national identity on constitutive appeals to loyal and disciplined bodies as opposed to "national traitors" and the "fifth column". On the other hand, unlike most of other promoters and sympathisers of biopower in the West, Russia is a special case in at least two regards. First, Russian sovereign power extends biopolitical instruments of control and regulation to necropolitical repressions within the country and external force projection. Secondly, Russian biopolitical sovereignty is an instrument for empire-rebuilding, as opposed to nation-building which is the case of countries of Central and Eastern Europe.

As this analysis demonstrated, the Russian case not only can be effectively studied using key biopolitical categories, but it also can contribute to advancing biopolitical theory. The concept of Russian Corporeal Order introduced in this chapter is an example of such a contribution. Unlike the notion of biopower with its grounding in the Foucauldian traditions of treating relations of power as productive and stimulating, the idea of Corporeal Order is bent on multiple zones of indistinction between the spheres of bio- and necro-politics. The intentionally blurred borderline between life and death creates an ample space for sovereign exceptionalism, interventionism and decisionism.

This chapter was meant to join the academic voices who wish to make biopolitical conceptualisations less Western-centric and less Western-skeptic. In this regard, I am supportive of projecting the concepts of *homo sacer*, bare life, the camp and *zoepolitics* to non-Western regimes, as opposed to their attribution to the liberal West. Russian Corporeal Order implies the grounding of politics in the laws of wild nature and physical force, with a strong subculture of prison, from GULAG to incarceration of the anti-Putin opposition. This Order is grounded in "natural", "organic" understanding of life, and legitimises possessive management, administration, and control over human bodies. Being religion- and family-centric, Russian Corporeal Order glorifies the "collective body" of the empire, normalises sexuality and reproductive behavior, and expresses explicitly illiberal disdain for lifestyles, feminism, and gender emancipation. At the same time, "care" and "protection" of "compatriots" beyond Russian borders is an indispensable part of this order as well.

A zone of indistinction between bio- and necropolitics is an important characteristic and a comprehensive feature of the new Russian Corporeal Order. It vacillates between politics of care and politics of death exemplified by the so-called "filtration camps" for Ukrainian refugees on the Russian border, where civilians are exposed to humiliating practices of inspection which includes their undressing and examination of their bodies. There are apparent parallels between physical violence inside the Russian repressive apparatus (police terror, regular torture of suspects and prisoners), and Russian war crimes in Ukraine: the necropolitics of the repressive regime have been exported on the body of the neighboring nation and has led to the ideas of inferiority of Ukrainians, and the glorification of violence. The biopolitical concept of the "Russian world", with its own "blood and soil", has led to the de facto ethnic cleansing of the Ukrainians, denying them their national sovereignty, identity, history, language and finally, physical existence. All this makes studies in necropolitics increasingly topical for the disciplines of security studies, foreign policy analysis, international relations and comparative politics.

References

Activatica. 2022. V Tbilisi 'Partiya Miortvykh' provela antivoennuyu aktsiyu. YouTube, May 11, https://www.youtube.com/watch?v=FPUU_HUjRBg

Agamben, G. 1997. The Camp as the Nomos of the Modern. *Violence, identity and self-determination*, 106–118.

Agamben, G. 2007. Il Regno e la Gloria. Per una genealogia teologica dell'economia e del governo. Vicenza: Neri Pozza.

Agamben, G. 2024. Qualche notizia sull'Ucraina, August 2, https://www.quodlibet.it/giorgio-agamben-qualche-notizia-sull-u2019ucraina?fbclid=IwY2xjawE_CThleHRuA2FlbQIxMAABHceIELoNyoWEKTuYelnoFsH4wMXM4pvovmUK_g3_CNZvx-dRGk5sCs9gyw_aem_aPb2LpdGSrVD-xQE5Od8ZQ

Alexeeva-Karnevaly, O. 2022. Konflikt na Ukraine: biopolitika i strategiya, Russian Council on Foreign Affairs, April 5. https://russiancouncil.ru/analytics-and-comments/columns/sanctions/konflikt-na-ukraine-biopolitika-i-strategiya/

Barbour, Ch. 2010. Sovereign Times: Acts of Creation, *Law, Culture and the Humanities* 6 (2), 143–152.

Bignall, S. and Svirsky, M. 2012. Introduction: Agamben and Colonialism, in Marcelo Svirsky and Simone Bignall (Eds.), *Agamben and Colonialism*. Edinburgh University Press.

Binkley, S. 2018. Biopolitical Metaphor: Habitualized Embodiment between Discourse and Affect. *Body & Society*, 24 (3), 95–124.

Bol'she vsekh nado. 2022. Kak voina delaet iz liudei zhivotnykh / Ilya Kolmanovskiy. YouTube, November 25. https://www.youtube.com/watch?v=yocp-L6ptSo

Bussolini, J. 2010. What Is a Dispositive? *Foucault's Studies* 10, November, 85–107.

Cairo, H. 2006. The Duty of the Benevolent Master: From Sovereignty to Suzerainty and the Biopolitics of Intervention, *Alternatives* 31, 285–311.

Deranty, J.-P. 2008. Witnessing the Inhuman: Agamben and Merleau-Ponty, *South Atlantic Quarterly*, winter, 165–186.

Diplomatrutube 2024. Vladimir Putin prerval Alekseya Smirnova vo vremia rasskaza o proniknovenii VSU v Kurskuyu oblast'. YouTube, August 12, https://www.youtube.com/watch?v=7SVB9LhtMI8

Dizzar. 2024. Fidgital Igry Buduschego 2024. Tseremonia otkrytia. YouTube, February 22, https://www.youtube.com/watch?v=FYB-tPcdBjY

Downey, A. 2009. Zones of Indistinction: Giorgio Agamben's 'Bare Life' and the Politics of Aesthetics. *Third Text*, *23* (2), 109–125.

Elden, S. 2010. Strategies for Waging Peace: Foucault as *Collaborateur*, in Michael Dillon and Andrew Neal (Eds.), *Foucault on Politics, Security and War*. Basingstoke and New York: Palgrave Macmillan, 21–42.

Evans, B. 2010. Foucault's Legacy: Security, War and Violence in the 21st Century, *Security Dialogue* 41 (4), 413–433.

Flohr, M. 2014. The Problem of Civil War in Agamben's Thought. *Semikolon* (Aarhus C), 15 (27), 61–68.

FREEDOM. 2023. Voina i faktor evoliutsii. Otvetstvennost' za moral'niy vybor. Interview s Kolmanovskim. YouTube, April 18. https://www.youtube.com/watch?v=vNOagxaNERA

Journeyman Pictures. 2017. Pavlensky: Life Naked / Trailer / Available Now. YouTube. March 2. https://www.youtube.com/watch?v=ETyDlNmaeCw&t=8s

Haskaj, F. 2018. From biopower to necroeconomies: Neoliberalism, biopower and death economies, *Philosophy and Social Criticism* 44 (10). 1148–1168.

Huang, Han-yu. 2014. "The Wound Is Healed Only by the Spear That Smote You": The Undead, Biopolitics and Psychoanalysis, *NTU Studies in Language and Literature* 39 (31), June, 39–63.

Gercek, S.E. 2020. From body to flesh: Lefort, Merleau-Ponty, and democratic indeterminacy, *European Journal of Political Theory* 19 (4), 571–592.

Gratton, P. 2006. A 'Retro-version' of Power: Agamben via Foucault on Sovereignty, *Critical Review of International Social and Political Philosophy* 9 (3), 445–459.

Grozny: the Chechen State Broadcasting Company. 2022. Ramzan Kadyrov rezko osudil podniatie flaga LGBT odnoi iz mechetei Germanii. July 3, https://grozny.tv/index.php/news/main/49179

Independent Institute of Philosophy. 2023. Reproductive Coercion and Biopolitical Control: Russian and Belarusian Cases and Trends. YouTube, July 3, https://www.youtube.com/watch?v=KcBv_VH3s-U

Kak teper'. Proyekt OVD_Info. 2023. Kak teper' voina, propaganda, emigratsiya vliyaut na mozg i nashi resheniya / Irina Yakutenko. YouTube, September 28, https://www.youtube.com/watch?v=kBZdDP5itzc

Kiersey, N. 2009. Scale, Security, and Political Economy: Debating the Biopolitics of the Global War on Terror, *New Political Science* 31 (1), 27–47.

Komsomol'skaya Pravda. 2024a. Vsemirniy festival molodiozhi—2024 v 'Siriuse'. YouTube, March 2. https://www.youtube.com/watch?v=tZuCzRtDEZ0

Komsomol'skaya Pravda. 2024b. Glava RT Margarita Simonyan vystupaet na Vsemirnom festival molodioshi. YouTube, March 3. https://www.youtube.com/watch?v=Ma4xCZskEXc

Komsomol'skaya Pravda. 2024c. Gubernator Zaporozhskoi oblasti Evgeniy Balitskiy na VFM-2024: priamaya transliatsiya. YouTube, March 6. https://www.youtube.com/watch?v=dIBhHisKFmA

Komsomol'skaya Pravda. 2024d. Glava DNR Denis Pushylin vystupaet na VFM—2024: pryamaya translyatsiya. YouTube, March 5. https://www.youtube.com/watch?v=ilbFioEQcUg

Komsomol'skaya Pravda. 2024e. Maria L'vova-Belova vystupaet na VFM—2024: pryamaya translyatsiya. YouTube, March 5. https://www.youtube.com/watch?v=vhwmDxFvZYg

Komsomol'skaya Pravda. 2024f. Vsemirnyi Festival Molodiozhi 2024 v Siriuse. YouTube. March 2. https://www.youtube.com/watch?v=tZuCzRtDEZo

Kotrikadze Dziadko. 2022. Bol'shinstvo zaimiot passivnuyu pozitsiyu. Kolmanovskiy o tom, pochemu liudi viyuyuy i veriat propagande. YouTube, November 21. https://www.youtube.com/watch?v=BtgSef8CkhI

Luka Ebkov. 2024. Nastya Ivleeva: anti-vecherinka zashla slishkom daleko. YouTube, May 3. https://www.youtube.com/watch?v=OhXb1WN3llk

Mbembe, A. 2003. Necropolitics, Public *Culture* 15 (1), winter, 11–40.

McLoughlin, D. 2012. Giorgio Agamben on Security, Government and the Crisis of Law, *Griffith Law Review*, 21 (3), 680–707.

Mishamsk. 2024. Tseremoniya otkrytiya vsemirnogo festivalia molodiozhi 2024. YouTube, March 2. https://www.youtube.com/watch?v=rX8bGmtX1Jw

Morozov, A. 2017. Agamben i 'Kirovles', INLIBERTY, February 12. https://old.inliberty.ru/blog/2503-Agamben-i-Kirovles

Murray, Stuart. 2016. Affirming the Human? The Question of Biopolitics, *Law, Culture and the Humanities* 12 (3) 485–495.

Naterstad, T.B. 2023. The Reproduction of Nationalism and the Nationalism of Reproduction: Putin's Biopolitics of Defending Tradition, 2012–2021. *Nationalities Papers*, 1–20. Doi:10.1017/nps.2023.85

Oksala, J. 2010. Violence and the Biopolitics of Modernity, *Foucault Studies* 10, November, 23–43.

Pavlovich, V. and Borovikhin, A. 2021. Upravlenie kak faktor razvitiya biopolitiki (biopoliticheskikh protsessov) v Rossii, *Context and Reflection: Philosophy of the World and Human Being* 10 (2A), 43–37.

Płonowska Ziarek, E. 2008. Bare Life on Strike: Notes on the Biopolitics of Race and Gender, *South Atlantic Quarterly* 107 (1), Winter.

Populiarnaya politika. 2023a. Putin schitaet rossiyan myasom / Dmitriy Bykov. YouTube, September 16. https://www.youtube.com/watch?v=koLLe8ObLwo

Populiarnaya politika. 2023b. Kolmanovskiy pro propagandu, evoliutsiyu i ChatGPT / Chestnoe slovo s Ilioy Kolmanovskim. YouTube, December 23. https://www.youtube.com/watch?v=QmyQA1b27ME&t=49s

Prodolzhenie sleduet. 2022. Biolog Ilya Kolmanovskiy: chto zastavliaet liudei nenavidet' i voevat'. YouTube, October 26. https://www.youtube.com/watch?v=TixJOfLMxM4

Pussy Riot 2019a. Pussy Riot—1937 (Official Music Video). YouTube, September 12. https://www.youtube.com/watch?v=eCA8pjbtcCM

Pussy Riot 2019b. CHORNYE SNEZHKI / BLACK SNOW feat. MARA 37 (Official Music Video). YouTube, July 9. https://www.youtube.com/watch?v=ss3ONoKKMN8

Pussy Riot 2020. Nozh—Knife (Official Music Video). YouTube, May 13. https://www.youtube.com/watch?v=dX8ePP2ungo

Pussy Riot 2021. SEXIST—feat. Hoffmanita (Official Music Video). YouTube, March 4. https://www.youtube.com/watch?v=_fjoa77sKUM

Pussy Riot 2022a. PLASTIC—Pussy Riot fi iLoveMakonnen (Official Music Video). YouTube, July 8. https://www.youtube.com/watch?v=UhdSQdXEdqo

Pussy Riot 2022b. MAMA, DON'T WATCH TV / MAMA, NE SMOTRI TELEVIZOR (ANTI-WAR SONG). YouTube, December 24. https://www.youtube.com/watch?v=zroGN2llJaY

Pussy Riot 2022c. Pussy Riot feat. Slayyter—HATEFUCK (Official Music Video). YouTube, April 13. https://www.youtube.com/watch?v=Z48fNpTEPqg

Pussy Riot 2023. Putin's Ashes (Official Short Film). YouTube, January 23. https://www.youtube.com/watch?v=ni_CRPAw_5Q

Radio Svoboda. 2023a. Gosudarstvo prisvaivaet tela liudei. Proyekt Sergeya Medvedeva. YouTube, November 12. https://www.youtube.com/watch?v=LDloZOiEgwA&t=1s

Radio Svoboda. 2023b. Ilya Kolmanovskiy: kak propaganda ispol'zuet nesovershenstvo mozga dlia ubiystva liudei / *Grani vremeni*. YouTube, June 25. https://www.youtube.com/watch?v=1v2TAA3bWjI&t=1851s

Radio Svoboda. 2024. Voina v Ukraine—rak, dayuschiy metastazy po miru? Proyekt Sergeya Medvedeva. YouTube, February 4. https://www.youtube.com/watch?v=YtdnELuzxgo

Radio Van. 2023. Ilya Kolmanovskiy ob issledovanii predvoennykh nastroeniy rossiyan i o tom, chto budet posle Putina. YouTube, November 17. https://www.youtube.com/watch?v=wgycDlaX4dU

Reid, J. 2004. War, liberalism, and modernity: the biopolitical provocations of 'Empire', *Cambridge Review of International Affairs*, 17 (1), 63–79.

Reid, J. 2010. Life Struggles: War, Discipline and Biopolitics in the Thought of Michel Foucault, in Michael Dillon and Andrew Neal (Eds.), *Foucault on Politics, Security and War*. Basingstoke and New York: Palgrave Macmillan, 65–92.

Rigi, J. 2007. The War in Chechnya. The Chaotic Mode of Domination, Violence and Bare Life in the Post-Soviet Context, *Critique of Anthropology* 27 (1), 37–62.

Rosenberg, G. 2022. Civilization and Auschwitz are not a contradiction in terms. *Rosenberg Ideas*, April 26 https://engelsbergideas.com/essays/civilisation-and-auschwitz-are-not-a-contradiction-in-terms/?fbclid=IwAR0rsEvPs-q_Up4iintD_3cF_YB5CF4mFrJFdlG1yl9K9Mxnw95n9_hwzQc

Schlosser, K. 2007. The bio-politics of bodies politic: nature and intertextuality in classic US geopolitical discourse. *GeoJournal* 69, 199–210. https://doi.org/10.1007/s10708-007-9046-3

Sem' na sem'. 2024. Nemyslimye repressii stanovyatsa rutinoi. Chto takoe normalizatsiya—mekhanizm kotorym rossiyskaya vlast' podchinyaet grazhdan. June 4. https://semnasem.org/articles/2024/06/04/nemyslimye-repressii-stanovyatsya-rutinoj-chto-takoe-normalizaciya-mehanizm-kotorym-rossijskaya-vlast-podchinyaet-grazhdan

Short, J. 2005. Life and Law: Agamben and Foucault on Governmentality and Sovereignty, *Journal for the Arts, Sciences, and Technology* 03 (01).

Shulman. E. 2023. Programma "Znanie—Sila". 1 vypusk. Gost'—Irina Yakutenko, molekuliarniy biolog i nauchniy zhurnalist. YouTube, April 23. https://www.youtube.com/watch?v=jOkKEU1xrbE

Spaila Media. 2021. Online priyomnaya Partii Miortvykh. YouTube, June 8. https://www.youtube.com/watch?v=W4SrWetuBpQ

Spieker, J. 2011. Foucault and Hobbes on Politics, Security, and War, *Alternatives: Global, Local, Political* 36 (3), 187–199.

Stumer, J. 2018. The Dead Are Coming: Border Politics and Necropower in Europe, *Cultural Politics* 14 (1), March, 20–39.

Terry, G.S. 2024. Martial necropolitics: The position of the combatant in the Russo–Ukrainian war. *New Perspectives* 32 (3), 307–324.

The Insider 2022. Samomogilizatsiya: Halloween Special ot Partii Miortvykh. YouTube, October 28. https://www.youtube.com/watch?v=l2Xdoxtfiag&t=2s

Troncota, M. 2017. Interventionism and Biopolitics in the Ukrainian Crisis: The Disputed Claims of Humanitarian Intervention in Russia's Legitimizing Narratives for the Annexation of Crimea, in Miruna Troncota and Vasile Rotaru (Eds.). *Russia and the Former Soviet Space: Instrumentalizing Security, Legitimizing Intervention*. Cambridge Scholars Publishing.

YV2 2018. Interview s osnovatelem dvizheniya Partii Miortvykh. YouTube, February 27. https://www.youtube.com/watch?v=3esh0YwiDJg

TV2. 2022. Izlechim li massoviy psykhoz rossiyan? Psykhiatr s 50-letnim stazhem Alexandr Pravdin v 'Ochevidtsakh'. YouTube, December 19. https://www.youtube.com/watch?v=KOLUPObphLk&t=670s

TV2. 202. Voina obnazhila dno russkoi kul'tury i istorii. YouTube, November 13. https://www.youtube.com/watch?v=w6wF4jYUChw&t=495s

Urzhanov, A. 2024. Vstanem! Kennan Institute, Wilson Center, April 1. https://www.wilsoncenter.org/blog-post/vstanem?fbclid=IwAR3qhjHpoB_3mDrJ9hPV1m1BSlMM1ihaFLckhpB5s-Lep3FmYIQQpcv5wOs_aem_AUHIddUE9fNw5DixSaWcWo8BHDbrXxS4DwDxMJOLylGF-jH2M4sWLLHW9As0aufp__Ec4eAImU2bTN83c7SU3AVc

Valverde, M. 2010. Law versus History: Foucault's Genealogy of Modern Sovereignty, in Michael Dillon and Andrew Neal (Eds.), *Foucault on Politics, Security and War*. Basingstoke and New York: Palgrave Macmillan, pp. 135–150.

Vlast' vs Vlaschenko. 2023. Diagnoz vsplyl posle lodki 'Kursk', kto ubival v techenie XX veka. Ilya Kolmanovskiy. YouTube, October 5. https://www.youtube.com/watch?v=TE-foK3h3ao

Wichum, R. 2013. Security as Dispositif: Michel Foucault in the Field of Security, *Foucault Studies*, February 15, 164–171.

Whyte, J. 2013. *Catastrophe and Redemption. The Political Thought of Giorgio Agamben*. Albany: State University of New York Press.

Yakutenko, I. 2023. Pochemu diktatory budut obyazatel'no regulirovat' vashu intimnuyu zhizn'. YouTube, August 24. https://www.youtube.com/watch?v=pNg6xcVPizA

CHAPTER 6

Narratives on Civilization and Barbarity During the War in Ukraine

Yuliia Kurnyshova

1 Introduction

In this chapter I discuss how the interwoven concepts of civilization and barbarity can be used as analytical categories for understanding the multiple political layers of Russia's war against Ukraine and its repercussions for international security. I argue that the mass-scale violence committed by Russia's occupational forces in Ukraine sharpened the academic necessity to find proper approaches to analyze the war and its malign effects on the system of International Relations. The war re-actualized the old opposition between civilization and barbarity, and modified it according to the current state of academic debates in security studies. The contrasting juxtaposition of these two concepts is instrumental not for reproducing simplistic political binaries, but rather for a more nuanced and contextual analysis of the emerging security discourses. Thus, in the case of Russia, my approach helps to critically deconstruct the Kremlin's strategic narrative of civilizational state and reveal its barbaric underside. When it comes to Ukraine, its civilizational discourse vacillates between asserting this country's belonging to Western normative space and appeals to universal or cosmopolitan norms distinguishing civilization from its opposite. As for the West, its discourses shift between limited attempts to exclude Russia from the purview of the civilized international society and adhering to the principle of an ordered status quo, even at the expense of the otherwise cherished justice.

Unlike many previous studies whose authors looked at the concepts of civilization and barbarity through identity lens (Frear and Mazepus 2021), I approach them as practices that define relations of inclusion and exclusion. I argue that civilizational discourse may be used as a camouflage for barbarian practices; yet the recognition of barbarity may lead to discursive closure that minimizes chances for a negotiated settlement of a conflict or crisis. Therefore, barbarity mostly functions as an ethical construct, but does not automatically translate into legal and politically consequential definitions.

My methodological approach is based on delineation between ontological and epistemological categories. In line with recent civilizational scholarship

 | DOI:10.1163/9789004747364_007

(Hall 2007), I question essentialist claims about civilization while acknowledging the influence these claims have in social and political practices. I share an interpretivist epistemology skeptical toward "assigning any ontological reality to civilizations beyond the understandings and practices of the social actors that constitute and instantiate them" (Bettiza 2014, 3). By the same token, I treat civilization as a discursive marker that defines the choices faced by actors directly and indirectly involved in the war. These choices determine the political function of civilization discourse: it is up to Russia to build its international role as a civilization (and thus integrate itself in the structures of international society) or as a barbaric state (and therefore to dissociate itself from it). By the same token, it is up to Ukraine to define its belonging to the Euro-Atlantic civilization or to the universal norms shaping world / global / cosmopolitan society. The Euro-Atlantic West, in its turn, faces the choice between excluding Russia from the normative and institutional spaces of world politics defined by democratic values, and finding a compromise with Putin's regime in a hope of its ultimate democratic transformation.

Two additional methodological moves are important at this juncture. First, for the purpose of my research, civilizational analysis has to be extended from issues of cultural sociology, ethnography and anthropology to the domains of geopolitics and security studies. In my previous publications I applied the nascent concept of civilizational security (Kurnyshova 2024a) for studying Ukraine's discourses during the war, and in this paper will continue projecting and developing it further on.

Second, the research agenda on civilizations is analysed through the lens of how references to civilizational identity are employed in political discourses: "How do representations of civilization/s impact upon and influence political perceptions and interaction?" (O'Hagan 2007, 16). On this basis, discourse analysis is helpful for differentiating civilization in the singular (as a marker distinguishing the civilized from the barbaric), and in the plural (as a term describing distinct macro-cultural communities with dissimilar ethnic, linguistic and historical backgrounds) (Katzenstein 2010, 1–40). Discourse analysis is instrumental for understanding how the notions of civilization and barbarism construct speaking positions of the involved actors and build different types of agency within international society. In other words, the discursive approach shifts attention from the question of *what* civilization *is* to the question of *how* it is politically operationalized, instrumentalized and signified. Jacinta O'Hagan (2020) has convincingly applied discourse analysis as an interpretivist method to unpack the multiple linkages between civilization as a rhetorical tool and key concepts of the English school, including the idea of international society, and I will follow this reasoning bearing in mind drastic changes incurred by the war in Ukraine.

The chapter contains four sections. I start with a theoretical framing of the opposition between civilization and barbarity and relate it to the current discussions about Euro-Atlantic international society. In the next section I discuss the idea of multiple civilizations in the Russian strategic narrative, followed by a contrasting example of Ukraine's discourse on civilization and barbarity. In the remaining part I turn to dilemmas and challenges faced by the Euro-Atlantic security community when dealing with Russia's voluntary digression from norms and principles defining civilized politics. For source material, I draw on Russian, Ukrainian and Western strategic documents and speeches by prominent officials after 24 February 2022.

2 Theorizing Exclusions: A Conceptual Conundrum

There are different ways of conceptualizing exclusion in contemporary political theory. In this section I assemble a series of exclusionary concepts as elements of a discursive chain, and then focus more specifically on the concept of barbarity and its political function of establishing criteria and standards of inclusion in and exclusion from international society. Notably, all these concepts are genealogically inherent to Western discourses and connote a certain degree of securitizable deviation from the socially established norms by the West.

This chain might be started with *Orientalization*, a concept known as part of post-colonial theorizing. The specificity of Ukraine's positioning within this discourse is that it finds itself as an object of different forms of Orientalization by Russia and the West, and meanwhile, it itself orientalises Russia. Indeed, on the one hand, Russian colonialism has for centuries represented Ukraine as its exotic, semi-wild, archaic southern borderland. And the collective West has easily adopted the same perspective of Ukraine and Ukrainian people as a *fait accompli* without even attempting to question its logic and validity (Kotlyuk 2023). On the other hand, an important component of Ukraine's agency is not simply distancing from the space of the *Russian world*, but a policy of orientalizing Russia as a despotic, brutal and tyrannic country with a heavy legacy of coercive subjugation of its smaller neighbours.

Another way of thinking about relations of exclusion is the idea of *dark power* which was applied to the most radical norms deniers challenging the normative supremacy of the Euro-Atlantic international society (Maier 2007). The projection of this approach to Russia points to such *dark* features of

Putin's regime as moral nihilism, "military expansionism, minority hate, a brutal repression of dissent, and the demonization of opponents" (Kuzio, 2024), the mafia-style mentality and disregard for human sufferings, all ultimately leading to self-dissociation of Russia from the space of global norms. Dark power can also be applied to criminal states (Kan 2019), as well as *gangster states* and *mafia states*. These labels criminalize governments disobeying international rules and disrespecting principles of transparency, accountability and openness. *Rogue states* fall into the same category which implies different forms of international pressure, including sanctions, embargoes and other restrictive measures.

One of the political philosophers who was interested in mechanisms of social exclusion was Michel Foucault whose concept of *monstrosity* connotes a radical breach of law that entails disturbing confusion in society. In multiple interpretations of this approach monstrosity was discussed as an unusual abnormality incarnating "the impossible and the forbidden" (Sharpe 2007, 386). The metaphorical figure of a monster symbolizes exception and "an absolute alterity exceeding and violating all pre-existing norms, rules, jurisdiction or jurisprudence" (Mercier 2019, 109). In other words, the concept of monstrosity points to transgressions: "the monster should appear beyond all models or patterns. It should instantiate, in its irruption, its rule, its own model" (Mercier 2019, 110). Importantly, monstrosity not only challenges law, but disables it, thus exposing a situation of undecidability on the side of norm producers.

The concept of evil has religious roots and brings a strong component of the ethics in constructing relations of radical alterity and exclusion. The interest in the phenomenon of evil in international relations was boosted by such security-ridden events as the Holocaust, 9/11 and cases of mass-scale genocide in Rwanda and Bosnia (Rengger and Jeffery 2005, 3–15). Arguably, if the enemy or threat is successfully framed as evil, it can change the character of the conflict to become apocalyptic, in turn increasing the legitimacy of "by any means necessary" (Sheikh 2014). The ensuing debates were instrumental for differentiating between norms and exceptions, and legitimizing extraordinary measures in exceptional cases: If the adversary is depicted as evil, negotiations are foreclosed. Such an actor does not represent an ordinary enemy (Geis and Wunderlich 2014).

All these discourses have a common element—a sense of radical otherness, crisis and exception. In this vein, they can be easily embraced by the civilizational language as it is used to define who is included within the boundaries of a community (O'Hagan 2007, 21). At the same time, the discourse of civilization has a dramatic amplification effect, and intends to mobilize support for military action against barbaric enemies (Linklater 2021, 34).

2.1 *Standard of Barbarity*

Barbarity is a part of the chain of—metaphorically speaking—"red cards" meant to expel violators of rules and laws from the international stage. In academic literature the phenomenon of barbarity is described as imposed on certain ethno-religious groups who are considered inhuman and cast out by the hegemonic core of world politics. As seen from a constructivist perspective, the key is a separation of a civilized interior from a barbaric exterior.

The political West, being recently the main object of decolonization of knowledge, engages with the concepts of civilization and barbarity quite cautiously and selectively, since it was lambasted by many authors as allegedly having racial and even white supremacist underpinnings. Since the end of the Cold War the Western liberal discourse embraced the cult of inclusiveness and the end-of-history paradigm which either excluded the concept of barbarity from the analytical purview, or blurred distinctions between civilization and barbarity.

Together with barbarity, the conception of evil was relativized, assuming that evil could be reintegrated simply by the attractiveness of the good. The multiple forms of barbarity were pushed out by the omnipresent *otherness*, which de facto legitimized any alterity and accepted any distinction and claim for a peculiar identity. Within this universal and politically correct semiotic system, nothing is beyond the civilizational patchwork, and each deviation is seen as just another sign of cultural specificity, ontologically equal to other identities.

Against this background, barbarism became invisible (Al-Azmeh 2001, 89), but it did not disappear as an inherent part of international relations. Beyond the perception of inferiority remains a strong linkage between the barbarity and organized ardent violence which is in contrast with whatever we might define as civilization. Barbarity is a term associated with particular forms of violence, certain types of behaviour such as murder, massacre, torture, rape, slavery, systematic beatings, planned and organized starvation or infection (Krieken 2003, 298).

The references to barbarity pop up in times of such critical junctures as 9/11 (Kassimeris 2006) and the war on terror, and with the emergence of violent extremist groups as Islamic State, Al Qaeda, Boko Haram and Hamas that have been roundly condemned both for their level of violence and its indiscriminate nature of attacks. Their conduct has been consensually denounced as cruel, brutal, despicable, and therefore barbaric.

Since more and more academic and political voices claim that "history is back", is barbarity returning as well to our analytical purview? I tend to answer this question affirmatively. Barbarism can't be defeated and erased once and forever from the international scenery: as a set of practices, it reappears under new circumstances in different versions, all based on the domination of physical and muscular force over norms and rules. In this sense, if the concept of a revamped *standard of civilization* is still used to define what practices might be defined as "civilized", it is logically to assume the possibility of re-barbarization as a specific set of norms or actions that define de-civilized behaviour, sort of *standard of barbarity*.

A standard of barbarity would denote practices that are consensually condemned for their violation of basic human rights, such as indiscriminate violence, disregard for international law, systemic dehumanization of opponents, and the use of force as a primary means of governance or conflict resolution. Just as the standard of civilization seeks to elevate norms of diplomacy, governance, and humanity, the standard of barbarity would categorize behaviours that regress towards imperial violence, ethnic cleansing, or the subjugation of entire populations through brutality and domination.

In advocating for bringing barbarism back to political analysis, I rely on the liberal interpretation of international political and legal institutions as barriers for violence, yet in the meantime, I am more critical of the erasure of barbarism as the opposite to civilization from the liberal discourse. By combining civilizational process of Norbert Elias and core arguments of the English School in IR, I hold that barbarity includes two major components reinforcing each other and giving a cumulative effect.

First, barbarity implies direct physical violence and substitutes institutional politics and legal norms with a direct physical/muscular action. It is expedient to treat barbarity and violence as synonyms and identify civilized society with non-violent forms in the settlement of conflicts at domestic, international, and transnational levels.

Most civilizations have established war conventions which recognize the rights of noncombatants to be protected from unnecessary suffering. Furthermore, the world's major religions uphold the ethical principle of non-harm to civilians. This ethos of non-violence and the principle of self-restraining from causing harm to others serve as a basic standard of conduct and civilization agreed upon by different societies. Despite varying lifestyles, moral and cultural differences, societies seek to achieve peaceful coexistence by committing to this fundamental principle of non-harm. Here civilizations in *the plural* overlap with civilization in the *singular*, as all countries that subscribed to the UN charter, Geneva Conventions and Additional Protocols to it commit to the universal norms of unnecessary harm avoidance.

Second, barbarity necessarily includes self-detachment and dissociation from the pivotal norms and principles of international society. This corresponds to Eric Hobsbawm's definition of barbarity as "the disruption and breakdown of the systems of rules and moral behaviour by which *all* societies regulate the relations among their members and, to a lesser extent, between their members and those of other societies". This interpretation points to breakdown rather than to contestation of norms and ethical principles (Bettiza et al. 2023). In a similar vein, Mark Salter (2007, 84–85) claimed that the discourse on civilization and barbarity constructs the enemy as outside the boundaries of normal politics, which leads to different forms of exceptionalism applied to those termed "barbarians".

Two important theoretical caveats need to be added to this discussion. First, civilization and barbarity, in my interpretation, are applicable not to societies or states, but to practices that might exist anywhere and take different forms. Social groups and governments might voluntarily accept and appropriate these practices and build their policies or strategies on this basis. This points to a processual analysis of civilizations and barbarism pioneered by Norbert Elias who discussed the civilization—barbarism nexus as a complex arrangement of habits, principles, and actions.

Second, civilization and barbarity is not a dichotomy but rather a frame that denotes a spectrum of attitudes and narratives practiced by many actors. Neither concept is static—there are plenty of shades of both civilization and barbarity as dynamic processes (hence re-barbarization, or de-civilization) that might overlap and hybridize. One may argue that "barbarity and civilization emerge as mutually constitutive components of dialectical contradiction" (Offe 1996, 355) embedded in modern societies facing multiple insecurities. At the same time, civilization is not immune to degrading into a barbaric state, which has been proved by plethora of historical examples.

3 "Civilization" a-la Russe and Its Barbaric Underside

The Russia's war on Ukraine exemplifies the construction and practical application of the civilization-barbarity nexus. From an ontological standpoint, civilizational elements should not factor into this conflict, as it occurs between two states that lack racial or significant religious-cultural differences. Contrary to the views of Toynbee and Huntington, which are currently heavily contested, countries that are civilizationally similar have found themselves on opposite sides of the front line. This supports constructivist view that we are "living in a world in which civilisation is fast becoming the currency of international politics" (Coker 2019). By invoking civilizational tropes, Russia taps into national megalomania, emphasizing notions of greatness, self-sufficiency, distinction from the West, and spheres of influence.

The usage of the concept of civilization in Russian discourse since 1991 is illustrative of the overall trajectory of Russia's political regime. In the 1990s the liberal economic and political reforms undertaken by the Yeltsin administration were often characterized as a "return to civilization" (Neumann 2017, 79–80). The reference to civilization in the singular was indicative of the state of mind among the first post-Soviet generation of the Kremlin rulers: civilization was symbolically associated with Europe and, in a broader sense, with the Euro-Atlantic West, and did not necessitate a plural form.

Claims for a separate civilizational niche became part of the official discourse with Putin's ascension to power as an integral part of the mutually complementary and reinforcing doctrines of the *Russian world* and *Eurasianism*. Putin subscribed to a view of world politics defined by inter-civilizational relations with Russia as the epicentre of a broad pan-Slavic, Orthodox or Eurasian civilization (Bettiza et al. 2023, 3). These claims reached their apex in Putin's insinuations about the so-called unity of Russian and Ukrainian peoples and lamentations about the deliberate separation of Ukraine from the Russian civilization by the West. According to this reasoning, the West allegedly leaves Russia no choice but to regain control over the Ukrainian state. Russia's strategic narrative framed the so-called "special military operation" in Ukraine as an effect of a civilizational collision between Russia and the Western liberal order. Clearly, Russia's intent was to rationalize its military intervention in Ukraine using a civilizational mythology and camouflage its imperial project with a less belligerent rhetoric.

Not less important is how it was done. As a Ukrainian observer put it, "where European space ends, another civilisation does not begin. Where European space begins, war and legal uncertainty end" (Portnikov 2023). Indeed, the Russian military offensive revealed the barbaric side of Russia's civilizational specificity which exposed itself in the intentional and overt shelling of civilian populations, and turning the targeting cities into places unsuitable for living. Practices like war atrocities, tortures, and forcible deportation Ukrainian children to Russia are explained by civilizational differences and enables Russia, in its own eyes, to "whitewash" its violence. In other words, the civilizational argument for Russia is a sort of "insurance policy" helping to normalize violence and barbarity, a manifestation of its alleged innocence. Parallels with the war in Chechnya are illustrative: as Julia Wilhelmsen argued, "the entire Chechen people was convicted in the eyes of Russian popular opinion, and any action against them, no matter how barbaric, was seen as morally justified" (Wilhelmsen 2017, 79).

In the case of aggression against Ukraine it is Putin himself who openly integrates barbaric tropes into its mainstream discourse: "We have heard many times that the West wants to fight us to the last Ukrainian. This is a tragedy for the Ukrainian people, but it seems that everything is heading towards this." (Putin, June 2022 in Reuters 2022). In the words of the spokesperson of the Russian Foreign Ministry, "The Ukraine that you and I had known, within the borders that used to be, no longer exists, and will never exist again. This is evident" (Zakharova 2022 in TASS 2022).

The references above indicate that the paradigm of "the silence of barbarians" has been fundamentally reversed. It is Russia that is employing mass-scale violence in Ukraine while invoking a civilizational narrative as justification and legitimization for its actions—a feigned innocence that ultimately descends into barbarity. If Russia's civilizational argument is accepted more broadly, the next step would be to absolve Russia of moral blame and responsibility for its actions, both past and ongoing ones.

This new barbarism is also different from savages' unawareness or illiteracy. Russian artillery and missiles have used advanced technology to kill civilian populations. Mercenaries of the Wagner group do not shy away from their brutality; on the contrary, they have elevated it to a military cult.

Specifically, the narrative of the civilizational state not only permits Russia to expand beyond the 1991 borders (according to Putin, Russia is a country without borders at all), but also legitimizes its unrestricted actions within these borders. Russia's aggression against Ukraine stands in clear contradiction to international political norms, including those produced by the Conference on Security and Cooperation in Europe, such as the 1975 Helsinki Final Act and the 1990 Charter of Paris for a New Europe. Civilizational discourse in this respect conflates with spheres of influence while also being premised on notions of self-sufficiency and impunity. The practical implication of this is Russia's detachment from international or universal law, as evidenced by the violation of the UN-based international legal order in general and, more recently, by Putin's blatant visit to Mongolia in September 2024 in defiance of the International Criminal Court's arrest warrant.

Another cognitive trick is Russia's conflation of the concepts of civilization and empire. From the standpoint of its war objectives—establishing dominance and, at worst, the large-scale destruction of Ukraine's population as a form of punishment for perceived disloyalty—the conflict exhibits all the characteristics of imperial violence against a subjugated people. Here Kremlin leverages civilization (in the plural) together with anti-colonial narratives and effectively plays on the existing anti-Western sentiments in the global South.

Moscow's civilizational tropes also echo the European colonial past, implying that Russia is simply following the same path once taken by European powers in their colonies. The significance of this justification should not be underestimated, as highlighting European colonial crimes is closely linked with a culture of shame that can morally disarm counterarguments. This dynamic weakens the ability to effectively restrain Russia's civilizational narrative.

One may argue that Russia presents a complex—geopolitical, military, as well as civilizational—challenge to Europe, because of normative disagreement regarding the rules and boundaries of the international society (Bettiza et al. 2023). Indeed, Russia's perception of acceptable domestic and international behaviour, along with visions of what is legitimate, starkly contrasts with Europe's standards, and goes far beyond norms contestation. The annexation of Crimea, nuclear threats from a permanent member of the UN Security Council, denial of Ukraine's statehood, and blatant battlefield violence against civilians all contradicts UN, OSCE and other international norms and principles, and represents a fundamental digression from liberal universalism and the realpolitik of authoritarian statism.

Incorporating all these arguments into the analysis of Russia's use of the civilizational narrative reveals that, in essence, the primary aim of this narrative is to obscure the central issue—that Russia is acting as a barbarian by contemporary standards of barbarity. What is clearly an illegal act of aggression by one state against another, as defined by international law, becomes layered with complexity and ambiguity at the behest of Putin and his civilizational ideologists. This complexity, cloaked in spirituality and civilizational distinctiveness, effectively removes Russia from the normative and even rational domain: everything becomes possible, and nothing is excluded under the guise of a civilizational state. A critical step is to call things by their proper names and deconstruct this so-called "complexity," a task that is crucial yet challenging, particularly for audiences in the global South.

4 Ukraine's Civilizational Path

The concept of the Euro-Atlantic international society is rooted in a strong civilizational background, with civilizational ideals driving the EU and NATO enlargement. It enabled the framing of NATO's expansion not only in military terms but as an extension of the Western cultural realm (Kuus 2007). In the case of the EU, the concept of normative power rests on a modern form of "civilizing mission" (Zielonka 2013). The prominence of this civilizational framing in the EU and NATO's international roles provides a relevant context for Ukraine's pursuit of membership in the Euro-Atlantic international society, aligning with Western normative standards.

The Russian aggression has challenged the idea of a stable, protected Western world, yet at the same time it made clear that Ukraine's future critically depends on its recognition by the EU and NATO as a full-fledged European country, which is impossible without internalization of Western civilizational norms. In this respect the normative agency of Ukraine is a civilizational construct embedded in Ukraine's bids for membership in the EU and, in the long run, NATO.

This duly corresponds with the idea of civilizational security (Adler 2010; Bowden, 2020; Lewicki, 2023) which presumes that nations derive part of their security from belonging to a (geo)cultural zone with shared and embedded norms and values. Ukraine is seeking long-term, solid and formal guarantees for its sovereignty and security which would differ from such previous non-binding assurances as the 1994 Budapest memorandum. Against this backdrop, Ukraine's bid for EU and NATO membership is defined as the most indispensable instrument to insure the nation's long-term strategic security. Coming from the vision of the EU and NATO as sources of civilizational power, Ukraine seeks civilizational belonging to Euro-Atlantic civilization.

Moreover, Russian aggression produced far reaching spillovers into other spaces beyond hard security: energy, refugees, food and environment became highly securitized policy domains. In all of them, in addition to military cooperation, Ukraine together with the EU and US strives to build web of interdependencies to foster its formal accession into Euro-Atlantic institutions and bind Ukraine further into the military, economic and political orbit of the West.

Apart from considerations of security, the war reinscribed Ukraine's sense of identity in civilizational terms. If Huntington is correct that civilizations are "the biggest 'we' within which we feel culturally at home as distinguished from the other 'thems' out there" (Huntington, 1996: 43), Ukraine sees itself "returning to its native European home"—the trope which is frequently emphasized by Ukrainian officials both domestically and abroad. A hallmark of the European civilizational belonging of Ukrainian discourse is its strong normative component: "Europe is the space of freedom, democracy, social development and fundamental values that unite *us all*" (Zelenskyy, We must … 2022). The Copenhagen criteria of democracy are perceived in Ukraine as a contemporary version of the *standard of civilization*, and the EU accession negotiations as a practical mechanism of its practical implementation.

In line with the discursive turn in civilizational analysis (Alker 2007), Kordan (2022, 164) refers to civilizational narratives as stories and argues that the Ukrainian story is firmly rooted in European values, while the Russian story follows an alternative trajectory revolving around authoritarian rule and centralized power. The symbolic tension is understood to entail both a spatial—"we might be dealing with people from another universe" (Marples, 2022, p. 209) and a temporal dimension—"the Russia story is linked to a distant, pre-modern past" (Kordan, 2022, 164). In sum, "these two stories, being in tension, help explain the historical origins and genesis of the present conflict" (Kordan, 2022, 164).

For years before the war, Ukrainian society identified itself with the civilized norms of the Western world, where peace and the value of human life are the key normative foundations. But on geopolitical map Ukraine remained in a civilizational "grey zone", which was one of the reasons that made the Russian invasion possible. Currently Ukraine's civilizational reference is deemed to overcome its international marginalization. In this sense, for Ukraine civilizational narrative is a way to avoid what Norbert Elias dubbed "group disgrace" and differentiate itself from the Russian empire by condemning its aggression and seeking recognition from the West. In the words of Ukraine's former Minister of Foreign Affairs Dmytro Kuleba, "we need a wall between civilisation and barbarians" (Ukrainska Pravda 2022), which chimes with President Volodymyr Zelensky's characterization of the Russian invasion as a war between civilization and non-civilization (Zelenskyy 2022, We would not ...).

As evidenced by the quotes above, Ukraine's civilizational discourse, revitalized by the war, is deeply rooted in emotional underpinnings. Symbols such as peace, hope, stability, prosperity, and human dignity are strongly associated with the Western civilization. Invoking these symbols also serves to bolster emotional identification with Europe. In contrast, Russia is portrayed as the embodiment of evil and barbarism.: "a source of absolute evil for generations ... (with) the most barbaric and inhuman army in the world" (Zelenskyy 2022, If Our Partners ...).

Contrary to Russia, Ukraine often uses the concept of civilization in the singular, claiming that Ukraine deters and holds back contemporary barbarians. The representation of a group as barbarians is an exercise of power that implies superiority, in this case morally grounded. Ukrainian mainstream discourse deems that Russia's rampant violence is causing their moral decline. Historically it was imperial conquerors who treated the colonized / indigenous peoples as being civilizationally inferior (Bowden 2007, 8); but in the case of the Russia–Ukraine war, it is the victim of the aggression who sees its colonizer as barbarian. Narratives that frame the conflict as a battle between 'barbarians' and 'civilized people,' democratic Ukraine against authoritarian Russia, modern against archaic forces are intended to fall into this category. For foreign audience, they warn about the threat posed by Russia that requires constant vigilance and control, and call for exceptional measures that otherwise might not be deemed legitimate or acceptable.

Two discourses—civilizational security and the moral blame of Russia—exist in parallel and reinforce one another. These narratives have helped Ukraine frame its struggle as both a defence of the European civilization and a moral stance against the Russian aggression. However, these discourses can also create tensions. The notion of the "European way of life" and its civilizational foundations, with which Ukraine strongly identifies itself, could lead to a tendency of carving out a "separate space for themselves" (Glencross 2021, 341). This separation could influence how Ukraine perceives its place in the broader international community.

From a practical standpoint, the conflation of these two discourses provides the foundation for distinct foreign policy strategies for Ukraine. One strategy prioritizes integration with the Euro-Atlantic West, reinforcing Ukraine's civilizational alignment with Europe. The other adopts an anti-colonial stance, positioning Ukraine as part of the global society (Kurnyshova 2024b). When strategically employed, these approaches can complement one another, showcasing Ukraine's unique potential to redefine what it means to be European.

5 What to Do: Western Dilemmas

As many historians deem, the idea of European civilization has a problematic history, associated as it is with colonial ideas of racial superiority and moral progress (Henry 2010, 270). Nevertheless, the inception and subsequent enlargement of the EU have been underpinned by a civilizational idea, and the criteria for accession remain a revamped standard of civilization. Europe's normative power extends well beyond its formal borders, as exemplified by Ukraine's pursuit of EU membership. However, the war in Ukraine has shifted the EU's focus from transforming its neighbouring states to addressing the pressing challenge of how to resist the hostile Russia. What was until February 2022 a political boundary with Russia has now been redefined as a cultural frontier. The war is not merely a geopolitical rupture; it is also laden with values that both Russia and Ukraine infuse in it. This intertwining of geopolitics and culture generates a clash similar to the Cold War. Nevertheless, the EU remains cautious about adopting a narrative of civilization versus barbarity, as it inherently suggests a global cultural conflict and might require a contemporary version of crusade. This trope not only negates the possibility of nonviolent solutions, but also precludes any form of negotiation.

Instead of labelling Russia's war against Ukraine as a barbarian act—the move which would require radical measures—the EU prefers to employ traditional legal mechanisms. This approach allows for a measured response, determining the extent to which Russia can be isolated or sanctioned without resorting to its full exclusion from the international political community. In this regard discourse analysis is helpful in identifying several contexts through which the civilization-barbarity narrative can be linked with politically and legally instrumentalizable concepts.

The first one—the narrative of a terrorist state—holds significant strategic importance in strengthening the anti-Russian sanctions regime and intensifying political and economic isolation of Moscow. The extant logical nexus between barbarity and terrorism was established by the 9/11 attack on the US, which provided a precedent of translation of the narrative on barbarism into practical actions. Russia's attack on Ukraine is of different nature, yet Russia was recognised as a state sponsor of terrorism by the Parliamentary Assembly of the Council of Europe (PACE), the European Parliament, and the NATO Parliamentary Assembly. Specifically, in its resolution (November 2022), the European Parliament called on EU member states to impose comprehensive international isolation on Russia, support investigations into its crimes, and establish a special international tribunal (European Parliament 2022). This document is an expression of a political stance with political consequences, but it has legal implications too, as it creates a ground for further sanctioning, freezing and confiscation of Russian assets, as well as prosecution of criminals. However, the designation of Russia as a state sponsor of terrorism by the United States, with potential major political and economic implications is still pending, demonstrating that the White House is not ready to go that far with Russia as it was the case of Islamic terrorist groups. Some practices of exclusion are underway, but they are moderate, cautious and self-restrained.

In 1933, Raphael Lemkin proposed an international convention to classify the extermination of human groups as a crime, initially labelling it as "Acts of Barbarity". A decade later, he coined the term *genocide*, still rooted in the concept of barbarity, to describe systematic extermination targeting ethnic, religious, or social groups through massacres, economic destruction, and violations of human dignity. Taken separately, all these acts are punishable in the respective codes; considered together, they constitute offenses against the law of nations, which is a clear reference not only to the direct victim of the assault, but also to the international order. In 2023 the International Criminal Court (ICC) issued arrest warrants for Vladimir Putin and Children's Ombudsperson Maria Lvova-Belova for the illegal deportation of children from Ukraine to Russia. Subsequently the Parliamentary Assembly of the Council of Europe adopted a resolution recognizing the deportation and forced transfer of Ukrainian children to Russia as genocide. Russia was stripped of its membership in the

UN Human Rights Council, expelled from the Council of Europe, removed from the chairmanship of UNESCO's World Heritage Committee, and excluded from several other international organizations.

In the academic literature one may find further examples of interconnections between barbarity and genocide. For example, the genocidal nature of Nazism has been widely discussed as a "reversion from civilization to barbarity" (Freeman 1996, 210). As a similar approach suggests, barbarism might be ingrained in civilization and may manifest itself through "cultural genocide", particularly practiced by imperial powers (Kriecken 1999). Moreover, colonial practices of ethnic and racial genocide might make civilization and barbarity indistinguishable from one another—hence the hybrid concept of "barbaric civilization" (Powel 2011).

The application of the civilization-barbarity dichotomy in contemporary international relations is hindered by what Norbert Elias described as the self-restraint of civilizations and their avoidance of violence. After World War II, Europe moved toward the elimination of various forms of violence by embracing rule-based conflict resolution and imposing self-restraint on violent behaviour (Linklater, p. 229). This contrasts with Russia's *normalization of violence* (Makarychev and Nizhnikau 2023) as a conflation of two paradigms that operate within different ontological frameworks and are difficult to reconcile. Against this backdrop, "Ukrainians view Europe with both admiration and apprehension: they admire it as a civilization that seeks to minimize the space for violence, yet they are horrified by its vulnerability and perceived helplessness in the face of violence from Russia and other proponents of the emerging anti-liberal order" (Yermolenko, 2019).

Indeed, the solution to barbaric violence along the *mission civilisatrice*, once used by European powers to civilize barbaric actors, is no longer viable due to the ethical rejection of colonial practices and civilizing offensives. Despite voices claiming that society can be protected only by defeating the state—violator (Dunn 2022), the idea of peacebuilding or military intervention against Russia, akin to the Western actions in Bosnia or Afghanistan, appears equally untenable, due to the structural limitations imposed by Russia's status as a nuclear power.

The West has largely resorted to symbolically placing Russia outside of the boundaries of *normal politics*, treating it as a *barbaric civilization* state that threatens the civilized international order. This strategy of isolation and deterrence underscores the limitations of traditional responses to barbaric violence. What it does demonstrate, though, is that civilizational discourse is closely related to assertion of legitimacy within international society: it acts to define the membership in this society with references to moral and political yardsticks of civilized behaviour, and to identify the interests of members of society with a common purpose of civilized conduct. In practise it involves defining a degree of institutional, economic, legal and other forms of exclusion of Russia.

Admittance at the level of discourse Russia's barbarism might be extended into political moves and practices of radical exclusion, but this translation doesn't come easy, especially considering that the very idea of European civilization turned into a "free-floating term ... consisting of many languages, civilizational models and political projects" (Delanty 2003, 21). Notably, Ukraine's civilizational discourse is perceived as being too radical for many Western countries who are guided by a paradigm of avoiding infliction of a major harm to the perpetrator. By expecting that all conflicts ought to be concluded at the negotiation table through compromises and even concessions, the West, in the words of one observer, tests the dilemma of Goethe's Faust: whether and how to make deals with evil (Yermolenko 2019). Or, in other words, how to agree on security guarantees with a state sticking to a minimal level of law obedience and trust.

Conclusion

In this chapter I juxtaposed civilization and barbarity as discursive practices that create speaking positions and therefore sustain agential characteristics of all parties involved in the current Russia's war in Ukraine. My analysis shows that these discourses facilitate either inclusive (Ukraine and the West) or exclusive (Russia and the West) policies shaping the contours of the Euro-Atlantic security order and its contenders. It also underscores how discursive practices in conflict shape not only the perception of legitimacy but also the broader geopolitical dynamics that influence security and policy decisions.

Russia transformed its failure to be accepted as a part(ner) of the liberal West into an idea of its own civilizational specificity with the subsequent civilizational self-closure. The concept of the *state-civilization* adds imperial meanings to Russia's perception of itself as a self-sufficient geo-cultural entity whose territorial borders should ultimately coincide with the arbitrarily established spheres of linguistic, religious or other forms of Russia's presence, influence and impact on its neighbours. It is the logic of congruence between territorial state and its civilizational underpinnings that justified the Kremlin's attitude to Ukraine as Russia's subaltern, an unreliable and treacherous part of the imagined *Russian world* that should be controlled and subjugated by force.

Moscow's voluntary break with the liberal international society and disdain for its normative basis was however seen in Ukraine not as a proof of Russia's civilizational peculiarity, but rather as self-barbarization manifested in the evidence of multiple atrocities and war crimes committed by the invading army in the occupied territories. By the same token, Ukraine responded to Russian attempts to Orientalize and marginalize this country by counter-Orientalization, reciprocal securitization and ultimately barbarization of Russia itself, thus asserting civilizational and moral superiority of the victim over the aggressor. Ukraine's alignment with European values and its emphasis on Russia's barbaric conduct serve to strengthen its claims for support and legitimize its aspirations for the EU and NATO membership. This civilizational positioning underscores moral and political standards, identifying Ukraine with the liberal international society committed to the principles of civilized conduct.

Ukraine's reliance on a civilizational discourse centred on Western values could potentially alienate non-Western actors. The exclusionary nature of such framing may seem limiting the universality of its appeal. However, Ukraine tries to convey the idea that Russian invasion should be viewed as an assault on universal norms because all countries (including the global South) are signatories of the UN Statute that determines the standards of the expected international behaviour.

This discourse also contributes to the ethical debates about the current war and underscores the necessity to distinguish civilization from barbarity even further. The current generation of European leaders instrumentalized the idea of civilization as inherently tied to values of moderation, restraint and dialogue, with the ensuing adherence to conflict resolution through negotiation and legal mechanisms established by the international community. However, for Russia civilization provides a licence for waging imperial war *against* this very standard of Western civilization, which considered and widely promoted as a spiritual bogey inside Russia and a building stone of Russian anti-colonialism in the global South. This duality in the concept of civilization reflects a foundational tension in global politics and reveals how the idea of civilization can both promote peace and become a tool for power politics, shaping alliances, legitimizing interventions, justify war.

Given the global dimension of the war, it is important to keep in mind the logical connections between the concept of barbarity, on the one hand, and the legal definitions of terrorism and genocide with their practical effects, on the other. The question of whether the border between Norway, Finland, Estonia, Latvia, Lithuania and Poland and Russia demarcates two conflicting civilizations, or delineates civilization from barbarity, remains open. Arguably, it is exactly the answer to this question that will largely define the degree of exclusion or isolation of Russia from international institutions created by the collective West, as well as legal procedures to be undertaken against the aggressor after the war ends. Two conceptual models—a *dialogue of civilizations* and exclusion of barbarism from international politics—are sharply competing with each other as never before since the Second World War.

References

Adler, Emmanuel. 2010. "Europe as a Civilizational Community of Practice." In *Civilizations in World Politics: Plural and Pluralist Perspectives*, edited by Peter Katzenstein. Routledge.

Al-Azmeh, Aziz. 2001. "Civilization, Culture and the New Barbarians." *International Sociology* 6 (1): 75–93.

Alker, Hayward. 2007. "Discussion: On the Discursive Turn in Civilizational Analysis." In *Civilizational Identity: The Production and Reproduction of 'Civilizations' in International Relations*, edited by Martin Hall and Patrick Thaddeus Jackson. Palgrave Macmillan.

Arnason, Johann. 2010. "Introduction: Domain and Perspectives of Civilizational Analysis." *European Journal of Social Theory* 1 (23): 5–13.

Bettiza, Gregorio. 2014. "Civilizational Analysis in International Relations: Mapping the Field and Advancing a "Civilizational Politics" Line of Research." *International Studies Review* 16: 1–28.

Bettiza, Gregorio, Derek Bolton, and David Lewis. 2023. "Civilizationism and the Ideological Contestation of the Liberal International Order." *International Studies Review* https://doi.org/10.1093/isr/viad006.

Bowden, Brett. 2007. "Civilization and Savagery in the Crucible of War." *Global Change, Peace & Security* 19 (1): 3–16.

Bowden, Brett. 2010. "Civilizational Security." In *The Routledge Handbook of New Security Studies*, edited by J.P. Burgess, 7–16. Routledge.

Coker, Christopher. 2019. *The Rise of the Civilizational State*. Polity Press.

Delanty, Gerard. 2003. "The Making of a Post-Western Europe: A Civilizational Analysis." *Thesis Eleven* 72 (February): 8–25.

Dunn, Kyle. 2022. "Society Can Only Be Protected by Defeating the Violator State." Defense One. https://www.defenseone.com/ideas/2022/05/society-can-only-be-protected-defeating-violator-state/367074/

Elias, Norbert. 1996. *The Germans*. Cambridge: Polity Press.

European Parliament. 2022. "Resolution on Recognising the Russian Federation as a State Sponsor of Terrorism." Legislative Observatory, November 23. https://oeil.secure.europarl.europa.eu/oeil/popups/ficheprocedure.do?lang=en&reference=2022/2896(RSP)

Fermor, Ben. 2021. "This Barbaric Terrorist Organization": Orientalism and Barack Obama's Language on ISIS." *Critical Studies on Terrorism* 14 (3): 312–334. https://doi.org/10.1080/17539153.2021.1932750.

Foster, John Bellamy, and Brett Clark. 2004. "Empire of Barbarism." *Monthly Review: An Independent Socialist Magazine*, March 1–15.

Frear, Matthew, and Honorata Mazepus. 2021. "Security, Civilisation and Modernisation: Continuity and Change in the Russian Foreign Policy Discourse." *Europe-Asia Studies* 73 (7): 1215–1235.

Freeman, Michael. 1996. "Genocide, Civilization, Modernity." *The British Journal of Sociology* 46 (2): 207–223.

Geis, Anna, and Carmen Wunderlich. 2014. "The Good, the Bad, and the Ugly: Comparing the Notions of 'Rogue' and 'Evil' in International Politics." *International Politics* 51: 458–474.

Glencross, Andrew. 2021. "The EU and the Temptation to Become a Civilizational State." *European Foreign Affairs Review* 26 (2): 331–348.

Hall, Martin, and Patrick Thaddeus Jackson, eds. 2007. *Civilizational Identity: The Production and Reproduction of "Civilizations" in International Relations*. Palgrave Macmillan.

Henry, Nicholas. 2010. "Politics Beyond the State: Europe as Civilization and as Empire." *Comparative European Politics* 8 (2): 262–280.

Huntington, Samuel P. 1996. *The Clash of Civilizations and the Remaking of World Order*. Simon and Schuster.

Kan, Paul. 2019. "Dark International Relations: When Crime Is the 'Dime'." *War Room: Online Journal*, July 24. https://warroom.armywarcollege.edu/articles/dark-international-relations-when-crime-is-the-dime/.

Kassimeris, George. 2006. *The Barbarization of Warfare*. New York University Press.

Katzenstein, Peter. 2010. "A World of Plural and Pluralist Civilizations: Multiple Actors, Traditions, and Practices." In *Civilizations in World Politics: Plural and Pluralist Perspectives*, edited by Peter Katzenstein, 1–40. Routledge.

Kordan, Bohdan S. 2022. *Strategic Frames: Europe, Russia, and Minority Inclusion in Estonia and Latvia*. University of Toronto Press.

Kotlyuk, Galina. 2023. "Colonization of Minds: Ukraine between Russian Colonialism and Western Orientalism." *Frontiers in Sociology* 8, October 30. https://www.frontiersin.org/journals/sociology/articles/10.3389/fsoc.2023.1206320/full.

Krieken, Robert van. 1999. "The Barbarism of Civilization: Cultural Genocide and the 'Stolen Generation'." *British Journal of Sociology* 50 (2): 297–305.

Kurnyshova, Yuliia. 2024a. "Civilizational Security and Transformative Power: Two Facets of Ukraine's Integration with the West." *Ukrainian Analytical Digest* 4. https://css.ethz.ch/content/dam/ethz/special-interest/gess/cis/center-for-securitiesstudies/pdfs/UkrainianAnalyticalDigest004.pdf

Kurnyshova, Yuliia. 2024b. "Ukraine and the Global South: The Ambiguities of a Postcolonial Perspective." *New Perspectives* 32 (3): 272–289.

Kuzio, Taras. 2023. "Russia: A Confused and Embittered Nation." Center for European Policy Analysis, September 11. https://cepa.org/article/russia-a-confused-and-embittered-nation/.

Kuus, Merje. 2007. *Geopolitics Reframed: Security and Identity in Europe's Eastern Enlargement*. Palgrave Macmillan.

Lewicki, Gregory. 2023. "Civilizational Security: Why the Russian Invasion of Ukraine Shows That 'National Security' Is Not Enough to Understand Geopolitics." *Comparative Civilizations Review* 89 (89). https://scholarsarchive.byu.edu/cgi/viewcontent.cgi?article=2284&context=ccr

Linklater, Andrew. 2021. *The Idea of Civilization in World Politics*. Routledge.

Maier, Charles. 2007. "Dark Power: Globalization, Inequality, and Conflict." *Harvard International Review* 29 (1).

Makarychev, Andrey, and Ryhor Nizhnikau. 2023. "Normalize and Rationalize: Intellectuals of Statecraft and Russia's War in Ukraine." *Journal of International Relations and Development* 26: 632–642. https://doi.org/10.1057/s41268-023-00299-x

Marples, David R. 2022. "The Economic Factors Behind the Russian Invasion." In *Russia's War on Ukraine: A Canadian Perspective*, edited by Bohdan S. Kordan, 205–214. McGill-Queen's University Press.

Mercier, Thomas Clément. 2019. "Resisting the Present: Biopower in the Face of the Event (Some Notes on Monstrous Lives)." *CR: The New Centennial Review* 19 (3): 99–127.

Neumann, Iver. 2017. "Russia's Return as 'True Europe'." *Conflict and Society: Advances in Research* 3: 78–91. https://doi.org/10.3167/arcs.2017.030107.

O'Hagan, Jacinta. 2007. "Discourses of Civilizational Identity." In *Civilizational Identity: The Production and Reproduction of 'Civilizations' in International Relations*, edited by Martin Hall and Patrick Thaddeus Jackson, 15–32. New York and Basingstoke: Palgrave Macmillan.

O'Hagan, Jacinta. 2020. "Shape Shifting: Civilizational Discourse and the Analysis of Cross-Cultural Interaction in the Constitution of International Society." *Journal of International Political Theory* 16 (2): 190–209. https://doi.org/10.1177/1755088220905039.

Offe, Claus. 1996. "Modern 'Barbarity': A Micro-State of Nature?" *Constellations* 2 (3): 354–377.

Portnikov, Vitaly. 2023. "How Ukraine Can Get into the EU." *Visegrad Insight*, August 23. https://visegradinsight.eu/how-ukraine-can-get-into-the-eu/.

Powell, Christopher. 2011. *Barbaric Civilization: A Critical Sociology of Genocide*. McGill-Queen's University Press.

Rengger, Nicholas, and Renee Jeffery. 2005. "Moral Evil and International Relations." *SAIS Review of International Affairs* 25 (1): 3–16.

Reuters. 2022. "Putin Says Russia Just Starting in Ukraine, Peace Talks Will Get Harder." Reuters, July 7. https://www.reuters.com/world/europe/russias-putin-if-west-wants-beat-us-battlefield-let-them-try-2022-07-07/.

Salter, Mark. 2007. "Not Waiting for the Barbarians." In *Civilizational Identity: The Production and Reproduction of 'Civilizations' in International Relations*, edited by Martin Hall and Patrick Thaddeus Jackson, 81–94. New York and Basingstoke: Palgrave Macmillan.

Sharpe, Andrew. 2007. "Foucault's Monsters, the Abnormal Individual and the Challenge of English Law." *Journal of Historical Sociology* 20 (3): 384–403.

Sheikh, Mona Kanwal. 2014. "Appointing Evil in International Relations." *International Politics* 51: 492–507. https://doi.org/10.1057/ip.2014.22.

Silverstein, Paul. 2005. "The New Barbarians: Piracy and Terrorism on the North African Frontier." *CR: The New Centennial Review* 5 (1): 179–212.

TASS. 2022. "Ukraine in Its Previous Borders No Longer Possible—Diplomat." TASS, June 18. https://tass.com/politics/1467871

Ukrainska Pravda. 2022. "Kuleba on Missile Strike on Odesa: Russia Should Be Recognised as a State Sponsor of Terrorism." April 23. https://www.pravda.com.ua/eng/news/2022/04/23/7341612/.

Wilhelmsen, Julie. 2017. *Russia's Securitization of Chechnya: How War Became Acceptable*. Routledge.

Yermolenko, Volodymyr. 2019. "Violence and Anti-Violence: Ukraine between Russia and Europe." Ukraine World. https://ukraineworld.org/en/articles/opinions/ukrayina-yevropa-i-nasillya.

Zelenskyy, Volodymyr. 2022a. "We Must Never Stop Defending Freedom and Democracy—Speech of the President of Ukraine at the European Council." Official Website of the President of Ukraine. https://www.president.gov.ua/en/news/mi-nikoli-ne-mayemo-zupinyatisya-v-zahisti-svobodi-j-demokra-78353

Zelenskyy, Volodymyr. 2022b. "I Would Not Like a Story About Ukraine as About 300 Spartans—I Wish Peace to My Country." Official Website of the President of Ukraine. https://www.president.gov.ua/en/news/volodimir-zelenskij-ya-ne-hotiv-bi-istoriyu-pro-ukrayinu-yak-73317

Zelenskyy, Volodymyr. 2024. "If Our Partners Have the Weapons Ukraine Needs, Their Duty Is to Help Protect Freedom and the Lives of People." Official Website of the President of Ukraine. https://www.president.gov.ua/en/news/yaksho-nashi-partneri-mayut-potribnu-ukrayini-zbroyu-yihnij-74417.

Zielonka, Jan. 2013. "Europe's New Civilising Missions: The EU's Normative Power Discourse." *Journal of Political Ideologies* 18 (1): 35–55. https://doi.org/10.1080/13569317.2013.750172.

CHAPTER 7

Towards Decentring Comparisons: Overcoming the *Uniqueness* Myth in Russian Studies

Aliaksei Kazharski

1 Introduction

The chapter examines the potential of developing comparative perspectives in Russian studies for the purposes of decentring Russia as a subject of academic analysis and policy discourse. The crux of its argument is that the Russian case has been studied in too much isolation from other peripheral Central and Eastern European (CEE) cases. As a result, there has often been a shortage of knowledge about the broader region on the part of Russianists, both Western and those coming from Russia itself. In turn this has contributed to strengthening certain false tropes, such as the trope of Russia's alleged "uniqueness" as well as the implied notions about the necessity to "understand" Russia on its own terms and often by implication the inapplicability of general rules and standards to its development. This mythology of uniqueness corresponds to the well-known ideologemes such as those of Russia being a "Eurasian" power or a "unique civilization". In the Western academic and policy circles theses ideologemes were often gladly—or at least uncritically—received, in part because they rhymed well with the charms of a romanticist-orientalist exoticization. The discourse in uniqueness has also traditionally been rather popular in Russia as a form of exceptionalist collective narcissism and has been used by its rulers as well as by its sympathizers internationally to defend, legitimize, and normalize Russian imperialistic exceptionalism, Russian transgressions of regional and international orders, and Russian privilege as a great power. This type of discourse has thus operated in parallel and sometimes even in synergy with those IR-realist discourses that also defended Russia's great power privilege

The chapter therefore argues that the myth of uniqueness in its various guises can and should be debunked by running systematic comparisons between Russia and other countries in Central and Eastern Europe (CEE) so as to demonstrate that exceptionalist claims to uniqueness are actually rather widespread in the region and, therefore it is, in fact, quite *typical* to be "unique." Furthermore, these claims stem from very similar experiences and concerns

 | DOI:10.1163/9789004747364_008

shaped by the trajectory of *peripheral* development of which Russia has historically been part, alongside multiple other countries. Here, I interpret CEE peripherality as a spectrum rather in terms of clear divides, e.g. into "Western," "Central," and "Eastern" Europe, so the effects of peripherality can of course vary in their strength, scope, and guise. Nevertheless, this *transitive peripherality* is a regional common denominator and its effects can be grasped through comparative work. Below I propose several ideas as to how these comparisons can be developed, in particular with respect to how peripheral nations develop their identity building strategies and position themselves on the core-periphery spectrum as well as with respect to their neighbours.

Assuming this perspective implies that—contra those who would altogether exclude Russia from CEE—it should also be analysed as its part. Arguably, the exclusion of Russia from CEE and from Europe at large also plays into "the uniqueness" ideologeme by supporting the Russians' own exceptionalist *self-Orientalizing* nationalistic mythology which as such has few historical or factual grounds. In analytical terms Russia needs to be recognized for what it is: one of many peripheral European cases that should not be entitled to any special treatment or privilege either because of its geographical size, military capabilities, or because of the international status stemming from the generous recognition by the international community of Moscow as the post-Cold War successor to the USSR.

2 On the Three Fallacies in Analysing Russia

Over the years there emerged several families of misconceptions in the academic, policy expert, and political discourse on Russia which underpinned various strategies of explaining *the sources of Russian conduct*. Some of these explanations have recently come under critical scrutiny as being justifications or normalizations in disguise (see Makarychev and Nizhnikau 2023). Among these particularly visible is the IR structural realist narrative that, at least since 2014, has been a regular irritant for both area studies specialists and scholars working in the social constructivist tradition. This heavily criticized narrative presents Russia as a strategic actor that is security-rational and is "defending itself" from NATO expansion, while ignoring a plethora of domestic factors such as regime type, ideologies, and identities. Though usually associated with a controversial op-ed published in 2014 by one of the prominent structural realists (see Smith and Dawson 2022; Specter 2022), this explanation-as-justification obviously predates the beginning of the first phase of the war. At some point, these or related narratives also became entrenched in some

Russia-leaning Central European EU member-states. As Katalin Miklóssy discusses in her chapter, in Hungary they became an ideological component of a repressive mechanism, meant to enforce conformity with the Russian official storyline.

The second broad misconception that has been, perhaps, somewhat less visible is also rooted in utility-rational models of state behaviour in IR. It fits with the ideologeme of "liberal interdependence" that, for obvious reasons, became popular after the end of the Cold War. Among other things, it has been used to justify increased economic integration with Russia in the apparent hope—whether genuine or faked—that gradual socialization through economic means can steer Russia's post-Soviet transition in the right direction. Inconvenient truths such as the ensuing asymmetric dependencies of Western liberal democracies on an increasingly authoritarian and internationally aggressive regime in Moscow were largely ignored. For many, the liberal interdependence paradigm, with its assumptions of utility-rational behaviour was also a source of confidence prior to February 24, 2022, helping sustain the argument that a big war is impossible because it would be too *irrational.*

Finally, the third fallacy is immediately connected to what one usually considers the domain of social constructivist approaches in IR, though, of course, the way it operates intellectually is in direct opposition to the premises of a scientific social constructivism. Here, I am talking about identity: collective self-perceptions, and perceptions of/by the Other. The crux of the fallacy is the trope of the so-called "uniqueness" of Russia, its supposedly "non-European" or "Eurasian" character which has been repeated in many different guises both in Russia itself and by external observers. This trope often goes hand in hand with a heavy romanticization of the alleged "uniqueness". One of its immediate implications is exceptionalism, whereby Russia may not/should not be judged by general standards and cannot be expected to behave/develop in accordance with them. As Viacheslav Nikonov, a prominent member of Putin's foreign policy establishment and a proud grandson to Viacheslav Molotov, once said Russia was "too big and too Russian," and this is why it was impossible to integrate it into the EU or NATO (Nikonov 2003; see also Curanović 2021).

This form of self-isolationist exceptionalism practiced in many guises provides an easy rationalization for the growing gap between Russia and the West, in fact, caused by Russia's democratic backsliding and the growing propensity for aggressive neo-imperialist behaviour towards its neighbours—with the two trends being obviously connected. As a cultural and political trope this romanticization of the alleged Russian uniqueness or irrationality is anything but new and is extremely well-known since the 19th century, from Fyodor Tyutchev's

"Russia cannot be known by the mind, nor measured by the common mile" to Churchill's "riddle wrapped in a mystery inside an enigma" and to other iconic clichés familiar to anyone who has so much as briefly touched Russian studies.

On a purely political level, the discourse of uniqueness can help normalize various transgressions of the commonly accepted standards of behaviour—from democratic backsliding to open military aggression. In a sense, it is a very strange relative of the realist "explanations"-cum-justifications of Russian aggression that frame Moscow's neoimperialist policies as a result of "merely reacting" or "having been left no choice." Its academic counterpart is the well-known grossly mechanistic neorealist image of international politics (see Specter 2022). The ultimate similarity derives from the fact that both normalizing tropes operate through a depoliticized essentialization. In the realist trope, domestic politics and decision-making are essentialized as being a mere function of the *structure of the international system*. The "uniqueness" trope essentializes politics, culture, and history as unchanging constants, suggesting that there is nothing to be done about how Russia *is* or "has always been", the best one can hope is to *understand* it. Understanding here thus tacitly implies accepting and justifying. In the words of a Russian literary classic, "*tout comprendre c'est tout pardoner*"—to understand everything is to forgive everything.[1]

Apart from this obvious effect, however, there are other caveats that pertain to knowledge production and regional studies. At this point, it has become commonplace to observe that the study of either CEE or "the post-Soviet space" has been unacceptably Russocentric and distorted by the toxic spells of "imperial knowledge" (Riabchuk 2024) both in the sense of being disproportionately tilted towards Russia in terms of the distribution of attention and—what is probably even more harmful—in terms of seeing other countries through the Russian cognitive lens. If the first aspect could be partially explained by the relative political, economic, or military weight of the countries in question, the second one can hardly be justified in any sensible way. It has nevertheless been present and has led to serious distortions in knowledge production, and beyond, i.e. in ways the agendas were set by journalists or policymakers, some trained at the very same universities where this distorted knowledge production was happening.

What could be perhaps better reflected is the role that the "uniqueness" trope probably played in this infamous distortion. Arguably, apart from justify-

---

1 I am grateful to James Sherr for leading me to this aphorism during our informal discussion on Western Russianists in Riga.

ing behaviour, supposed "uniqueness" can also be used to normalize *epistemic injustice*. Thus, an allegedly "unique" country logically needs more researcher attention and more funding, while the rest of region can then be dealt with as a box of more or less "standard" "Eastern European" cases. However, the illusion of "uniqueness" can often stem from the lack of knowledge of other cases—as, it seems, it could have well happened to some Western Russianists who are not thoroughly familiar with the rest of CEE.

I would venture further to argue that this also has to do with the *positionalities* of researchers, especially those who come from outside the region. Here, the charm of exoticization or a genuine scholarly love for the object of research can no doubt play an important role. In this context, elevating one's object of research to the status of "unique" is a natural tendency but it leaves too many dangerous blind spots. In a slightly different manner, this also concerns those Russianists who come *from Russia*, but often lack knowledge of other peripheral countries. The latter is also not an uncommon case as I could myself witness time and again. Contra this, I would propose a comparative research agenda that would use systematic comparison as a way of decentering Russia in CEE and potentially overcoming the harmful impact of the uniqueness myth that manifest itself on many discursive plains.

2.1 *Background: The Uniqueness Myth and Self-Orientalizations*

Arguably, the background to this is formed by two broader historically trends. One of them is the othering or orientalization of some European peripheries, apparently originating in the "invention of 'Eastern Europe'" during the Enlightenment period (Wolff 1994). The other is self-orientalization (or auto-othering), which could involve reappropriation of foreign ideologies, tropes, and images. The two trends were thus also connected, and that is hardly surprising, as, being a periphery of Europe, Russia naturally shared a dialogical intellectual space with it. In this context, it is telling that even its most recognizable and broadly used animalistic symbol seems to have been gradually adopted by Russia from the West, having its origins in British political cartoons and the early modern imagology of European peripheries (see Khrustalyov 2011; Platoff 2012).

Bears aside, Russian attempts to romanticize pseudo-uniqueness have European ideological origins and this certainly has not escaped the researchers' attention. As Iver Neumann (1996) points out in his classical study of Russian identity and international relations, the 19th century Slavophiles with their notions of the "organic unity" of the nation and "spiritual superiority" to the "ossified" West were under direct influence of German romanticism. Here, it goes without saying that the Romantic movement has inspired many a

European nationalism. As a reaction to the rationalism of the Enlightenment, it put forth a cult of irrational intuitivism (cf. "cannot be known by mind, nor measured by the common mile"). The most ardent 19th century seekers for Russia's "authenticity" coming from its educated class were very well versed in European languages, philosophy, and literature, so naturally, the ideological means they used to emphasize that alleged uniqueness and non-Westerness turned out to be anything but unique or non-Western.

The 21st century presented us with a kind of a parody of this 19th century paradoxical situation. Thus, some of the internationally medialized Russian icons of the anti-Western so-called "neo-Eurasianism" were exposed as massive borrowers and compilators of Western far-right thinkers (Laruelle 2008, 107–141). That is both ironic and telling in this context, particularly in light of the attention they draw from the international publics who imagine them as the geopolitical "masterminds" behind Russia's anti-Western drive. There is thus a recurrent tendency in Russia towards constructing pseudo-authenticity or *pseudo-uniqueness*, but there also tends to be an international audience for that.

The reasons for the latter appear to be manifold but, for sure, one element of it is the rather standard charm of (fantasized) exoticism which has always been part of Orientalist imaginations. Orientalization of Russia does not stop at the bear nor at the Matryoshka doll. The intellectual excitement about "a riddle wrapped in a mystery", the Romantic appeal of "cannot be known by mind," the never-ending search for the supposedly "enigmatic Russian soul," and other similar tropes act as cognitive lures ensuring that the myth of "uniqueness" continues to be reproduced both in and outside academic circles.

Contra that, there are two analytical tasks that work towards debunking the myth. The first one is exposing the empirical unsustainability of the claims to uniqueness by highlighting their historical and intellectual roots. We can find expiring examples of that in the classical work of critical scholars like Iver Neumann cited above. The second task, which has perhaps been given somewhat less attention is putting the Russian case in a comparative perspective in order to demonstrate *inter alia* that the strategies of self-orientalization, exceptionalism, claims to "uniqueness," and othering of the West are by no means endemic to Russia. They are part of the standard repertoire of peripheral nations who often suffer from very similar, well recognizable complexes—collectively held complexes that can, under certain circumstances, evolve into extremely toxic nationalistic and imperialistic ideologies.

3 Towards Decentring Comparisons: Distant Others

It would, of course, be unfair to assume that no work has previously been carried out in this direction. For example, authors like Neumann (1996), Zarakol (2011) or Morozov and Rumelili (2012) tried to put Russia into a comparative perspective with Turkey—another peripheral presence in Europe. A 2010 volume edited by Emil Pain shed light on the striking similarities between Russian and German exceptionalisms, embodied in the ideology of a *Sonderweg—osobyi put'*. These are some examples, but it is more than likely that, through my own ignorance, I have not mentioned a number of serious comparative analyses here.

It is not difficult to understand why the comparisons with Germany resonate after 2022 and were quickly drawn, among other, by established Russianists (e.g. see Iver Neumann's interview to Tavberidze 2022). Notions like "Weimar Russia," first coined in the 1990's (see Hanson and Kopstein 1997) sadly became even more relevant since then. Russia's 2022 invasion of Ukraine has had clear indications of an attempted genocide underpinned by public discourses that deny whole nations their very right to exist, thus evoking horrifying parallels with World War II. The comparisons are understandable on the emotional level. However, I would venture to argue that they are more than simple figures of speech and that similarities between Russia and pre-WW2 Germany should be attributed to their historically defined position of *peripherality* in Europe. Here, notwithstanding the important differences between the two cases, a number of shared boxes can be checked, including a *relatively* late "catch-up" modernization trajectory and a deeply ingrained sense of grievance and inferiority vis-a-vis the "West". The "West" comes to be both coveted—as a typical role-model—*and* hated—as a perceived adversary or rival, who is seen as culturally, technologically, and politically superior. This has traditionally been reflected in Russia's (in)famous virtually incessant quest for recognition and the status of a great power. Thus, Ivan Gomza's chapter in this volume builds on earlier observations by Neumann (see Neumann 1996; 2015) to demonstrate how the Russian "Great Power ideology" is historically a product of this gap between ambition and limited capacity.

In Europe, the Eastern periphery is of course also a *relative* term. On the one hand, it goes without saying that Central and Eastern Europe is anything but homogenous and one should be always careful when drawing comparisons. On the other hand, the region can nevertheless be seen as deeply shaped by a *transitive peripherality*, understood as a condition in which the East-West differences form a geographically distributed spectrum, stretching across the Eastern and South-Eastern parts of the continent, rather than a clear dividing line unambiguously separating the "East" from the "West." Consequently, countries like Austria or Germany which are these days placed in "Western Europe" without a second thought, can also be seen as having been shaped by peripheral experiences over the long course of their history. Russia is of course also a part of this spectrum of transitive peripherality.

Transitive peripherality manifests itself in similar and easily recognizable patterns of self-positioning and othering in national identity building strategies. The standard repertoire here includes the tropes pertaining to uniqueness, messianism, exceptionalism, "the shield of Europe" or "bulwark of Christianity" (*Antemurale Christianitatis*), and simultaneously (!) also self-orientalization. Apparently, much of this has to do with compensatory mechanisms employed by peripheral polities. Alexander Kiossev's concept of self-colonizing cultures emphasizes this aspect precisely: the postcolonial inversion that claims spiritual superiority over the Western Other is itself a product of a long-term collectively held sense of inferiority. Put in more mundane terms: if you feel that you are not able stand on par with the West, then you have a powerful political and psychological stimulus to justify it, above all, for your collective self. As pointed out above, there are several ideologemes that can be instrumental here. The "bulwark of Europe" is one of them, while "uniqueness" is yet another. The trope of "uniqueness" logically implies a relativization of general standards, giving the collective Self a chance to escape from the feeling of inferiority towards its "Western" or "European" Other.

As any Russia expert will easily guess here, the Russian versions of the "bulwark" and messianic myths respectively are the themes of "shielding" Europe from the Mongols and delivering it from Nazism—and this is not to mention the nearly seventy years of the Marxist-Leninist eschatology. Eurasianism, on the other hand is a form of self-orientalization that works with a construct of an imaginary "Easterness." Unsurprisingly, self-orientalization is a common peripheral move that has been employed by regional discourses on national identity on more than one occasion. It easily becomes entangled with messianism, as Poland's romanticist myth of Sarmatian origin demonstrates, although arguably, a more accurate analogy here would involve Hungarian Eurasianism or Turanism (see Balogh 2015; 2022).

In both cases we are dealing with ideologies produced in social and intellectual environments that are deeply shaped by European intellectual and cultural influences even if situated somewhat peripherally vis-a-vis the European "core." Similarly, both are built on a romanticization of the alleged non-Europeanness, of a (Euro)Asiatic "nature" or "destiny." It is hardly a coincidence that both Hungarian and Russian discourses on "Eurasia" have had strong inclination towards authoritarianism and fascism and involved a resentment towards the Western "liberalism". Unsurprisingly, in the more recent period, the "Eastern origins" trope has been used by Hungary's PM Viktor Orbán to rhetorically support the so-called *Eastern opening* policy—a doctrine of hearty relations with various authoritarian states in Eurasia.

In the neighbouring Slovakia a similar grammar of self-orientalization can be observed in national identity building strategies of the Russophile pan-Slavists (see Marušiak 2023). Frustrated by their marginalized political position in the Habsburg empire, the 19th century Slovak revivalist felt rather disappointed by the West and its liberal political culture and were looking up to Russia as a Slavic "alternative" looming in the "East." Bátora's analysis shows how this resulted in "an erroneous narrative placing Slovakia on the intersection between 'the East' represented by an imaginary Slavic community led by Russia and the West with its feudal and later liberal democratic institutions," whilst "Slovakia de facto never was at such a civilizational intersection as it always belonged to the West" (Bátora 2017, 139). Furthermore, here, again the "Eastern" imaginary has been used to normalize authoritarianism. As the populist authoritarian PM Vladimír Mečiar said in response to the Western criticism of democratic backsliding in Slovakia, "if they do not want us in the West, we will turn to the East" (Duleba 2005).

4 More Decentring Comparisons: Proximate Others and the Logic of Genocide

Briefly discussed above were merely some sporadic examples from the regional neighbourhood. The intention was not to provide a comprehensive catalogue nor to develop the comparisons in a systematic matter. The latter calls for a comprehensive research project of its own. The point was rather that to illustrate that self-orientalization is generally speaking, a common move for peripheral identities. It is closely related to the claims of "uniqueness" though it is not the only strategy associated with it, as messianism and other forms of exceptionalism usually go hand in hand. Peripheral identities tend to have a menu of strategies that political and intellectual elites adopt to their specific identitary needs as they choose to see them. Their routine work here is turning marginality into centrality, defeat into victory, and lagging into "spiritual" superiority.

Here, another very disturbing analogy comes to mind. As Moscow unleashed its full-scale invasion and the world became witness to terrifying images from Bucha, Irpen, and other massacre sites, there arose the question of possible comparisons to the Yugoslav wars with their dark history of genocide and use of rape as a weapon of war. At the time of writing, the Russian war of aggression still rages, and we are nowhere near the phase of transitional justice. Yet, at this point it is clear that both the public rhetoric of some Russian media, politicians and public figures and Russia's multiple post-2022 practices clearly suggest genocidal motives, as some genocide scholars have also publicly pointed out (Stavljanin 2024; see also Riabchuk 2022).

Systematic securitization and dehumanization of Ukrainians in Russia had been going on for years prior to 2022 (see Gaufman 2017 for a comprehensive study) but with the war of conquest this kind of rhetoric came to justify the actual atrocities. Following the attack on the Ohmatdyt children hospital in July 2024, commentators from the Orthodox fundamentalist Tsargrad TV, owned by the Kremlin-linked oligarch Konstantin Malofeev, gave a candid account of their view. According to them Russian missiles were not killing any people, "not a single one" simply because on the other side "there were no people" (Kapnik 2024). Shocking as it may be in its blunt genocidal sincerity, this quote is but a tiny tip of the iceberg, with multiple calls during the course of the war to bring an end to the independent Ukrainian nation, to destroy its cities, to burn and drown children (BBC 2022), or to set up concentration camps in Ukraine for the purposes of "de-Ukrainization."

Of course, the comparativist has to tread very carefully bearing in mind the vast differences between the post-Soviet and post-Yugoslav contexts. However, it seems that, at the very least, in light of the recent developments, in particular including their most dark and depressing moments, this academic discussion cannot be avoided. At most, illuminating similarities could potentially be discovered in how various peripheral identity discourses mutate into an antagonistic, genocidal discourse, denying the neighbouring Other their very right to exist. Here, despite what some might naively assume, cultural proximity (which in itself must always be understood in strictly relative terms) does not necessarily have to be a factor restraining hatred and violence but, on the contrary, one that can aggravate it under certain circumstances. As Adam Bence Balazs (2020, 12) observes, writing in the context of Yugoslav conflicts, the proximity of the "almost-other" can sharpen and reinforce difference and, in the end, "alikeness might be more 'frightening' than difference."

Serbian nationalism, which had been a major—but, of course, far from the only—perpetrator of genocide in the post-Yugoslav area, is well-known for its tendency towards peripheral inversions discussed above. The European periphery is characterized by multiple strategies of re-narrating inferiority as superiority or marginality as *centrality*—hence, for example, also the regional prominence of the geopolitical concept of "Central Europe." The Kosovo myth (see Ejdus 2020), which has been pivotal for modern Serbian nationalism since its inception, works exactly along these lines: (alleged) military defeat becomes a "spiritual" victory and strongly cultivated victimhood is rendered a moral triumph of the Serbian "Heavenly Kingdom" (Judah 2009). This proneness towards self-victimization survived well into the 21st century, with the dominant Serbian narrative remaining that of "a victimized nation, of people struggling, in vain, for national survival, or engaged in an honourable, but futile struggle to defend themselves against one of the myriad and always more powerful enemies" (Subotić 2013, 309).

It should be noted that a rather common element of the peripheral condition is some unfortunate combination of self-victimization and imperialist or expansionist attitude towards neighbours. In fact, the former typically serves as justification for the latter with aggressors infrequently styling themselves as victims. Notably, this stands true, for what is—up to date—the most toxic mutation of CEE nationalisms, as German Nazism is also known to have drawn heavily on images of a wronged and victimized Germany. The present Russian case is obviously no exception either, for as its rulers said following the 2022 full-scale attack, throughout its history Russia, in fact, "never attacked anybody" but has always only defended itself (Sibir.Realii 2022).

Self-victimization and the stemming revanchism are, of course, always carefully cultivated, albeit there are also some shared structural factors in peripheral development that can be seen as stimulating the rise of this type of nationalism. Arguably, they pertain to the late nation formation pattern and incongruence—at least as proclaimed by the nationalists themselves—between the political and cultural borders. Here, again the (in)famous 19th century "German question" in CEE is exemplary. In some cases, there is an additional element at play, which is the competition between different identitary projects and the resulting fuzzy and contested nature of borders between identities. In the Serbian case this seems to apply to Yugoslavism, which, in certain periods, could be viewed by some as a vehicle for the Greater Serbian project (Bulatović 2004) but it could also equally apply to alternative definitions of Serbianness based on religious or linguistic markers (Judah 2009).

Needless to say, this multi-layered confusion is particularly conductive for the self-cultivation of insecurities. In the Serbian case this cultivated insecurity was persistently reinforced—strongly but once again far from exclusively if seen from the broader regional perspective—by the trope of betrayal. The betrayal trope runs through the nationalist mythology of the 1389 "defeat" on the Kosovo field and shapes the attitude towards the neighbouring Others such as Bosniak Muslims. Thus, the figure of the Kosovo traitor Vuk Branković "represents the Slavs who converted to Islam under the Ottomans and any Serb who would live with them or tolerate them'" (Carmichael 2002, 64).

Consequently, one of the most influential classics of Serbian literature, *The Mountain Wreath*, tells the story of a massacre, perpetrated on purely identitary grounds, of Muslim Montenegrins. Their "betrayal" of "Serbianness," defined here along religious lines, is presented an intolerable abomination, though at the same time they are still seen as some form of blood relations (Wachtel 2004), thus making the declared "betrayal" an internal one. This sets a model for a genocidal, antagonistic identity discourse which denies the neighbouring Other the very right to independent existence.

Similar patterns could also be produced in relation to the Croatian Other. Of interest here are the ideological creations of the Serbian far-right, for example, those belonging to Vojislav Šešelj. A deputy prime minister in Slobodan Milošević's government and a convicted war criminal, Šešelj (2007) also happens to be the author of a massive treatise with a mostly self-explanatory title: *The Roman Catholic Criminal Project of the Artificial Croatian Nation*. Taking many hundreds of pages, Šešelj lays out a meticulously detailed conspiracy narrative, according to which the modern Croatian nation is a product of a long-term anti-Serbian Western plot, spearheaded by the Vatican and involving "artificial Croatization", and delusion of the "Catholic Serbs" who are skilfully manipulated into believing themselves to be Croats. The narrative spans several centuries and involves a broad variety of perpetrators, from the Communist Josip Broz Tito to Pope John Paul II who is said to be the ultimate culprit behind the 1999 NATO bombings of Yugoslavia.

For anyone familiar with post-Soviet politics it is not too hard to see the parallels to conspiracy narratives that long took a deep root in Russia. According to them Belarusian and Ukrainian nations are "artificial" creations, that either do not exist at all or exist in some marginal form, e.g., as a Western Ukrainian ("Galician") language and identity that have been "forced" onto the "Russians" living in Central and East Ukraine. The narrative also typically involves portraying Belarusian and Ukrainian nation-building as tools of external forces (e.g., "the West", "Austrians", "Poles") and also draws on the trope of betrayal, e.g., the "traitor" Hetman Mazepa in the case of Ukraine (see Grob 2007).

These narratives pre-existed the rule of Putin by a long time (Basenko 2018) but were increasingly activated by him and his ideologists. For example, shortly prior to the full-scale war, in his new role as amateur historian, the Russian president tried to persuade the public that Ukraine was both an artificial Soviet creation and an "anti-Russia project" of Western authorship (President of Russia 2021), while Russian media and establishment referred to Ukrainians as "Banderite" traitors who betrayed their Russian legacy.

At a glance, it may seem bizarre to compare Russia to much smaller East or South European states, but, in fact, this is exactly the kind of decentring comparisons that we need, and as I argue below, we may need them for more than one reason. To begin with, categories like "smallness" are a matter of social fact: they have a socially constructed and intersubjective nature and do not simply refer us to some material characteristics (see Abulof 2009). Insecurities and self-victimization can easily be cultivated by great powers, sometimes with catastrophic effects as—once again—the historical example of Nazism demonstrates. Consequently, trauma narratives about the "geopolitical catastrophe" and "a divided people" (Laruelle 2015) were also a very important part of Putin's "Weimar Russia".

The post-Soviet ideologues of Russian ethnonationalism tried to reframe the Russian people as the main victim of the Soviet experiment, having been most disadvantaged in comparison to the non-Russians (Krylov 2012). Notably, this is reminiscent of the way that, acting in parallel, both Serbian and Croatian nationalisms tended to other the Yugoslav project as detrimental to the interest of their respective nation and serving the rival nationalisms. In the early 2010's, some observers were optimistic enough to think that Russian ethnonationalism might become an ally to Russia's democratisation because it was a "defensive" rather than imperialist-expansionist nationalism (Popescu 2012). In that respect, the figure of Alexey Navalny and his political evolution were also emblematic (see Laruelle 2014). However, as we know now, almost a decade and a half later, Putinism successfully co-opted primordialist Russian ethnonationalism through its "Russian world" concept with ideologemes like "divided people" underpinning the annexation of the neighbouring countries' territories (see Kazharski 2019).

Putin's ethnonationalism was a step away from the legacy of the Soviet official doctrines which at least formally recognized the existence of independent Belarusian and Ukrainian peoples—even if they were bound to be in a perpetual "brotherly" union with Russia. The Putinist discourse on Ukraine—and by a mostly silent implication also on Belarus—as "artificial" states created by Lenin and the Bolsheviks, denied them even this limited form of existence. Sherr and Gretskiy (2023) make an interesting point about "a Russian imperial practice that survives to the present day: the incorporation of other peoples into the identity of Russia itself." While I would not dispute this line of reasoning in general, I feel like the word "incorporation" sounds a little too neutral here. Perhaps, cultural appropriation, "kidnapping" or "trophy-taking" would be more suitable metaphors.

Thus, for the Russian imperialist discourse, Belarusian and Ukrainian identities are tolerable merely as regional or ethnographic ones, imagined by the imperial gaze along the ethnostereotypes that date back to Nikolai Gogol's literary work and other cultural foundations of Russia's grand imperial narrative.

Any transgression by the Other of these boundaries that are set by the imperialistically articulated Self is met with hostility and identified as "treason." Naturally, the Others may not wish to be thus "incorporated" and this is where the identitary antagonism stems from, as well as the ensuing *genocidal* mode of the imperialistic Self: those who dare not be(come) us must be annihilated.

In principle, genocide can be seen as *one* mode of the imperialistically articulated Self, while the relationship to its proximate Others remains both dynamic and dialogical. At a glance, it can contain seeming paradoxes. For example, Putin's denial of Ukraine as an "artificial state" is coupled with the fact that his amateur historian articles on the Russo-Ukrainian "unity" were published in *both* Russian and Ukrainian. We do not need to be deceived by this seeming partial recognition, as we know all too well what the amateur historian did next. Nevertheless, it is useful to remember that the relationship between imperialistically articulated Selves and those Others who become targets of their kidnapping attempts is not a static one: it is dialogical and, to some extent, *negotiated*. Apparently, similar examples could also be found in the post-Yugoslav contexts: situational recognition can co-exist with assimilation or annihilation attempts (c.f. Biondich 2004: 60–64; Judah 2009 61–62).

Comparisons between Russia's full-scale invasion and the Yugoslav wars are sure to be emotionally very difficult work. However, they seem necessary in light of the gruesome parallels in the use of practices such as torture, mutilation, and rape as *weapons of war*—practices that apparently take the war objectives beyond the physical destruction of the enemy soldiers and military infrastructure and onto the level of a symbolic triumph over the Other. This grim triumph can involve a sadistic performance of utter humiliation or a complete erasure of the Other's cultural presence and a systematic stripping of individual and collective dignity. As Carmichael (2002, 70) points out for example "rape has become an integral part of ethnic cleansing in the Balkans (and elsewhere) because it is directed against civilian populations, intended to humiliate, traumatize and punish them for their continued existence". Furthermore, "some scholars have also argued that in nationalist discourse, since women come to represent the entire nation, acts of sexual violence are also the 'symbolic rape of the that community'" (Carmichael 2002, 68). In other instances, the genocidal logic of erasure could involve the destruction of cultural heritage, such as the demolition of historical mosques in Bosnia (Carmichael 2002, 86).

Russia's deliberate brutality in Ukraine on so many fronts has also involved a broad range of similar practices. Thus, the use rape as a weapon of war has been well documented. In one particularly well-known instance, a Russian soldier was encouraged by his own wife "to rape the Ukrainian bimbos" (Kulick 2022). The raped Ukrainians in Bucha were promised to be raped till they would bear no more Ukrainian children and vilified as "Nazi prostitutes" (Limaye 2022). Russia's use of the "Nazi" label is part of the dehumanization strategy, which, as some observers remarked, is akin to the "big lie" about a Jewish conspiracy which underpinned Nazism itself (Hardoš 2022). In the post-Yugoslav context, similar dehumanization techniques were used as Croatian or Serbian populations could receive the stigmas of "Ustashas" or "Chetniks" respectively, thus also drawing on the bloody images of the Second World War in order to justify murder.

These and many other shocking parallels can be subsumed under the horrifying puzzle of genocidal violence perpetrated towards people that, as a matter of fact, often do not stand too far apart in geographical or cultural terms, i.e. towards *proximate Others*. The empirical questions here naturally pertain to the techniques of dehumanization but also its deeper mechanisms and roots. The archaization of war with deliberate brutality becoming a *spectacle* of cruelty and humiliation suggests that identities are involved here in some sinister way. In the aforementioned *Mountain Wreath*, which some go as far as to call "an ethnic cleansing manual" (Pavlović 2001), the existence of the Other is presented as intolerable based on identity grounds: Muslim Slavs are viewed as "traitors" to the Orthodox Serbia, apostates who are "worse than the Turks" (Judah 2009, 76). The Russian imperialistic Self has also been systematically articulated in such a manner as to stigmatize any Ukrainian (or Belarusian) independent existence as an offense to that Self, a "treachery" that must be (ex)terminated.

The grim comparison I very briefly outlined is instructive in many ways. Obviously, it runs along the lines of what I argued above with respect to decentring Russia and exposing it as one of many European peripheries with similar dark experiences—rather than a "unique civilization" or a "Eurasian power" that must be understood on its own terms. It also goes without saying that the abyss of genocidal violence that is underpinned by systematically and deliberately constructed identity antagonisms demonstrates the weakness of any arguments made in the realist tradition. This tradition is fond of portraying Russia as an actor driven by a cold strategic, security-seeking rationality, and these arguments continue to be presented despite the multiple public acknowledgements made by top Russian officials of the actual goals of the 2022 invasion as being about putting an end to an independent Ukrainian nation.

Finally, a somewhat less obvious corollary to this has to do with how Russia has framed itself and Ukraine post-2014. The so-called "Ukraine crisis" played all too well into the carefully promoted Russian narrative on Ukraine as "a weak," "corrupt" or "failing state" (see Kazharski 2017; Tyushka 2023). This narrative externalized a Balkanesque image of violence, "crisis", "ethnic conflict," or "civil war" onto Ukraine while presenting Russia as a responsible global player and a great power who was participating in the "crisis management" through the Minsk Agreements together with other European powers. This aspect is discussed in detail in the chapter by Anselm Schmidt and Sanshiro Hosaka, who scrutinize ways in which the "civil war" narrative also managed to deceive some members of the international academia due to their methodological shortsightedness.

Considering that Russia was actually the perpetrator all along, having used the domestic political crisis in Ukraine in order to annex Crimea and start a proxy war in the Donbas, this narrative—on which so many in the West picked up between 2014 and 2022—was part and parcel of the ongoing international normalization of Russian behaviour, and of presenting the 2014 Russian aggression against Ukraine as a "crisis". Consequently, exposing Russia's own identity issues in a comparative-decentring perspective also exposes the grand deception performed by this narrative. It demonstrates that all this time the problem—or "crisis"—has not been in Ukraine. It has been in Russia who was growing pathologically obsessed with Ukraine, in the Russian failure to build a consolidated democratic nation-state post-1991, and in the instrumentalization by the Russian authoritarian regime of national identity issues and genocidal identity politics.

Conclusion

What I attempted above is a very brief and sketchy survey of some possibilities for comparative work that would situate Russian studies in the broader region of CEE—a region defined by the common denominator of European peripherality. Some of these possibilities have to do with how peripheral nations constitute its relations with the European "core"—whether real or imaginary, while others have to do with intra-regional othering and relations with immediate neighbours or what I termed proximate others.

I am aware that some of these suggestions are likely to be met with scepticism and criticism, yet I believe that it is important to at least initiate these discussions. For too long Russian studies have been dominated by an exclusive focus on Russia, underpinned, whether explicitly or tacitly, by notions of uniqueness and of Russia's ensuing special status. Consequently, it is important to bear in mind that the exceptionalist "uniqueness" myths discussed above are not just a form of collective narcissism which inhibits the domestic post-imperial transition and prevents an honest reckoning with Russia's murderous imperial-colonial past—and—at the time of writing—present. They also affect knowledge production and other types of discourses outside of Russia, including the Western (and global) expert and policy-maker discourse.

Therefore, I suggest that comparisons are important not only for better understanding—but also a decentring instrument. Systematic comparisons with other CEE cases should demonstrate that Russia is far from being "unique", but on the contrary, in many respects it is very typically peripheral. It is one of many peripheral cases in CEE, which is, in turn, itself historically but one of multiple forms of European peripherality. Here, I fully concur with the point Vladimir Gel'man makes in his respective chapter about "a serious need to put [the Russian] case into a broader theoretical and comparative-historical framework in a systematic way." The author also seems to worry that the proposed *decolonisation* of knowledge production would divert resources from studying Russia and thus result in a serious shortage of expertise on the strategic adversary. However, this seems to me like a false dilemma or, indeed, an unnecessarily narrow (mis)reading of "decolonisation" (alternatively: "de-imperialisation") Introducing the comparative perspective would not mean ceasing to study Russia altogether, only decentring it according to the logic I outlined above.

Likewise, my argument also runs in parallel with the suggestion Andrey Makarychev makes in his respective chapter about comparing Russian authoritarian biopolitical practices with other countries in the region, although I am not firmly convinced about the sharp distinction being drawn between Russian "empire-rebuilding" and what is identified here as CEE "nation-building." On the one hand, it is true that the broader regional neighbourhood (fortunately) does not seem to harbour any cases with a comparable scale of global ambition or military and economic resources. On the other hand, some of the regional nationalisms clearly combine a sense of resentment with territorial revisionism and irredentism (cf. Hungary, Serbia). In any case, the comparative perspective here could help us better understand ways in which democracy failure in peripheral European polities yields anti-liberal/ authoritarian biopolitical regimes strongly fixated on issues of gender and sexuality.

In sum, Russia does not deserve being "understood" on its own terms any more than any other country in the region and any disproportionate attention to its alleged idiosyncrasies is both analytically and normatively unjustified. The exceptionalist discourse of Russia's revisionist rulers and exoticization of Russia in Western academic and policy circles have been mutually sustaining. They have worked jointly to normalize Russian transgressions and to indirectly justify a great power privilege. It is time to end that privilege.

References

Abulof, Uriel. 2009. "'Small Peoples': The Existential Uncertainty of Ethnonational Communities." *International Studies Quarterly* 53: 227–248, https://doi.org/10.1111/j.1468-2478.2008.01530.x.

Balogh, Péter. 2015. "Returning to Eurasia from the heart of Europe? Geographical metanarratives in Hungary and beyond." In *Beyond Transition? Memory and Identity Narratives in Eastern and Central Europe*, edited by Bernsand, Niklas. Narvselius, Eleonora., and Törnquist-Plewa, Barbara. The Centre for European Studies (CFE) at Lund University.

Balogh, Péter. 2022. "Clashing geopolitical self-images? The strange coexistence of Christian bulwark and Eurasianism (Turanism) in Hungary." *Eurasian Geography and Economics* 63: 726–752, https://doi.org/10.1080/15387216.2020.1779772.

Balazs, Adam B. "Introduction—All Happy Regions are Alike." In *The Visegrad Four and the Western Balkans*, edited by Adam Bence Balazs and Christina Griessler, Nomos Verlagsgesellschaft—Andrássy Universität Budapest.

Basenko, Ivan. 2018. "'The German Intrigue' as an Element of the Anti-Ukrainian Campaign: A Case Study of Kyiv's Russian Language Press, 1914–1918." *East/West: Journal of Ukrainian Studies* V(2): 149–173. https://doi.org/10.21226/ewjus422.

Bátora, Jozef. 2017. "Slovakia's layered entry into international society and the possibilities of its exit." In *Memories of Empire and Entry into International Society*, edited by Filip Ejdus, Routledge.

BBC. 2022. "Krasovsky: Russia bans 'burn Ukrainian kids' TV presenter", October 24, https://www.bbc.com/news/world-europe-63378613.

Biondich, Mark. 2004. "'We Were Defending the State': Nationalism, Myth, and Memory in Twentieth-Century Croatia." In *Ideologies and National Identities. The Case of Twentieth-Century Southeastern Europe*, edited by John. R. Lampe and Mark Mazower. CEUPress.

Bulatović, Marko. 2004. "Struggling with Yugoslavism: Dilemmas of Interwar Serb Political Thought." In *Ideologies and National Identities. The Case of Twentieth-Century Southeastern Europe*, edited by John. R. Lampe and Mark Mazower. CEUPress.

Carmichael, Cathie. 2002. *Ethnic Cleansing in the Balkans. Nationalism and the destruction of tradition*, Routledge.

Curanović, Alicja. 2021. *The Sense of Mission in Russian Foreign Policy. Destined for Greatness!* Routledge.

Gaufman, Elizaveta. 2017. *Security Threats and Public Perception: Digital Russia and The Ukraine Crisis*, Palgrave Macmillan.

Grob, Thomas. 2008. "'Mazepa' as a symbolic figure of Ukrainian autonomy." In: *Democracy and Myth in Russia and Eastern Europe*, edited by Wöll, Alexander and Wydra, Harald, Routledge.

Duleba, Alexander. 2005. "Slovakia's Relations with Russia." *Studia Politicae Universitatis Silesiensis* 1: 325–340.

Ejdus, Filip. 2020. *Crisis and Ontological Insecurity. Serbia's Anxiety over Kosovo's Secession*. Palgrave Macmillan.

Hanson, Stephen E. and Kopstein, Jeffrey. S. 1997. "The Weimar/Russia Comparison." *Post-Soviet Affairs* 13(3): 252–283.

Hardoš, Pavol. 2022. "Kremeľ našiel svoju veľkú lož." *SME*, March 4. https://komentare.sme.sk/c/22853616/kremel-nasiel-svoju-velku-loz-pise-pavol-hardos.html.

Judah, Tim. 2009. *The Serbs. History, Myth and the Destruction of Yugoslavia*. Yale University Press. (New Haven and London: 2009).

Kapnik, Elena. 2024. "V Rossii posle udara po "Ohmatdetu" prizvali k novym ubijstvam graždanskih." *TSN*, July 13. https://tsn.ua/ru/ukrayina/v-rossii-posle-udara-po-ohmatdetu-prizvali-k-novym-grazhdanskim-ubiystvam-2619396.html.

Kazharski, Aliaksei. 2017. "From 'colony' to 'failing state'? Ukrainian sovereignty in the gaze of Russian foreign policy discourses." In *Vocabularies of International Relations after the Crisis in Ukraine*, edited by Andrey Makarychev and Aleksandra Yatsyk, Routledge.

Kazharski, Aliaksei. 2019. *Eurasian Integration and the Russian World. Regionalism as an Identitary Enterprise*. CEUPress.

Khrustalyov, D. 2011. "Proishoždenie 'russkogo medvedja.'" *Novoe literaturnoe obozrenie* 107 (1/2011), https://www.nlobooks.ru/magazines/novoe_literaturnoe_obozrenie/107_nlo_1_2011/article/13029/.

Krylov, Konstantin. 2012. "Ot RSFSR—k Evrazijskomu sojuzu." *Voprosy nacionalizma* 4(12): 28–36.

Kulick, Orysia. 2022. "Gender and violence in Ukraine: changing how we bear witness to war." *Canadian Slavonic Papers* 64 (2–3): 190–206. https://doi.org/10.1080/00085006.2022.2108264.

Laruelle, Marlene. 2008. *Russian Eurasianism. An Ideology of Empire*, Woodrow Wilson Center Press.

Laruelle, Marlene. 2014. "Alexei Navalny and challenges in reconciling 'nationalism' and 'liberalism.'" *Post-Soviet Affairs* 30(4): 276–297. https://doi.org/10.1080/1060586X.2013.872453.

Laruelle, Marlene. 2015. "Russia as a 'Divided Nation,' from Compatriots to Crimea: A Contribution to the Discussion on Nationalism and Foreign Policy." *Problems of Post-Communism* 62(2): 88–97. https://doi.org/10.1080/10758216.2015.1010902.

Limaye, Yogita. 2022 "Ukraine conflict: 'Russian soldiers raped me and killed my husband'", *BBC*, April 11, https://www.bbc.com/news/world-europe-61071243.

Makarychev, Andrey and Nizhnikau, Ryhor. 2023. "Normalize and rationalize: Intellectuals of statecraft and Russia's war in Ukraine," *Journal of International Relations and Development* 26: 632–642. https://doi.org/10.1057/s41268-023-00299-x.

Marušiak, Juraj. 2023. "Slovakia: Emergence of an Old-New Pseudo-Pan-Slavism in the Context of the Conflict Between Russia and Ukraine After 2014." In *Pan-Slavism and Slavophilia in Contemporary Central and Eastern Europe*, edited by Mikhail Suslov, Marek Čejka, Vladimir Đorđević, Palgrave Macmillan.

Morozov, Viatcheslav and Rumelili, Bahar. 2012. "The External Constitution of European Identity: Russia and Turkey as Europe-Makers," *Cooperation and Conflict* 47(1): 28–48. https://doi.org/10.1177/0010836711433124.

Neumann, Iver B. 1996. *Russia and the Idea of Europe: A Study in Identity and International Relations*. Routledge.

Neumann, Iver B. 2015. "I Remember When Russia Was a Great Power." *Journal of Regional Security* 10: 5–16, 10.11643/issn.2217–995X151SPN56.

Nikonov, Vjačeslav. 2003. "Soblazn osobogo puti." *Rossija v global'noj politike*, September 2. https://globalaffairs.ru/articles/soblazn-osobogo-puti/.

Pain, Emil. 2020. *Ideologija "osobogo puti" v Rossii i Germanii*, Tri Kvadrata.

Pavlović, Srdja. 2001. "The Mountain Wreath: Poetry or a Blueprint for the Final Solution?", *Spacesofidentity.Net* 1(4). https://doi.org/10.25071/1496-6778.8038.

Perla, Andrej. 2024. "Detskaja bol'nica v Kieve—ne slučajnost'. Pora èto priznat' i perestat' bojat'sja", Tsargrad, July 8, https://web.archive.org/web/20240713203657/https://tsargrad.tv/articles/detskaja-bolnica-v-kieve-ne-sluchajnost-pora-jeto-priznat-i-perestat-bojatsja_1025063.

Platoff, Anne M. 2012. "The "Forward Russia" Flag: Examining the Changing Use of the Bear as a Symbol of Russia", *Raven: A Journal of Vexillology*, 19: 99–126.

Popescu, Nicu. 2022. "The Strange Alliance of Democrats and Nationalists", *Journal of Democracy* 23(3): 46–54, 10.1353/jod.2012.0046.

President of Russia. 2021. "Article by Vladimir Putin 'On the Historical Unity of Russians and Ukrainians'", July 12. http://en.kremlin.ru/events/president/news/66181.

Riabchuk, Mykola. 2022. "Calling a Spade a Spade: How to Unwrap a Genocidal Essence from the Kremlin Anti-Ukrainian Rhetoric." *Eidos. A Journal for Philosophy of Culture* 6: 8–20, https://doi.org/10.14394/eidos.jpc.2022.0031.

Riabchuk, Mykola. 2024. "Imperial Knowledge and Challenges of Decolonization", *Warsaw East European Review* XIII: 11–12.

Sibir.Realii. 2022. "Počemu slova o tom, čto "Rossija nikogda ne načinala vojnu",—nepravda", May 5. https://www.sibreal.org/a/pochemu-slova-o-tom-chto-rossiya-nikogda-ne-nachinala-voynu-nepravda/31835365.html.

Šešelj, Vojislav. 2007. *The Roman Catholic Criminal Project of the Artificial Croatian Nation*, Serbian Radical Party.

Sherr, James and Gretskiy, Igor. 2023. "Why Russia Went to War. A Three Dimensional Perspective", International Centre for Defence and Security. https://icds.ee/wp-content/uploads/dlm_uploads/2023/01/ICDS_Analysis_Why-Russia_Went_to_War_James_Sherr_Igor_Gretskiy_January_2023_EDIT.pdf.

Specter, Matthew. 2022. "Realism after Ukraine: A Critique of Geopolitical Reason from Monroe to Mearsheimer," *Analyse & Kritik* 44(2): 243–267. https://doi.org/10.1515/auk-2022-2033.

Stavljanin, D. 2024 "Genocide Scholar: 'I Do Think That Russia's Violence In Ukraine Is Genocidal' ", *RadioLiberty*, April 2. https://www.rferl.org/a/russia-genocide-ukraine-scholar-war-crimes-prosecution-hague/32888386.html.

Subotić, Jelena. 2013. "Stories States Tell: Identity, Narrative, and Human Rights in the Balkans", *Slavic Review*, 72(2): 306–326.

Tavberidze, Vazha. 2022. "Prof. Iver B. Neumann on the Russian Mentality, Putin's Failing Regime, and What's Coming Next", *Georgia Today*, July 28. https://georgiatoday.ge/prof-iver-b-neumann-on-the-russian-mentality-putins-failing-regime-and-whats-coming-next.

Tyushka, Andriy. 2023. "In 'crisis' we trust? On (un)intentional knowledge distortion and the exigency of terminological clarity in academic and political discourses on Russia's war against Ukraine."*Journal of International Relations and Development* 26: 643–659, https://doi.org/10.1057/s41268-023-00313-2.

Wachtel, Andrew B. 2004. "How to Use a Classic: Petar Petrović Njegoš in the Twentieth Century." In *Ideologies and National Identities. The Case of Twentieth-Century Southeastern Europe*, edited by John. R. Lampe and Mark Mazower. CEUPress.

Zarakol, Ayşe. *After Defeat: How the East Learned to Live with the West*, Cambridge University Press (New York: 2011).

CHAPTER 8

Knowledge Production in Crisis

Constraints on Academic Freedom in Finland, Hungary, and Poland

Katalin Miklóssy

This article investigates how and why Russia's war induced changes and political limitations on the academic endeavour in Finland, Hungary, and Poland.[1] These countries were selected because of their spatial situatedness as *frontline terrain* and their different relatedness to the conflict, which began in 2014 with the occupation of Crimea and the breakout of war in the Donbas. There are some similarities in these countries' scholarly traditions in relation to the study of Russia, as well as divergencies regarding not only Russia but also Ukraine. The main question is how existential threats and the ongoing European security discourse are converted into a noticeable change in academia in these countries. What is interesting in these cases is that they all ended up, for different reasons, with a black-and-white academic environment, producing one-sided knowledge.

This problem becomes particularly relevant when we take into consideration the progressing polarisation we witness Europe-wide in epistemic communities due to the gradually advancing politicization of academia and academic knowledge. In the humanities and social sciences in particular, the often-close relation between universities and political decision-making has a bearing on the evaluation of research content. Being useful for the state by tackling social challenges brings attention to the long-standing dichotomy between the pursuit of academic freedom, on the one hand, and the will of serving common interests, on the other (Schmid-Petri, Bienzeisler and Beseler 2022, 45–63). What constitutes the common interest, however, and how 'well' it is served by research results is defined by political elites and their ever-changing value-based priorities. In the European academic environment this relationship is reflected in the dependence on funding from (often state) authorities. This development carries serious consequences on scientific inquiry, detectable in

1 Acknowledgement: The work presented in this paper is part of the ARENAS project. https://arenasproject.eu/

This project has received funding from the European Union's Horizon Europe research and innovation programme under grant agreement No:101094731.

 | DOI:10.1163/9789004747364_009

academia's politicization (Rekker 2021, 352–368). Furthermore, given the frequency of crises occurring in the world, there are situations which demand swift reactions from political powerholders, which bring about equally rapid changes in the evaluation of what constitutes 'useful' science. The media as a primary influencer of a political cause often passes judgments of the usefulness of research outcomes, which are then interpreted and circulated by politicians (Brüggemann, Lörcher and Walter 2020, 1–22; Gustafson and Rice 2020, 614–633).

Russian aggression against Ukraine brought light onto this politicalness of academic attitudes, revisiting once again how Eastern Europe should be studied. Eastern Europe has been a frequently contested area since the early 20th century, representing a space that has been defined and redefined in relation to how the West perceived the significance of Russia. Thus, *Eastern Europe* became a moving target, a narrative constantly structuring and restructuring the geopolitical space (Miklóssy 2016, 59–78; Korhonen 2010, 1–22; Palonen 1997, 239–260). Studying Eastern Europe, therefore, has been more prone to political changes than Russian studies, in comparison. While Russia's physical border has changed over time, scholarly focus was Russia-centred even during the era of Sovietology—as it was evidently one country.

Eastern European studies have always been conditioned by Europe's political positioning to Russia and the Soviet Union. The first discipline (established in 1915) at the London School of Slavonic and Eastern European Studies became an independent institute in 1935 in response to understanding newly emerging states after the disappearance of the Austro-Hungarian and Russian Empires. The importance of this area grew after the Bolshevik revolution and the appearance of a new communist empire. In 1919 Sir Harold Mackinder proposed even a supranational state, a Europe in-Between, which would create a buffer zone between Germany and Soviet Russia (Mackinder 1919/1981). Eastern Europe was to serve as the Western *Cordon Sanitaire*, a security belt against the Soviet Union. The buffer zone potential of Eastern Europe stuck since the interwar period. Eventually the area acquired a pendulum-motion: during the Cold War it became the Eastern Bloc, protecting the USSR from the West, and after the collapse of the communist system, it was the site of the Eastern Enlargement of the European community, separating Europe from the unstable, and thus dangerous, Russia. After Russia attacked Georgia in 2008, the EU expanded its buffer belt with countries situated between the EU and Russia through the introduction of the Eastern partnership policy, a Polish initiative. This area was now known as 'Eastern Europe', since all these states became the buffer zone for the Eastern flank of the European Union.

As such, the definition of who belonged to 'Eastern Europe' in academic terms changed according to political interests. Russian and Eastern European studies look back to a dubious history of politically coloured arrogancy and oversimplifications, as well as inaccurate knowledge-production that modified Cold War understandings. New academic enthusiasm occurred in studying Eastern Europe after the end of the Cold War, but scholarship of the post-communist democratisation process was still dominated by a Western paternalistic attitude, which became a new brand of scientific endeavour, *transitology* (Schmitter 2014, 71–86; Markwick 1996, 255–276). The Russian attack on Ukraine brought back the prescriptive stance and value-loaded demand of fast knowledge of a maximum area for political purposes (Miklóssy 2020, 245–259). This period has even been called Cold War 2.0 (e.g., Verma and Brossig 2024, 794–801; Sakwa 2023; Sakwa 2025; Libman 2022, 148–159). The conflict induced an urge to take sides, but this moral ground has politicized and thus blurred academic inquiry on both Russia and Ukraine. This open political positioning of academia was encouraged by its institutional framework.

1 Institutions in Focus

Crises always underline the importance of how political decision-makers frame national interest with a direct impact on the expectations on 'relevant' research. Russian studies and Eastern European studies are embedded in an intertwined relationship between space and time. Areas which are seen as vitally important to study holistically and interrelatedly revealed the prevailing political atmosphere and attitudes of the time. Spatiality also referred to coexisting levels of the international, regional, national and local institutions and agency. In our case, academia and epistemic communities display this multilayered spatiality. They are internationally interlinked as science inevitably necessitates the exchange of ideas and networking, yet scientists are anchored to their home institutions and disciplinary communities. In addition, the academic subsystem is dependent on the ever-changing societal environment as well.

Knowledge production can be explored by applying the theoretical framework of historical institutionalism, which emphasizes the interaction of agency, structures, and temporal context. Institutions are assumed to be relatively stable entities, because the operational culture of institutions, in other words, the formal and informal rules, norms, and practices, sustain their durability. This relative stability is the guarantee of trust in the institutions of any political system. Alterations are supposed to be slow and gradual, taking place from within the institutions in the form of *endogenous evolution*. Institutions change suddenly and dramatically only by *exogeneous shocks*. These radical and instantaneous transformations mark a turning point and hence a critical stage where uncertainty increases (Thelen and Conran 2016, 51–70; Capoccia and Keleman 2007, 341–369; Soifer 2012, 1572–1597).

Historical institutionalism offers a special theoretical toolkit to understand one form of modern authoritarian governance, which relies on gradual evolution and non-violent means to tighten its grip over society (Haber 2008, 694–707). This policy of small steps and the subtle calibration of institutional frameworks have been a feature of two of our target countries, Hungary and Poland, in the 2010s. Yet, it can be argued that the war in Ukraine created a general crisis also in the European arena as a whole, regardless of the governance model. The security threat caused by Russian aggression launched an exogeneous shock and all-encompassing institutional transformation, especially in the sphere of academia. Knowledge production has changed because of the war. Although there is always a discursive relation between academic knowledge producers and societal knowledge appliers, nevertheless, area studies have been evaluated by how its research results were able to answer practical challenges; in other words, how useful scholarship has been for politicians. This is particularly true for countries where authoritarian tendencies emerged in the 2010s and where the freedom of academia was openly questioned. Yet, increasing polarization became apparent in European societies in general, creating black-and-white attitudes to questions that have been framed as nationally important matters by the political elites.

2 The Academic Watershed: War in Ukraine

How academic institutions reacted to Russian aggression against Ukraine depended on their geopolitical, regional and national context, and the perceptions of insecurities linked to spatiality. The three countries in this respect differed a great deal in their relation to understanding threats and the domestic political relevance of them. Finland and Poland experienced the war and Russia's actions as a direct existential fear, partly because the war reactivated historical traumas. Hungary, however, while referring to a different kind of trauma, did not see Russia as a threat, but rather, a possibility.

In all three cases, the role and significance of academic inquiry has been affected by the narrative turns and discursive reconfigurations of national interest. This fluctuating national frame bore consequences on academic priorities, partly due to the changing emphasis of resourcing certain contents in the scientific endeavour. Hence, the evaluation of scholarly results, as well as the political and public receptions of them, depended on how well core values of national identity and loyalties to political goals were addressed in times of insecurity. This new focus on knowledge production directly affected the re- or even deconstructions of the institutional bases, transforming disciplinary requirements and sometimes even restricted leverage for scholarly interpretations.

2.1 *Liberal Traditions and the Finnish Academic Infrastructure*

From the Finnish perspective, Russian aggression against Ukraine was a reminder of the Stalinist attack on the country and the Soviet invasion of the Eastern Karelian territories in the Second World War. The current war introduced a new discourse on security, quickly leading to an unprecedented consensus of joining NATO. This relatively rare mutual understanding of the Finnish political elite was seen also in the enthusiasm to align with EU sanctions against Russia and in the full support of Ukraine through military aid. This self-evident and consistent political decision was reflected in the institutional arrangements of academia. As in many Western countries, the war was a wake-up call to repolish Finland's run-down and previously neglected Ukrainian studies by channelling massive resources.

Ukrainian studies were originally established at the University of Helsinki in 2008, coordinated by the Aleksanteri Institute (Finnish Center for Russian, Eurasian and Eastern European Studies). The initiative came from the university itself, arguing that Ukraine is acquiring international attention, and thus it was important to increase Finnish expertise systematically (Opetusministeriö 2007). The new program was constructed as a minor field, partly consisting of language courses and some area studies lectures, and was to serve as extra competence in the major fields of Russian studies or Eastern European studies. The long-term problem was that Finland severely lacked in-depth area expertise, so the University of Helsinki established a partnership with the National University of Kyiv-Mohyla Academy. After ten years, however, the Ukrainian studies program still relied on foreign exchange because Finnish academia was unable to produce satisfactory Finnish expertise due to low interest (Salmivaara—Into 2019). Due to diminishing budgetary resources, by 2020 the program lost the language option and began to fade away, regardless of the upward trend of student enthusiasm on Ukraine, especially after Russia's illegal annexation of Crimea (Ukrainan opintokokonaisuuden tilastoja 2008–2019; Ukrainian Studies 2020). It was decided that the program was to be reduced even further in the fall 2021, targeting the next academic year 2022–2023 (Onninen 2022). The full-scale invasion beginning in February 2022 saved the finances of the program, which could now continue offering courses for a massively rising student demand. Interestingly, the University of Helsinki was the only Finnish institution where a complex area studies program was offered about Ukraine (Posio—Kuikka 2022).

Finnish scholarly communities reacted immediately to the situation in Ukraine and began to collaborate to put bottom-up pressure on the authorities. The Finnish *Scholars at risk* academic network, Federation of the Finnish Learned Societies, Young Academy Finland, academic research projects, and individual scholars initiated dialogues with the Ministry of Education and the National Agency of Education to demand centralized actions (Lattu and Mäkinen-Rostedt 2022). Political decision-makers responded positively and efficiently. In May 2022, the Ministry of Education summoned the representatives of higher education institutions at the university and college levels to map out what needed to be done to receive Ukrainian refugee students. The Ministry allocated 5.5 million Euros to assist 21 projects in expanding their welcoming potential for increasing numbers of Ukrainians. In addition, a special supporting organization was established by the authorities to assist in scientific matters regarding Ukraine and Russia (Piironen 2022). Due to this significant investment and official advertisement activity, between 2022 and 2024, 564 Ukrainian students enrolled into Finnish higher education institutions. Interestingly, at the same period 3459 Russian students entered these institutions (Opetushallinnon tilastopalvelu, also Finnish National Agency for Education Studyinfo). The universities themselves took the initiative to inflict pressure on politicians. The country's biggest university, the University of Helsinki (2022) for example raised money to launch a Crisis Fund to support researchers and students fleeing from Ukraine to study and do research in Finland. In addition, Finnish decision-makers established extra funds to assist Ukrainian academics in coming to Finnish universities (Research Council of Finland 2023).

Finnish academia took a decisive stand regarding the war and aligned with the Finnish political elite and the EU line. In response to the declaration of Rectors of Russian universities supporting Putin's 'special military operation' in Ukraine, the European Commission (2022) instructed European universities and research institutes to cut all institutional contacts to Russian universities. This resulted an exceptional situation where decades of networks and ongoing collaborative projects had to be shut down. This type of total isolation of academic communities for political reasons was unknown even during the harshest periods of the Cold War when academic interaction was maintained against all odds over the Iron Curtain (Kunkel 2021, 473–484; Stensrud 2018).

From the Finnish perspective, this EU regulation was followed to the letter by the recommendation of the Ministry of Education (Opetus- ja kulttuuriministeriö 2022). This meant, for example, that universities' management should discourage researchers from maintaining any contacts with their colleagues, to have work related trips, to participate in conferences or do fieldwork in Russia or Belarus. Furthermore, all already funded projects came under scrutiny and the Academy of Finland, the main national funding agency decided to interrupt already awarded grants (Research Council of Finland 2022).

First, Russian studies researchers and others whose research depended on fieldwork or cooperation panicked. Scholars in Finno-Ugric or Arctic studies, or those who were losing project funding due to their cooperation with Russia, protested (Merimaa 2022). They criticised the policy as 'burning bridges' and demanded a more individualised attitude, pointing to a statement made by 8000 Russian scholars' opposing Russian aggression (T-Invariant 2022). The situation turned, however, gradually inflexible and stringent. When an ex-minister and scholar Tarja Cronberg participated in a conference in Kaliningrad, the academic community of the University of Eastern Finland published a press release where they cancelled Cronberg's honorary doctorate (University of Eastern Finland 2024, Teivainen 2024). Similarly, when professor emeritus of Arctic politics Lassi Heininen maintained his Russian contacts and attended conferences in Russia, his contract was terminated at the University of Lapland. He argued that Arctic research is vital for Finland and while the war will end eventually, academic contacts will be damaged for good (Siivikko 2024).

The polarization of the academic community began to show dogmatic features when the relevance of pre-2022 knowledge of Russia was not only questioned but accused on the basis that it relied on the wrong premises and naivety. Researchers were misled by their Russian informants, and they failed to realize that their fieldwork was manipulated. Criticism centred on the claim that behind the overtly benevolent interpretations of Russia laid researchers' close links with Finland's political establishment and scholars' willingness to serve the country's trade interests. The debate was public, and in the context of the war where the Finnish public opinion sided unanimously with Ukraine, the accused researchers had considerable difficulty to respond (Tynkkynen 2023; Bäckgren 2023; Hyttinen 2023).

In such a heated atmosphere, epistemic communities, especially in Russian area studies, were to revisit their previous modus operandi and reach for what was now considered as truth. While the war continues, the Finnish academic atmosphere became one-sided, silencing divergent opinions. This also applies to the evaluation of Ukraine. Criticism of Ukraine's democratic development or corruption is considered as politically incorrect scientific inquiry. While obviously there are enormous challenges embedded in carrying out robust research during wartime, nevertheless, politics should not define what is the right time and place, not to mention the topics to be studied. This attitude is alarming particularly for two reasons. On the one hand, it is necessary for Finland to properly evaluate its unstable and potentially dangerous neighbour, Russia, from a security perspective. The overpoliticization of how Russia should be understood brings back the old Sovietologist gaze, with oversimplification and arrogancy towards undesired details. Due to the official ban on fieldwork, Russian studies are relying mostly on sources accessible only on the internet (Heusala 2025). Secondly, and even more seriously, when the academic community becomes blind to its societal embeddedness and interconnectedness with the ever-changing political interest, it resembles despised authoritarian Russian practices.

What is 'true' knowledge was also the key question in Hungarian and Polish institutional restructuring.

2.2 *Culture War in the Hungarian and Polish Academic Field*

Academic freedom has been politically jeopardized since the mid 2010s in Hungary and Poland, creating extra pressure on scholarly conduct. This was the result of an underlying process in which right wing conservative political elites attempted to take over institutions, aiming to establish cultural hegemony. The idea of cultural hegemony was borrowed ironically from the Marxist philosopher Antonio Gramsci, who was well read by conservative theorists (e.g., Békés 2020). According to Gramsci, ideology and culture are the most efficient and enduring instruments for maintaining power. The creation of a consistent political system therefore requires a shared, homogeneous, and uniform culture that produces a collective will and common affinity of the people. To carry out a consistent systemic transformation, political powerholders must take over cultural institutions to cement their power for a long durée (Gramsci 1982, 27; Badino 2020).

In 2016, Hungarian Prime Minister Viktor Orbán and the leader of the Polish Law and Justice party, Jarosław Kaczyński declared that it was time to start a 'cultural counter-revolution' (Foy and Buckley 2016). According to them, this was necessary to offer an alternative to liberal traditions, multiculturalism, gender ideology, and human rights enthusiasm in the EU. To strike against institutions populated and overrepresented by liberal intellectuals was imperative in pursuit of establishing a new conservative era. Liberalism had to be rejected because in the national conservatives' understanding liberalism was closely linked to leftism (Kazharski 2018, 754–780; Legutko 2016 and 2019). As Orbán crystallized in 2018, an era is much more than a plain political system: it is a special cultural context, a collective consciousness, a way of thinking and behaving, a common taste and shared atmosphere (Orbán 2018).

The purpose of creating a uniform culture and congruent academia was to marginalize and downplay any uncontrolled activity and product that could challenge officially accepted values, worldview, and scientific interpretations from 'below'. In Hungary and Poland, to carry out the 'conservative revolution', the institutional take-over was a gradually expanding process and the new policy became apparent in various sectors of academia, such as in universities, research institutes, and think tanks simultaneously. While Hungary and Poland developed national responses to meet this task, nevertheless they also implemented common strategies. The most efficient ones were changing institutional structures and management strategies, in addition to calibrating resources.

In Hungary, scientific research has been traditionally carried out in three types of institutional arrangements: the Academy of Science, at universities, and in research institutes. The most important centre of research has been in discipline-based institutes of the Hungarian Academy of Science, where researchers could concentrate on doing research without teaching obligations like at universities. In 2019, the government decided to reorganize the Academy: to promote scientific innovations and competitiveness of Hungarian science, it was deemed necessary to concentrate disciplinary institutes under a new 'research network' (in Hung. Eötvös Loránd Kutatási Hálózat), where the Board would be populated by politicians of the government party Fidesz, and by loyal academics. The Board would make decisions on scientific resources and channel them into topics that were considered nationally important.

Organisational changes in universities started at about the same time, with the purpose of limiting 'liberal academia.' The first dubious sign was purging Central European University in 2018 as a 'too international' and 'too liberal institution.' From 2021, the government created foundations to manage universities. The foundations were populated by Fidesz politicians who had power over the university's budget, exercised employer's rights, and thus had a decisive influence on who was employed and what was worth teaching. Since the foundations had the ultimate access to European research grants which academic staff were eligible to acquire through competitions, the European Commission excluded all Hungarian universities which were run by foundations (European Commission 2023).

The first Orbán government (1998–2002) started to build up multiple research institutes that would promote its values. For this purpose, the administration established a centrally run foundation dedicated to study of Central and Eastern Europe (Magyar Közlöny 14/1999). Under this foundation, the government started creating new institutes. Two research institutes, that of 20th century (1999) and of 21st century (2020), were attached to the freshly constituted

House of Terror Museum (2002). When Orbán's Fidesz party returned to power in 2010, the new institutions started to mushroom: institutes of Communism (2011), of Post-Communist Transition (2013), of National Heritage (2013), Veritas (2014) and of Hungarian-ness (2019). The Foundation studying Central and Eastern Europe oversaw the activities of numerous research institutes and was led by Orbán's personal trustee and advisor Mária Schmidt. She had decisive influence on what sort of research was carried out and what kind of interpretations were allowed (Miklóssy 2024, 123–132).

In Poland, institutional restructuring started in the late 1990s when newly established institutions became part of the political struggle. Yet, while the Hungarian institutional approach was more extended, the Polish was in many ways more focused and more intensive. Whereas in Hungary, the leadership of institutions was decided by the governing party, in Poland, appointing the leadership of these institutions required the approval of the majority of the House of Representatives, the Sejm. This gave an extraordinarily strong mandate for these institutions. One of the most powerful institutions was the *Institute of National Remembrance*, that resembled a ministry in its leverage and which the left-wing opposed fiercely (Stola 2012, 45–58). The Institute had a dual task: on the one hand, it carried out historical research on the crimes against the Polish nation and emphasized the continuous struggle for freedom. On the other hand, the research results had a practical consequence. The personnel also included a commission of legal experts that launched charges ending up in courts and they also conducted lustration, i.e., investigating public figures holding public office (Act 1998/no. 155; also Nalepa 2010, 58–96). It became a huge institution with over 2000 employees with affiliations throughout the country.

The significance of the Institute increased in 2018, when the Law and Justice party issued a new law to protect the reputation of Poland and the Polish people. The purpose of the law was to fight against untruthful interpretations about Polish guilt during the Second World War. Individual researchers could be accused and punished with imprisonment if they published the 'wrong kind' of interpretations. According to the narrative, Poland was the victim of great powers and as such blameless (Dz.U.2018 poz. 368; also, the explanation of the legal initiative, see Druk nr. 806). The authorities left the interpretation of this law open for political interpretations and legal consequences. This induced self-censorship among scholars. While they could investigate the past, they could no longer popularize their research.

3 Hungary: Controlled Interpretations of the War

After the institutional take-over, academic accounts aligned with the official Hungarian stance on the war. The war in Ukraine did not launch similar shock or existential threat as in Finland or Poland. From the Finnish point of view, Hungary was its complete opposite, not only regarding the model of governance and academic freedom but also in relating to Ukraine. Hungarian leadership had long troubled relations with the Zelenskyy administration, and when the full-scale invasion began there was not much empathy for neighbouring Ukraine, nor any willingness to help. Official attitudes were affected by Ukrainian language laws in 2012, 2014 and 2018 which weakened, among others, the Hungarian minority's cultural rights in Transcarpathia (Fedinec and Csernicskó 2017, 81–110). For the national conservative Fidesz, Hungarian minorities in the neighbouring states belonged as much to the concept of the Hungarian nation as people living in Hungary. These minorities were granted double citizenship; hence they are able to vote in Hungarian parliamentary elections. In other words, to take care of this population is vitally important for the governing party. Bilateral relations between Hungary and Ukraine had deteriorated before the full-scale invasion (Miklóssy 2024, 188–190).

Russia, on the other hand, represented a resource in counterbalancing the strengthening pressure from the EU regarding the gradual erosion of the rule of law during the Orbán administration. Due to frequent clashes with the EU, the administration revisited Hungary's Western orientation. Admiration of Russia's governance was articulated already in 2014 when Viktor Orbán declared his intent to turn the country towards an illiberal direction (Orbán 2014). Since then, Hungarian powerholders consistently and stubbornly began to develop closer relations with Putinist Russia.

Political affinity is reflected in academic accounts regarding the evaluation of the war. Because of the highly centralised institutional framework there are key influencers promoting the officially accepted interpretation of the conflict. The works these academics publish receive wide media publicity because of the Fidesz media empire which controls over 80 percent of the Hungarian media environment (Reporters without Borders 2023). The aforementioned Mária Schmidt, who is closely embedded within the political elite, acquired a central role as leading national conservative theoretician. Her book analysing the war was published in October 2022 by the Foundation of Central and Eastern Europe, which she herself oversaw. And since the Foundation functioned as an umbrella organization of research institutes dealing with history and social sciences, the book sent a powerful message on how the international situation should be understood. The war was seen as a regional conflict of two Slavic nations, which grew out of proportion through the active engagement of the US-led West. Schmidt primarily accused the United States for triggering the war due to its imperialistic aims and secondarily the Western globalist elite in aligning with the US (Schmidt 2022; Terror Háza 2022).

Universities that were managed by foundations with Fidesz politicians on board were in a delicate situation not to question the governing perception that resembled Russian propaganda about the war. The decisive official line had great influence on public opinion yet individual scholars could engage in different angles that mostly remained within the walls of universities. Those who still were able to get published explored topics that did not challenge the official line but did not necessarily agree with it either. This was a tightrope walk on the last small drops of academic integrity. These researchers focused on economic matters, history, military strategy, business relevance, Ukrainian refugees and labourers, or the like (e.g., Huszák, Szirbik and Balogh 2024, 388–404; Ferenc and Koväly 2022, 47–77; Poór et al. 2024, 939–954). Other published works related to area studies, security or political science, on the other hand, reproduced the official narrative (e.g., Bernek at al. 2022; Jójárt 2022, 91–120; Holányi and Sebők 2023, 123–147). In these, the blame game focused on the NATO and the collective West, on the EU for accelerating economic disaster by introducing sanctions of Russia, and on Ukraine for attacking minority rights.

4 The Polish Consensus

Relating to Ukraine has been a complex issue that went through considerable transformations due to the war. Ukraine belonged to the Polish concept of Eastern Europe that relates to the Eastern neighbourhood of the EU. Interestingly, Poland has reflected on this Eastern area as its own buffer zone. Polish attitudes towards Ukraine have been always intertwined with Polish-Russian relations, and therefore Ukraine was a security issue for Warsaw. Competition between Poland and Russia over the possession of Ukraine between the 13th and the 18th centuries was a significant part of historical experience that influenced Polish political and strategic thinking. After Poland reappeared on the map in the interwar period, Ukraine was perceived as a vitally important buffer zone against Soviet-Russia. The sour point of common history concerns the massacre of 20,000 Poles by Ukrainian nationalists in 1943 and the ethnic cleansing of Polish minorities in Eastern Galicia and Polesie in 1945. Polish historians and the political elite made joint efforts to codify the official interpretation through the Institute of National Remembrance of the guilt of Ukrainians (see, 1943 Volhynia Massacre). The clash with those Ukrainian historians who cherish the memory of the Bandera-led nationalists as freedom fighters have been evident.

It can be argued that from the Polish side even this most difficult area of Polish-Ukrainian relations did not convert into clear political division of academics regarding Ukraine. Conflicting views have been downplayed, and the emphasis was moved to Ukraine's role on Polish security. Only when we leave the sphere of academic work and look at popular scientific publications can we see the divisions more sharply. Here one can find a whole range of martyrological publications which not only unequivocally draw the line between victims and executioners but also condemn the Polish state authorities for not making sufficient demands on the Ukrainian side regarding the policy of remembrance.

The historical perception of Russia as a threat is relatively clear-cut and has been analysed by multiple academic authors for a longer period of time (e.g., Zarycki 2004, 595–627; Taras 2014, 710–734). The history-driven fear of Russia was only fortified after the annexation of Crimea. The mistrust and outward hostility towards Russia have been manifest on manifold levels: in post-1989 history schoolbooks, in official pronunciations of politicians (especially of the Law and Justice governments), in right-wing and left-wing media outlets where also public intellectuals depict domestic political cleavages as due to Russia's malicious influence (Maresz 2013, 144). Despite official cracks between Warsaw and Brussels caused by the erosion of the rule of law and the PiS governments' anti-Western rhetoric, Russia still remains the main 'foe' towards which the public and political sphere position themselves. Russia and the 'East', on the other hand, are unanimously condemned by the government and the opposition alike. The Russophobic disposition of the Polish political and intellectual elite goes so far as to prompt accusations against their adversaries of actually serving Russian interests and even acting as Putin's proxies. Voices criticizing anti-Russian hysteria were isolated and often accused of blind Russophilia (Łagowski 2016).

For Polish strategic goals, Ukraine became vitally important in the 1990s and 2000s as means of weakening Russia. This new pro-Ukrainian foreign policy affected the Polish society and the academia alike (Jędraczczyk 2010). Before the war, the Polish academic community related to Ukraine with a certain distance, albeit after the invasion of Crimea more positively. Since 2022, however, the community fully identifies with Ukrainian affairs and show a clear and unequivocal siding with Ukraine in the conflict with Russia. Of course, individual scientists and research centers evaluate related phenomena and data in different ways, but nevertheless the whole academic environment was convinced that the efforts of the Ukrainians in repelling the Russian invasion must be supported.

The most visible change that can be observed between 2020 and 2024 is the avalanche of publications about the current military, political, social and economic situation triggered by Russia's February 2024 invasion. The primary focus of the academic community has been on the description, interpretation, and evaluation of current events and their impact on the reformulation of beliefs, assumptions, and future perspectives. According to the unanimous view of the academic community, Russia's full-scale invasion of Ukraine marks a turning point in theoretical thought, comparable only to the effects of the Second Wold War II (just to mention a few from the many: Diec et al. 2022; Stępień 2022; Olchowski 2023; Niedziółka and Próchniak 2023).

Criticism is targeted against actions that are considered to undermine Ukraine's chances in the current struggle. Hence, the traditionally good relations with Hungary and Hungarian academic circles, aligning with the Orbán-administration, came to a halt. Academics also engage in furthering national objectives to support Ukraine's position in all possible ways, recommending various policy actions. Hence, it can be argued that Polish academia actively support the political sphere that showed mutual understanding across the parties in their decisive stand against Russia.

It is also noteworthy that the participation of Ukrainian scholars in Polish academic life has been increasing since 2022. This can be also observed in the growing number of Ukrainians publishing in Polish and co-publishing in collaboration with Polish scholars. This process can be especially detected when browsing the archives of master's and bachelor's theses—in some universities, even every second thesis on Ukraine is written by students from Ukraine.

Conclusion: Normativity and Black-and-White Academia

Academic work is dependent on the institutional framework that provides a community, infrastructure and resources. There are political convergences and divergencies among academics in the three target countries, Finland, Hungary, and Poland, in relation to the interpretation of the ongoing crises, and their position on Ukraine and Russia. All these factors had a serious implication on the competition of producing viable and useful knowledge. The nationally required political stance from academic communities affected the selection of topics, the evaluation of scientific inquiry, and the social validity of knowledge. The national context on the other hand has been deeply marked by the different countries' historical trajectories and geopolitical insecurities that have been activated in times of crisis. In countries where the war launched an existential threat, as in Finland and Poland, the alignment with the EU-line as seeking for shelter against aggressive Russia translated into standardized academic approach to politically and morally 'sensitive' issues. In Hungary, there appeared an even more politically controlled scholarship. Here, the existential threat was perceived differently and cultivating friendly relations with Russia overshadowed institutional limitations on academia. Yet, in all cases the over-politicised context where researchers have been working in is gradually transforming the perception of knowledge.

At the time of writing this chapter, the expanded phase of the war has lasted three years, and the conflict in its entirety, beginning with the annexation of Crimea, 11 years. By this long period, European academia went through a profound change, reflecting and reacting to the political atmosphere. Because of the prolonged crisis, academia became increasingly rigid and inflexible. Confrontations and polarization have been emerging within the scholarly environment where students, academics, departments and universities took political stand and resist diverging ideas (Revers and Traunmüller 2020, 471–497; Alonso 2023). Limiting the scholarly gaze aligned with political interest and serving security considerations have had grave consequences on once cherished academic freedom in Europe. Hence, the strictly defined 'politically correct' interpretations, relying on community pressure and blame game, are ending up with black-and-white institutional knowledge production. And this does not represent anymore a firm juxtaposition towards authoritarian (i.e., Russian) academic practices but a mirror image.

Russian and Eastern European studies have been always investigating the *OTHER*. The question is how one relates to otherness. Is it a dialogical or dialectical phenomenon? In the dialogical approach the self and the other is perceived as a conceptual pair hence they are in a continuous interaction. In other words, the self is going through changes in relation to the other—so the concept of change integrates both. Regarding the dialectical attitude, the Self and the Other represent opposite values, they are negations to each other, exclude one another. In this understanding the other could be seen as threat for the self or what the self stands for. This results a normative perception of the Other, i.e., the object of investigation. Normativity most often becomes the prevailing stance in times of crisis and conflicts, when the Self is perceived to be threatened. Normativity on the other hand, can be more common in authoritarian regimes where the centralized power defines the core values to be followed in a society, and in academic institutions.

References

Act of 1998, no. 155, on the Institute of National Remembrance, Polish Journal of Laws issued 18 December 1998. https://www.legal-tools.org/doc/fc69d7/pdf/

Alonso, Johanna. 2023. "Shouting Down Speakers Who Offend." *Inside Higher Ed.* April 13. https://www.insidehighered.com/news/students/free-speech/2023/04/13/shouting-down-speakers-who-offend

Badino, Massimiliano. 2020. *Cultural Hegemony in a Scientific World.* Brill.

Békés, Márton. 2020. *Kultúrális Hadviselés.* Közép- és Kelet-Európai Történelem és Társadalom Kutatásáért Közalapítvány.

Bernek, Ágnes, Gábor Andrékó, Anita Faust and László Simon. 2022. *2022—A geopolitika által uralt világ.* Eurasia Center, Neumann LK.

Brüggemann, Michael, Ines Lörcher, and Stefanie Walter. 2020. "Post-normal science communication: exploring the blurring boundaries of science and journalism." *Journal of Science Communication* 19 (3): 1–22. https://doi.org/10.22323/2.19030202

Bäckgren, Noona. 2023. "'Pelkään, mitä Ukrainan tappio voisi tarkoittaa Suomen kannalta', sanoo Aleksanteri-instituutin entinen johtaja." *Helsingin Sanomat,* November 10. https://www.hs.fi/helsinki/art-2000009977112.html

Capoccia, Giovanni and Daniel Keleman. 2007. "The Study of Critical Junctures. Theory, Narrative and Counterfactuals in Historical Institutionalism." *World Politics* 59 (3): 341–369.

Diec, Joachim, Anna Jach and Krystian Pachucki-Włosek, eds. 2023. *Inwazja Rosji na Ukrainę 2022: źródła, przebieg, konsekwencje.* Wydawnictwo Księgarnia Akademicka.

Druk 2016. nr. 806, Sejm Archiwum, https://bit.ly/2Inj9uR

Dz.U. 2018. poz. 369, Sejm—ISAP: Internetowy System Aktów Prawnych, https://isap.sejm.gov.pl/isap.nsf/DocDetails.xsp?id=WDU20180000369

European Commission. 2022. *Commission suspends cooperation with Russia on research and innovation.* March 4. https://ec.europa.eu/commission/presscorner/detail/en/ip_22_1544

European Commission. 2023. *Joint statement by Commissioners Hahn and Gabriel on the application of Council Implementing Decision of 15 December 2022 in relation to Hungarian public interest trusts,* January 26. https://commission.europa.eu/news/joint-statement-commissioners-hahn-and-gabriel-application-council-implementing-decision-15-december-2023-01-26_en?s=09

Fedinec, Csilla and István Csernicskó. 2017. "Language Policy and National Feeling in Context Ukraine's Euromaidan, 2014–2016." *Central European Papers* V (1): 81–100.

Ferenc, Viktória and Katalin Kovály. 2022. "Kékgalléros ukrán munkavállalók Magyarországon: egy empirikus kutatás eredményei." *Kisebbségi Szemle* 7 (4): 47–77.

Finnish National Agency for Education. *Studyinfo: List of studies available for fleeing from Ukraine*. https://opintopolku.fi/konfo/en/sivu/list-of-studies-available-for-people-fleeing-from-ukraine

Foy, Henry and Neil Buckley. 2016. "Orban and Kaczynski vow 'cultural counter-revolution' to reform EU." *Financial Times*, September 7. https://www.ft.com/content/e825f7f4-74a3-11e6-bf48-b372cdb1043a

Gramsci, Antonio. 1982. *Vankilavihkot. Kokoelma, osa 2*. Translated by Martti Berger, Mikael Böök and Leena Talvio. Kansankulttuuri.

Gustafson, Abel and Ronald E. Rice. 2020. "A review of the effects of uncertainty in public science communication." *Public Understanding of Science* 29, (6): 614–633. DOI: https://doi.org/10.1177/0963662520942122

Haber, Stephen. 2008. "Authoritarian Government." In *The Oxford Handbook of Political Economy*, edited by Donald A. Wittman and Barry R. Weingast. Oxford University Press.

Heusala, Anna-Liisa. 2025. "Puheenvuoro: Venäjä-tutkimuksen suurin haaste on lyhytjänteisyys." *Ulkopolitiikka*, February 25. https://ulkopolitiikka.fi/venaja-tutkimuksen-suurin-haaste-on-lyhytjanteisyys/

Holányi, Ákos and Miklós Sebők. 2023. "Külpolitikai törésvonalak az Amerikai Egyesült Államokban az orosz-ukrán háború kapcsán." *Külügyi Szemle* 22 (4): 122–147. https://hiia.hu/wp-content/uploads/2024/02/8-Holanyi-Sebok.pdf

Huszák, Loretta, Miklós Szirbik and Laura Balogh. 2024. "Orosz-ukrán háború: vállalati reakciók és azon jogi keretei." *Magyar Tudomány* 185 (3): 388–404. https://doi.org/10.1556/2065.185.2024.3.9

Hyttinen, Tuomas. 2023. "'Keskustelu katkesi siihen'—Venäjä-asiantuntijat paljastavat, kuinka venäläiset yrittivät udella heiltä varuskunnista ja Ukrainasta." YLE, October 21. https://yle.fi/a/74-20056140

Jędraczczyk, Katarzyna. 2010. *Strategiczne partnerstwo ukraińsko-polskie. Polska w polityce niepodległej Ukrainy*. Wydawnictwo Poznańskie.

Jójárt, Krisztián. 2022. "Út a háborúba—A 2022-es orosz-ukrán háború diplomáciai és katonai előzményei." *Nemzet és Biztonság* 15 (4): 91–120.

Kazharski, Aliaksei. 2018. "The End of 'Central Europe'? The Rise of the Radical Right and the Contestation of Identities in Slovakia and the Visegrad Four." *Geopolitics* 23 (4): 754–780.

Korhonen, Pekka. 2010. "Naming Europe with the East." in *East and the Idea of Europe*, edited by Katalin Miklóssy and Pekka Korhonen. Cambridge Scholars.

Kunkel, Sönke. 2021. "Science diplomacy in the Twentieth Century." *Journal of Contemporary History* 56 (3): 473–484. https://journals.sagepub.com/doi/10.1177/00220094211006762

Łagowski, Bronisław. 2016. *Polska chora na Rosję*. Oratio Recta.

Lattu, Annina and Katri Mäkinen-Rostedt. 2022. "#Scienceforukraine, Suomi, Tiede ja Sota—aika reflektoida." *Tieteessä tapahtuu* 40 (5): 38–41. https://journal.fi/tt/article/view/122944/73179

Legutko, Ryszard. 2016. *The Demon in Democracy. Totalitarian Temptations in Free Societies*. Encounter Books.

Legutko, Ryszard. 2019. "International Institutions vs. National Independence." *Presentation at the National Conservatism Conference*, July 15. https://www.youtube.com/watch?v=UWuD3tSsbQI

Libman, Alexander. 2022. "A New Economic Cold War? The Future of the Global Economy after the War in Ukraine." *Horizons: Journal of International Relations and Sustainable Development* (21 Summer): 148–159. https://www.jstor.org/stable/e48510986

Mackinder, Harold. 1919/1981. *Democratic Ideals and Reality: A Study in the Politics of Reconstruction*. Greenwood Press.

Magyar Közlöny 1999. *1020/1999. (II.24.) számú kormányhatározata*. February 24. https://magyarkozlony.hu/dokumentumok/a8ac4183ef0a11d7d5cab075e823b2e0e3b14a1f/megtekintes

Maresz, Teresa. 2013. "Narracja i ikonografia historyczna współczesnych polskich podręczników do historii. Ich rola w kreowaniu wizerunku Rosji i Rosjan," *Sensus Historiae XI* (2): 135–147.

Markwick, Roger D. 1996. "A Discipline in transition? From Sovietology to 'Transitology'." *Journal of Communist Studies and Transition Politics* 12 (3): 255–276.

Merimaa, Juha. 2022. "Siperiassa ikirouta sulaa—mutta tietoa siitäkin voi olla vaikea saada, kun tiedeyhteisön yhteydet Venäjällä katkeavat." *Helsingin Sanomat*, May 11. https://www.hs.fi/tiede/art-2000008774979.html

Miklóssy, Katalin. 2016. "Russian and East European Studies with a Finnish Flavour." In *The Politics of East European Area Studies*, edited by Gareth Dale, Katalin Miklóssy, and Dieter Segert. Routledge.

Miklóssy, Katalin. 2020. "Itäisen Euroopan tutkimuksen poliittisuus ja Venäjä." In *Sandarmohista Skolkovoon*, edited by Kaarina Aitamurto, Elina Kahla and Jussi Lassila. Into.

Miklóssy, Katalin. 2024. *Demokratian rappio: Unkarin tiellä tulevaisuuteen*. Tammi.

Nalepa, Monika. 2010. *Skeletons in the Closet: Transitional Justice in Post-Communist Europe*. Cambridge University Press.

Niedziółka, Dorota and Mariusz Próchniak. 2023. *Ekonomia wojny: skutki społeczne, ekonomiczne i geopolityczne wojny w Ukrainie*. Oficyna Wydawnicza SGH.

Olchowski, Jakub. 2023. *Zbrodnie wojenne Federacji Rosyjskiej podczas agresji na Ukrainę*. Instytut Europy Środkowej.

Onninen, Oskari. 2022. "Nyt leikataan Ukrainasta: Koska rahaa ei ole, Helsingin yliopistossa ei voi tulevaisuudessa enää opiskella erityisiä Ukraina-opintoja." *Helsingin Sanomat*, February 27. https://www.hs.fi/feature/art-2000008593537.html

Opetushallinnon tilastopalvelu. *Korkeakoulujen hakeneet ja paikan vastaanottaneet*. https://vipunen.fi/fi-fi/_layouts/15/xlviewer.aspx?id=/fi-fi/Raportit/Haku%20ja%20valinta%20-%20korkeakoulu%20-%20live.xlsb

Opetusministeriö. 2007. *Määrärahan käyttöön asettaminen.* Viite: Slavistiikan ja baltologian laitoksen sekä Aleksanteri-instituutin hakemus. November 12.

Opetus- ja kulttuuriministeriö. 2022. *Suomi jäädyttää korkeakoulu- ja tutkimusyhteistyön Venäjän kanssa.* March 9. https://okm.fi/-/suomi-jaadyttaa-korkeakoulu-ja-tutkimusyhteistyon-venajan-kanssa-suomessa-olevia-ukrainalaisopiskelijoita-tuetaan

Orbán, Viktor. 2014. *Speech at the 25th Bálványos Summer Free University and Student camp.* Tusnádfürdö July 30. Prime Minister webpage. https://2010-2015.miniszterelnok.hu/in_english_article/_prime_minister_viktor_orban_s_speech_at_the_25th_balvanyos_summer_free_university_and_student_camp

Orbán, Viktor. 2018. *Speech at the 28th Bálványos Summer Free University and Student camp.* Tusnádfürdö, July 28. Prime Minister webpage. http://www.miniszterelnok.hu/orban-viktor-beszede-a-xxix-balvanyosi-nyari-szabadegyetem-es-diaktaborban/

Palonen, Kari. 1997. *"Kootut retoriikat. Esimerkkejä politiikan luennasta."* SoPhi 11. Jyväskylä University Press.

Piironen, Kaisu. 2022. "Ukrainalaiset korkeakouluissa—lyhyt ajankohtainen katsaus." *Opetus ja Kulttuuriministeriö.* May 16. https://www.oph.fi/sites/default/files/documents/Tiedonjako-%20ja%20verkostoitumiskokous%20korkeakoulujen%20Ukraina-toimista.pdf

Poór, József—Dajnoki, Krisztina—Pató Gáborné Szűcs Beáta—Szabó-Szentgróti, Gábor—Kőműves, Zsolt Sándor—Kun, András István—Kálmán, Botond Géza and Tóth, Arnold. 2024. "A bizonytalan gazdasági helyzet befolyása a vállalati menedzsmentre és a HR-re a koronavírus, a kilábalás és a háború hatásainak tükrében." *Magyar Tudomány* 185 (7): 939–954.

Posio, Matti and Joonas Kuikka. 2022. "Yliopistojen Ukarina-opetus on harvinaista." *Helsingin Sanomat*, August 21.

Rekker, Roderik. 2021. "The nature and origins of political polarization over science." *Public Understanding of Science* 30 (4): 352–368. DOI: 10.1177/0963662521989193

Reporters without Borders. 2023. *Statistics of the freedom of the Media.* https://rsf.org/en/index

Research Council of Finland. 2022. *Russian invasion of Ukraine and its effects on Academy of Finland operations.* April 7. https://www.aka.fi/en/about-us/whats-new/press-releases/2022/russian-invasion-of-ukraine-and-its-effects-on-academy-of-finland-operations/

Research Council of Finland. 2023. *Call for inviting researchers from Ukraine to Finland.* https://www.aka.fi/en/research-funding/apply-for-funding/calls-for-applications/apply-now2/call-for-inviting-researchers-from-ukraine-to-finland-autumn-2023/

Revers, Matthias and Richard Traunmüller. 2020. "Is free speech in danger on university campus?" *Köllner Zeitschrift für Soziologie und Sozialpsychologie* 72 (3): 471–497. https://link.springer.com/article/10.1007/s11577-020-00713-z

Sakwa, Richard. 2023. *The Lost Peace: How the West Failed to Prevent a Second Cold War*. Yale University Press.

Sakwa, Richard. 2025. *The Culture of the Second Cold War*. Anthem Press.

Salmivaara, Kati and Niina Into. 2019. *Kansainvälisyys on pienten alojen elinehto*. Helsingin yliopisto, Humanistinen Tiedekunta. https://www.helsinki.fi/fi/humanistinen-tiedekunta/ajankohtaista/kansainvalisyys-pienten-alojen-elinehto

Schmidt, Mária. 2022. *Látlelet az orosz-ukrán háborúról*. KKETTK Alapítvány.

Schmitter, Philippe C. 2014. "Reflections on 'Transitology' Before and After." In *Reflections on Uneven Democracies*, edited by Daniel Brinks, Marcelo Leiras and Scott Mainwaring. John Hopkins University Press.

Schmid-Petri, Hannah, Nils Bienzeisler and Arista Beseler. 2022. "Chapter Three—Effects of politicization on the practice of science." *Progress in Molecular Biology and Translational Science* 188 (1): 45–63. DOI: 10.1016/2021.11.005

Siivikko, Jarmo. 2024. "Lapin yliopisto painoi paniikkinappulaa, sanoo Venäjä-yhteyksien vuoksi sopimuksen menettänyt emeritusprofessori." *YLE*, May 8. https://yle.fi/a/74-20087844

Soifer, Hillel D. 2012. "The Causal Logic of Critical Junctures." *Comparative Political Studies* 45 (12): 1572–1597.

Stensrud, Ingeborg. 2018. *Europe not taken for granted: the Ford Foundation's exchange programs in Easter Europe in the 1950s and '60s*. PhD Dissertation, NTNU, Norway.

Stępień Stanisław, ed. 2022. *Wojna na Ukrainie: Kijów—Warszawa: razem do zwycięstwa: analizy—refleksje—kalendarium*. Fundacja Wolność i Demokracja.

Stola, Dariusz. 2012. "Poland's Institute of national Remembrance: A Ministry of Memory?" In *The Convolutions of Historical Politics*, edited by Aleksey Miller and Maria Lipman. Central University Press.

T-Invariant. 2022. "Open letter of Russian scientists and science journalists against the war in Ukraine." February 24. https://t-invariant.org/2022/02/we-are-against-war/?fbclid=IwAR3VvJUvVKQa2fCfcqR-giETm1BKeuvjRxQwwk-cUFMqzEl2gPyFQfOGI5w

Taras, Raymond. 2014. "Russia Resurgent, Russophobia in Decline? Polish Perceptions of Relations with the Russian Federation 2004–2012." *Europe-Asia Studies* 66 (5): 710–734.

Thelen, Kathleen and James Conran. 2016. "Institutional Change." In *Oxford Handbook of Historical Institutionalism*, edited by Orfeo Fioretos, Tulia Falleti & Adam Sheingate. Oxford University Press.

Teivainen, Aleksi. 2024. "University of Eastern Finland call off Cronberg's honorary promotion over appearances in Russia." *Helsinki Times*, May 15. https://www.helsinkitimes.fi/finland/finland-news/domestic/25214-university-of-eastern-finland-calls-off-cronberg-s-honorary-promotion-over-appearances-in-russia.html

Terror Háza. 2022. *Bemutatták Schmidt Mária: Látletet az orosz-ukrán háborúról című könyvét*. October 14. https://www.terrorhaza.hu/hu/programok-a-muzeumban/hirek/2022-10-14-bemutattak-schmidt-maria-latlelet-az-orosz-ukran-haborurol-cimu-konyvet

Tynkkynen, Veli-Pekka. 2023. "Venäjä-tutkimuksessa tehdyistä virheistä on otettava opiksi." *Helsingin Sanomat*, November 8. https://www.hs.fi/mielipide/art-2000009972858.html

Ukrainan opintokokonaisuuden tilastoja 2008–2019. 2019. Aleksanteri Institute, University of Helsinki, November 1.

Ukrainian Studies. 2020. *Correlation of study units*. Aleksanteri Institute, University of Helsinki. https://www.helsinki.fi/assets/drupal/s3fs-public/migrated-generic-group-long-pages/files/178954-ukraina-vastaavuustaulukko-ops-17-20-ja-ops20-23.pdf

University of Eastern Finland. 2024. *University of Eastern Finland cancels Tarja Cronberg's honorary doctorate*. Academic Community, May 7. https://www.uef.fi/en/article/university-of-eastern-finland-cancels-tarja-cronbergs-honorary-doctorate

University of Helsinki. 2022. *Ukraine appeal*. https://www.helsinki.fi/en/innovations-and-cooperation/support-us/ukraine-appeal

Verma, Raj and Malte Brosig. 2024. "The Russia-Ukraine war, the evolving global order, the Global South and the emergence of non-alignment 2.0." *Global Policy* 15 (4): 794–801.

Volhynia Massacre 1943: Truth and Remembrance. Institute of National Remembrance. https://volhyniamassacre.eu/zw2/history/173, What-were-the-Volhynian-Massacres.html

Zarycki, Tomasz. 2004. "Uses of Russia. The Role of Russia in the Modern Polish National Identity." *East European Politics and Societies* 18 (4): 595–627.

CHAPTER 9

Sociology Facing Wartime Russia

A Pragmatic Gaze on a Scientific Dilemma

Françoise Daucé

1 Introduction

In 1991, the fall of the USSR put an end to the episodes of great violence of the twentieth century (civil war, "great" famine, "great" terror, "great" patriotic war, holocaust, cold war, etc.) and usher in a period of democratisation and pacification. At that time, experts and social science researchers were surprised by the liberal wave that led to the demise of the Soviet Union in just a few years. In Russia, this liberalisation offered them new opportunities for undertaking investigations to document the transformations underway and to develop new social theories. This context has been conducive to the flourishing of sociological research in public and private spaces that opened to debate. Pragmatic sociology, or the sociology of critique, developed by French researchers, has taken its place in this new epistemological environment. After V. Putin came to power in 2000, the conditions for sociological research gradually deteriorated, as a result of the increasing constraints on public space but still allowed researchers to continue their investigations. After the massive aggression against Ukraine in February 2022, opportunities to conduct social science research in Russia have almost completely disappeared as a result of the political constraints placed on Russian society in wartime.

While Russia is gradually closing itself off to sociological research, other countries are more welcoming to scientific enquiry and debate. In societies faced with Russian expansionism, there is intense critical activity, both in the academic field and in society in general to understand the causes and issues of the war. By following the critiques that emerge, sociologists can take an interest in new questions and new spaces, which lead them to decenter their research in relation to the Russian centre. This demand for decentering is a response to the criticisms levelled at them. Russia's massive war against Ukraine is calling into question the centrality of 'Russian studies' in their field of research. In April 2022, Olesya Khromeychuk, director of the Ukrainian Institute in London, asked this uncomfortable question: "Where is Ukraine on the mental map of the academic community?". She addressed the scientific audience at the

 | DOI:10.1163/9789004747364_010

annual conference of the British Association for Slavonic and East European studies (BASEES) in Cambridge (Khromeychuk, 2022). Following her, calls for the decolonisation of knowledge have been made by researchers worried about the domination of Russian studies in post-soviet studies.

By remaining faithful to pragmatic sociology, the new debates and controversies that are emerging beyond Russia, driven by the actors involved in the war, have a heuristic virtue: controversy is an open-ended process that forces players to clarify their positions, make their evidence known, modify their proofs. It facilitates the work of investigators by making visible and comprehensible what the various black boxes usually contain (Meadel, 2015). Taking the controversies seriously can help guide researchers towards new research questions and shift the focus towards spaces and places that have hitherto received less attention.

To document the epistemic shifts underway, I first propose to assess the work in the sociology of contemporary Russia that have been done for thirty years in France, in cooperation with international scholars. This retrospective and reflexive gaze is necessary to underline the contributions of comprehensive sociology to the understanding of contemporary Russian society but also its limitations in a growing context of authoritarianism, violence and brutality. In this chapter, I present and discuss the contributions of French pragmatic sociology (Boltanski&Thévenot, 2006; Barthe & al., 2013; Quéré & Terzi 2014; Gauthier&Renault, 2020) which, despite its rich history, has until recently only garnered limited attention from English-speaking research communities (Holmqvist, Telling, 2024). In Russian studies, this sociological school was used in analysing protest mobilisations and political critics from below but also growing political domination and institutional constraints (Thévenot, Rousselet, Daucé, 2017; Clément, 2020; Gabowitsch, 2020; Zhuravlev, 2020).

After February 2022, I discuss new scientific questions arising from the impossibility of practising empirical sociology in Russia in wartime because of the threats to people involved in any form of critical thinking. These obstacles to the social sciences and constraints on the public sphere mean that researchers have to move into spaces that are conducive to public debates and critical thinking. The deep scientific displacement sparked by Russia's massive aggression against Ukraine led them to take an interest in societies seeking to emancipate themselves from Moscow's domination. In these new contexts, faithfulness to comprehensive sociology and ethnographic methods can help new issues to emerge. The displacement and decentring of research from the bottom is necessary to propose a new gaze on the sociology of former Soviet societies. The term "decentring" is used here in a broad sense, both in a geographical sense (linked to displacement and migration) and as a reflexive experience that leads to "taking an interest in something other than oneself" and "adopting points of view other than one's own".

This chapter is based on research experience and debates that emerged in France within a network of researchers specialising in the study of former Soviet countries. Many of them work within the CERCEC (Centre for Russian, Caucasian, East-European and Central Asian Studies / EHESS—CNRS), but they also network with colleagues from other French universities and research centres abroad. In a context of strong internationalisation of the social sciences and Europeanisation of university practices, this experience probably has points in common with the dynamics observed in many other European academic areas, and perhaps some specificities as well.

2 From Critical Sociology to Pragmatic Controversies

With perestroika, the so-called "triple transition" (Offe, 1994) favoured the deployment of sociological models, imported and applied to the case of former Soviet societies. In particular, they fed into the programmes for transition to the market and democracy, which were dominant in the 1990s. These models justified the structural reforms carried out in the early 1990s under the aegis of international organisations, to enable the transition from autocracy to political and social democratisation and economic development. The general nature of these models meant that the question of local cultures and configurations took a back seat. In the post-Soviet space, transition models were thought to be standardised in the different new independent states. At the end of the 1990's, these works have been heavily criticised for their teleological nature and their inability to conceive of multiple and divergent trajectories. Increasingly open criticism has been expressed against the domination and export of Western social sciences. The researchers were warned against using forward-looking models and preferred to avoid structural generalisations, preferring instead to document the cultural realities of the societies studied (Dufy, Thiriot, 2013). The concept of 'post-Western social sciences' has even been developed to show the limits of Western theoretical models and to advocate the production of situated knowledge in Eastern European countries (Koleva, 2020) as well as in non-European societies in general (Roulleau-Berger, 2017).

2.1 De La Justification *and Other Pragmatic References*

Criticism of transitological models has led to the emergence of other paradigms for studying the societies of the former USSR. Among them, French pragmatic sociology proposed to access to local debates and controversies in order to better understand the reconfiguration of politics in Russia. French pragmatic sociology appeared in France in the middle of the 1980's as an alternative between methodological individualism and the theory of social reproduction by Pierre Bourdieu. This sociology is also known as the "sociology of critique" (Boltanski, 2011). In English, several anthologies exposing this theoretical tradition are available (Diaz-Bone and de Larquier, 2024; Turner and Susen, 2014). According to Chateauraynaud (2015), contemporary pragmatic sociology has taken shape around three key trends, forming a triangle in which each vertex can be designated by a key concept: situated action (Quéré, 1999), the actor-network (Callon; Bruno Latour, 2005) and the justification test (Boltanski & Thévenot, 2006). In the last decades pragmatism has steadily and increasingly emerged as an innovative theoretical frame of reference for the study of daily politics. Departing from the dominant paradigms of rational choice theory, on the one hand, and critical theory on the other, sociologists have elaborated new and insightful ways of accounting for the moral dimension of social life (Silber, 2016).

French pragmatic sociology is inspired by the American pragmatic philosophy of the XXth century (Cefaî, 2023) and the concept of democracy developed by John Dewey (Dunn, 2022). Members of a society use "collective intelligence" in order to resolve the practical problems they face through an "inquiry" conducted in an experimental way. In keeping with the type of pragmatism championed by Latour, pragmatic sociology expands Dewey's notion of public to beings who are not only humans (animal species or natural things). In this approach, the action or decision is made in the course of an inquiry, whose conclusion is constitutively unknown and unexpected. A plurality of normative orders exists to which individuals refers in each aspect of social and political life (Ogien, Toby, 2014).

Cyril Lemieux (2018) describes some principles that can define pragmatic sociology: empirico-conceptualism, capabilities and relative indeterminacy. Each situation is original and social situations and logics of action must be considered in terms of what makes them irreducible. Actors raise complaints and criticisms and justify themselves: these operations cannot be reduced to covert or unconscious operations such as maximising self-interest or devising a strategy. All social agents have certain capabilities that evolve over time. The principle of capacity is a mean of combating the naturalization of incapacities. The sociologist must follow the actors rather than give them an external definition of their situation. This principle echoes the interactionist tradition in sociology and the ethnographic method of tracking. It gives a central place to the idea of process (action in the making). Human actions possess a form of regularity and predictability, but they cannot be absolute. Pragmatic sociologists attach great importance to the notion of resistance, i.e. the materiality that the world opposes to human action and discourse. Asymmetries exist in the social world, but scholars need to look at how they are produced. The status of the dominant and the dominated is not given in eternity. This involves considering the constitutive heterogeneity of human action in the very order of action.

In France, the seminal work by Luc Boltanski and Laurent Thévenot, *De la justification. Les économies de la grandeur*, was published in 1991, at the same time as the end of the USSR. This chronological coincidence and the impact of their work on social sciences have had a lasting influence on many fields of research in France, including Russian studies. L. Boltanski and L. Thévenot have developed the idea of "regimes of action", which is inspired by the notion of "multiple reality" proposed by Alfred Schutz, and is close to Goffman's "frameworks of experience". Faced with the principle of indeterminacy, it is necessary to give up the idea of predicting future developments or of offering foresight studies that are constantly contradicted by unexpected developments. This principle of indeterminacy leave all the doors to interpretation open, from the most obvious to the most inventive. In this spirit, the renewal of studies on former Soviet societies was based on listening carefully to the diversity of voices that were expressed in these societies, from the most civic to the most domestic, from the most liberal to the most conservative. Putting the concept of indeterminacy back at the heart of research on Russia was a key rule for avoiding peremptory assertions that are contradicted by unforeseen developments.

2.2 *Pragmatic Sociology Facing Political Domination*

The pragmatist paradigm suggests that power arises within and operates through modes of joint participation of human activity. Instead of looking for democracy only in the official sphere of professional politics, it examines the new conditions—technologies, interfaces, platforms, networks, and mediations that allow things to be made public. It considers that the old definition of politics is too narrow; there are many techniques of representation—in politics, science, and art—of which Parliaments and Congresses are only a part. Politics is not just an arena, a profession, or a system, but a concern for things brought to the attention of the fluid and expansive constituency of the public. There are many other kinds of assemblies, which are not political in the usual sense, that gather a public around things: scientific laboratories, supermarkets, churches, and disputes involving natural resources like rivers, landscapes, and air (Latour, Weibel, 2005).

The main critic addressed to pragmatic sociology was its inability to think domination, violence and coercion. Pragmatism stands accused of being naïve about power and presenting the spectre of an overly soft program for doing social science (Nash, 2014). However, the issue of political domination has been addressed by researchers working in this field. Pragmatic sociology's task is to understand how power devices *actually* work and to identify the concrete supports used in situation by those who manage to make others perform certain actions. It also analyses how those who try to challenge the dependency or domination they suffer from go about it, as well as the limitations they face in doing so (Barthe et al., 2013). According to Francis Chateauraynaud (2015), one of the rules of pragmatic method requires focusing on the actors themselves when undertaking to reveal, through work on controversies, everyday disputes and trials, the figures of power and domination. Disputing processes show how new arguments can appear and can be transformed in common features. The researcher needs to follow actors who have the competences to mobilize networks and tools, and who are able to produce discourses, testimonies and expertises.

Pragmatic investigation can shift its focus to arrive at a fresh critique of power relations, with their degree of visibility depending on actors' trajectories, seen within the logic of crisis and conflict. Chateauraynaud underlines that, since the 1980s, numerous studies have sought to characterize a form of 'soft domination'. But the 'softness' in question has not ceased to translate into new forms of violence, of which the notion of 'harassment' (*emprise*) is one legal expression (Chateauraynaud, 2025). The dominant players evoke the figure of the spider, weaving a multiplicity of webs, playing out countless relationships that are difficult to visualize. The hold is built up over time, exploiting the blind spots in the public space, and by the time the captor is identified it is often too late: the consequences of the asymmetries of lasting hold that it has managed to create or develop are cruelly felt by those who then appear as losers or victims. These considerations were important for understanding and theorizing the new forms of political constraint in Russia over the long period of the last twenty years.

3 Pragmatic Sociology in Authoritarian Russia

From the early 2000s onwards, Putin's experience and Russia's authoritarian trajectories showed that the developmentalist and democratic hypothesis of the 1990's did not hold. Russia has gradually become an emblematic case for new studies on autocratization. The old conceptual models of authoritarianism (as Juan Linz's influential 1964 description[1]) have been overtaken

1 He characterized authoritarian political systems by four principles: limited political plural-

by new conceptualisations of modern political constraints (Guriev, Treisman, 2022). Researchers have tried to test new models to understand the authoritarian trajectory in Russia. Sociologists and anthropologists developed renewed approaches of political domination sensitive to the phenomena of uncertainty, the double game of language, the questioning of the democratic narrative, and the plurality of people involved in change to analyse the progressive rise of authoritarianism. Pragmatic sociology has provided a heuristic framework for reconsidering the new forms of domination but also of resistance in Russia till 2022. Drawing in particular on the sociology of controversy (Callon, 2013) and conflict, it has shown the actors, the places and the justifications that have animated Russian society since the end of the USSR. Far from the spheres of institutions and official power, the investigations were conducted as close as possible to Russian citizens, studying their ordinary debates, their daily trials, and even their intimate doubts. These approaches have undoubtedly produced important results in terms of demonstrating the agentivity of people in Russia, their daily reflexivity and the productivity of their discontent. They also showed how difficult it is to talk about politics in Russia, particularly in ordinary interactions. This aversion to political debate is correlated with the rise of oppression. After Russia's massive aggression against Ukraine, the last spaces for public debate and discussion have disappeared, making it almost impossible to study from below the forms of loyalty or criticism of the war.

3.1 *From Contention to Coercion*

Pragmatic sociology has been particularly useful for documenting and analysing the new forms of mobilisation, critics and troubles that have emerged in Russia in the 2000s and 2010s. Laurent Thévenot (2020) was one of the main architects of the rapprochement between Russian studies and pragmatic sociology in France. He ran a research program on dispute and justification in such various regions as Western Europe, the United States, and Russia (Lamont, Thévenot, 2000). The collaborative French-Russian program "From close ties to public places", he coordinated, enabled the comparative study of learning politics in practice in both contexts (Thévenot, 2017). This paradigm inspired researchers active in the political sociology of Russia who have found analytical tools to account for the paradoxical agentivity and reflexivity of ordinary citizens in a context of monopolisation of political space by ruling party elites. These works have not avoided the political question and have proposed an

ism, legitimacy based on emotion, minimal social mobilization most often caused by constraints on the public, informally defined executive power with often vague and shifting powers.

analysis of the causes of populism in Russia. As explained Thévenot, "*is also from certain modalities of sharing close ties that it can turn into authoritarian populism in the hands of dominating rulers*" (*Thévenot, 2017*). *Familiarity thus constitutes a central tool in the toolkit of populism* (*Ylä-Antilla, 2017*).

Important research on the sociology of discontent and protest from below has been based on the propositions of pragmatic sociology. For example, researchers from the Public Sociology Laboratory (PSLab), an autonomous research group focusing on politics and society in Russia and post-Soviet region in comparative perspective since 1991, integrated different approaches such as pragmatic sociology, cultural sociology, eventful approach, and social movement studies to study the politicization of a-political activism stimulated by the "eventful protests" of 2011–2012 (Zhuravlev et al, 2020). Pragmatic sociology also made it possible to study ordinary forms of politicisation, as close as possible to the everyday lives of Russian citizens (Colin-Lebedev, 2013; Clément, 2015, Clément & Zhelnina, 2020). Beyond Russia, this theoretical model has been the basis for a number of important comparative studies (Kharkhordin & Alapuro, 2011; Gladarev & Lonkila, 2013; Luhtakallio and Thévenot, 2018). Research into pragmatic sociology over the last thirty years has undeniably led to significant advances in knowledge of political and moral justifications in debates from below in Russian society.

However, this politicisation from below has not led to the emergence of a civil society independent of Russian power. The study of the concrete mechanisms for enlisting people (militants, activists) and the mechanisms for supporting civil society (laws, subsidies, charity) showed the concrete appropriation of civic justifications and practices by power structures. The unorthodox use of the reference to civil society accommodated itself to the construction of an authoritarian state. Associations were invited to contract with the administration to fulfil different functions of general interest in return for public subsidies. The category 'civil society' became a tool for the depolitization of relations between society and the government and the transfer of general interest missions to nonprofit actors. The very notion of civil society (*grazhdanskoe obshchestvo*) was paradoxically appropriated as "a lieu commun" by a wide variety of actors and institutions even in Putin's Russia (Daucé, 2017). It has become a commonplace appropriated by both independent human rights organisations, state NGOs and radical right movements (Daucé, Favarel, 2019). These multiple appropriations have created tensions among activists, instilled doubt and misled citizens who were confronted with different uses and meanings of the very term 'civil society'. This uncertainty has favoured the gradual establishment of political control over association networks and activist groups, which have embarked on the creation of a loyal and a-critical civil society.

3.2 *The Tests of Pragmatic Sociology in Wartime Russia*

While pragmatic sociology offers theoretical support for investigating ordinary relationships with politics and the tensions that accompany them, it also requires solid empirical support. One of its stance's underlying principles is that the social world's *asymmetries* are all the better describable when observed from the standpoint of an epistemology of symmetry. This sociology places particular emphasis on the fact that each of the two poles of the relationship plays an active role in the evolution of the relation. No power can be unilaterally exerted since its exercise necessarily involves a reaction from the one who obeys or resists (Barthe et al. 2013). This type of approach requires ethnographic fieldwork and ethnographic methods to identify the moments of debate, controversy and conflict during which different social actors express and act in politics. It insists on the centrality of field research (ethnographic/qualitative) to get an intimate knowledge of society and the subjectivities of actors from below. The main idea is to insist on the importance of situational constraints and to take into account conflicts and controversies, which allow to discuss social order.

In many countries today, not just Russia, social science researchers are facing increasing difficulties (Aldrin et al. 2022). Attacks on scholars are even a symptom of despotic power. In Russia, the growing reinforcement of authoritarianism has been accompanied by harsh attacks on the autonomy of universities and researchers. In academic institutions, conferences and seminars, and in teaching in Russia, it has become increasingly difficult to disseminate critical knowledge or to discuss sensitive political issues (protest, demonstrations, politics, gender, corruption, inequality, rights violations, etc.). The most serious warnings have been the direct attacks on researchers (threats on researchers living in Russia forcing them into exile, young researchers being summoned to the police station, arrest and expulsion of Western colleagues) (Daucé & Rousselet, 2023).

Methodological and practical obstacles have also contributed to the flaws and fallacies revealed by Russia's invasion of Ukraine. Most obviously, sociologists have been faced with increasing institutional constraints. Whereas in the 1990s it was possible to conduct interviews in the State Duma, in central ministries or in regional administrations, these possibilities have gradually been reduced to nothing. This closure of access to official sources has left researchers half-blind, with only partial access to information transmitted by nongovernmental actors in Russia. The ministries "of force" (police, army, justice, security services) have become particularly opaque to any form of social science research. The lack of access to official institutions, at all levels of the state apparatus, could not be compensated for by access to archives, unavailable for the present day. But despite these constraints, the methodological imagination and cooperation of sociologists with independent organisations (such as the associations Memorial and OVD-Info, for example) have made it possible to collect sources and propose joint analyses.

In addition to institutional constraints, the most decisive factor for sociologists has been the decline of pluralist debate, the gradual strengthening of censorship and most of all the progressive disappearance of critiques. With colleagues, we have studied the genesis of digital authoritarianism in Russia between 2012 and 2022 (Daucé, Loveluck, Musiani, 2023). After 2022, researchers face the difficulty of carrying out sociological research in Russia, almost no dialogue, debate or critique is possible on the ground, either offline or online, it is too stressful for local people. Colleagues from Public Sociology Laboratory (PS Lab) still produce research on Russian society in wartime, using both quantitative and qualitative data, in-depth interviews and ethnographic observations (Erpyleva & Kappinen, 2024). They conclude, in December 2024, that, in Russia, "the formerly extraordinary nature of the war is giving way to normalisation: the war is gradually becoming something ordinary, another unremarkable part of the surrounding world" (Erpyleva & Kappinen, 2024). But is it possible to achieve different results in a country that is heavily controlled? Due to the increasingly extensive application of repressive laws, especially the law on foreign agents, any interaction between social actors in Russia and abroad can be detrimental to them (Daucé, 2023). In March 2024, the Russian Ministry of Justice designated the Public Sociology Laboratory as a "foreign agent." The vertical of power has been transformed into the vertical of fear (Favarel-Garrigues, 2023). If we remain faithful to the principles of inductive sociology and empirical-conceptualism, we must accept that it is now difficult to conduct political sociological research in Russia. At most, we can still offer remote monitoring of the main developments observed in the media and on social networks, without claiming to offer a detailed and nuanced analysis of the changes taking place in Russian society. The gradual closure of the public sphere, increasing censorship of the media and pressure on the academic world have limited the public emergence of critique and debate, reducing them to the language of Esope (Dubina, Arkhipova, 2023).

4 Following Critics to Open up New Perspectives

The return of military violence to the European continent is encouraging the dominance of geopolitical and culturalist interpretations of the current conflict at the expense of sociological approaches. What new research proposals in the social sciences should be put forward in a context of inter-state confrontation? In the face of war, following the critics of Russia that emerge on new fieldworks can give rise to new epistemic actors. It can also help to respond to the post-colonial criticisms levelled at Russian studies and open new perspectives for decentering research beyond Russia. The war gave rise to a major political and scientific controversy over Russian imperialism but also about the place of Russian studies in the field of ex-Soviet studies. Pragmatic sociology can contribute to this epistemological turn, by studying controversies, critics and the reflexivity of actors in wartime context. The main issue here is to study the transformative consequences of the current debates and controversies, on the assumption that, even if the war ends, the field of research on former Soviet societies will not return to the status quo ante. The context in which sociological knowledge is produced has changed profoundly.

4.1 *Daily Experiences of Wartime Dilemmas*

Sociologists can be at the forefront of reflection on the possibilities of producing scientific knowledge in countries affected by war. While war is closely correlated with the control of the public sphere in Russia, there are still areas for debate in societies fighting against Russian expansionism. It is to these societies that sociology can now turn its attention, thereby helping to readjust the academic balance. Historically, sociology was born in societies emancipating themselves from the dominations and injustices of the past. It can remain faithful to its founding principles if it is to accompany the critical movements that are today developing and formulating alternative ways of thinking to the civilisational and naturalistic narrative promoted by Russia.

Social scientists cannot reasonably work in situations where access to the field (archives, sociological surveys, observations, media spheres) has become almost impossible. In France, all cooperation programmes with Russian universities have been suspended. This impossibility is also an opportunity to shift our gaze towards societies that are more open to the social sciences. Guided by the major controversies and violent debates that accompany war, researchers can follow new questions and raise new issues, as close as possible to the people in the societies studied. In this new dynamic, pragmatic sociology can be useful in breaking out of the geopolitical, strategic and macro discourses on states in conflict and neo-imperialism. It can look at the ordinary dilemmas, provisional compromises and breakdowns that arise in everyday practices and the way in which players justify them.

The challenge is to shift the focus to contribute to the contextual study of societies grappling with Russian-centered domination and the historical and social realities of these regions. This means carrying out field work that listen to the subjectivities of the social actors involved in war debates. Pragmatic sociology considers reflexivity to be a core element of practical action and rejects the idea that qualities, ends or values can be determined without practical experience (Quéré, Terzi, 2014). It means taking an interest in the material and technical mechanisms which fuel social thinking on the current tensions. In societies open to debate, many new fields of study can be opened up to document and understand the ordinary reflexivity of citizens confronted with the question of Russia's imperial domination and war violence. These opportunities are correlated with the possibilities of public debate, online and offline, individually and collectively. In this situation, researchers need to pay particular attention to the conditions under which debates can take place in society, by focusing on modes of expression in the public sphere so that controversies or political disputes can arise. Empirical investigations can bring to light the multiple experiences of the (post) imperial and wartime condition.

To illustrate the new conditions of knowledge production in Ukraine, I propose here the concrete example of the differentiation of university contexts in wartime. In the territories occupied since 2014 in Donbass, Ukrainian academics face daily new realities of Russian domination. Some of them had to displace in Western Ukraine cities or to exile abroad to escape. They have been replaced by loyal university programmes and professors devoted to Russian official curricula. The dominant models of Russian higher education now apply in the illegally annexed Ukrainian territories. The stakes of this academic colonialism are particularly high in the social sciences. As a case study, Donetsk University works under new Russian rules since 2014. The university website is now only accessible in Russian. As explain on its main page: "Together with the Russian World Foundation, the university implements educational projects for students and teachers—advanced training in leading Russian universities, participation in the projects 'Professor of the Russian World' and 'Student of the Russian World'".[2] Since the early 2000s, the rise of civilizational rhetoric in Russia shows its prevalent role in current Russian geopolitical and aggressive imaginations around the notions of the 'Russian world' or 'Eurasianism' (Suslov, 2018; Mier & Turoma, 2020).

On the other side of the front line, in March 2022, missiles hit on the campus of V.N. Karazin Kharkiv National University. Russian missiles damaged its buildings, destroyed its Institute of Public Administration and exploded in the university's Rare Book Library. Despite these crimes, Kharkiv University is still

2 https://donnu.ru/about (accessed on 3 September 2024).

open and its website states that "The full-scale war has become a new chapter of heroism, courage, and bravery of Ukrainian defenders. And, unfortunately, a new chapter of grief and painful losses. Since the beginning of the full-scale invasion, many members of the Karazin University community have given their lives for Ukraine. Currently, the university is aware of 56 fallen defenders. Their names have been added to the "Memorial of Remembrance," located on the second floor of the main building of Karazin University".[3] This military violence is also an intellectual challenge to which Ukrainian academics are trying to respond. At Kharkiv's Faculty of history, in January 2024, scholars organised a public lecture on "The Problem of the Colonial Status of Ukrainian Lands within the Russian empire in the 'Long 19th Century' ".[4] Ukrainian scholars assert their interest in decolonial critical theories to study and analyse their experience.

4.2 *Decolonialism in Practical and Everyday Life*

While pragmatic sociology suggests studying the debates and controversies that arise in the societies studied, it also encourages scholars to reflect on the very conditions under which knowledge is produced in research centres specialising in the former Soviet countries. With the war, a huge academic controversy arose, regarding the domination of Russian studies in foreign universities, whether in the teaching of history, languages or Soviet and post-Soviet literature. The war is revealing the implicit effects of Russian domination in the academic field which had long been naturalised. Gorodnichenko, Sologoub and Deryugina (2023) collected data in course offerings in 2021/22 and 2022/23 academic years in top 13 private and public US universities. They show a disproportionate focus on Russia that "makes generations of US students largely unaware of non-Russia in Eastern Europe".

The very notion of post-Soviet studies became also controversial (Eggard, 2022). Criticizing the 'post-Soviet' label means getting rid of deep-rooted colonial mentalities, which neglect Eastern European perspectives in favor of those constructed in Moscow since the beginning of the 20th century. The international community of researchers working in the field of Russian and post-Soviet studies is concerned by the wartime controversy. Some scholars consider that "Western debates on Russia and Ukraine are shaped by interimperiality" (Hendl et al, 2024). Indeed, in the field of post-Soviet studies, publications and aca-

---

3 https://karazin.ua/en/news/den-pamiati-zahyblykh-zakhysnykiv-ukrainy/ (accessed on 3 September 2024).

4 https://karazin.ua/en/news/laquo-problema-kolonialnoho-statusu-ukrainskykh-zemel-u-nbs/

demic curricula are heavily dominated by studies on Russia, in the Russian language, at the expense of the other states and societies that have emancipated themselves from the USSR.

Facing this disbalance, Ukrainian researchers are inviting their foreign colleagues to reflect further on the structure of post-Soviet studies in Western universities. Pragmatic sociology cannot remain aloof from these debates and can follow critics, which take place in many different places, in Ukraine of course, but also in other former Soviet states, in Eastern Europe, in the Caucasus or in Central Asia, wherever citizens retain the opportunity to express themselves, debate and exchange views publicly. This sociology can contribute to grasp how decolonialism is lived and practised by people confronted with war, in the light of their own experience and their own investigations. The aim here is to study ordinary decolonial critique, which emerges from practical tests which concern all areas of everyday life: domestic relationships, cultural consumption, economic choices, civic engagements, technical uses or religious beliefs.

This is a particularly difficult task: it requires familiarity with the local historical, social and political contexts, and above all the acquisition of new linguistic skills in order to understand the local discourses. To remain faithful to the principles of empirical conceptualism, it is necessary to travel to explore other fields of investigation. It's not a question of finding new fieldworks 'by default', because of a lack of access to Russia. It is not a question of using the Soviet archives in Lithuania or Armenia to make the history of the USSR. It is not a question of collecting interviews with Russian exiles in the Caucasus or Central Asia to find out what is happening in Russia. The aim is to grasp, from below, how social actors seize on practical issues to raise broad political questions about the post-imperial and war conditions.

Conclusion

The massive war in Ukraine is a powerful reflexive force for all academic communities engaged in the study of States and societies emerging from the USSR. A decentering movement is already underway to better consider the trajectories of the various societies emancipated from the USSR. In this vast undertaking, which concerns all researchers, the discussion concerns both the spaces studied and the theoretical and methodological frameworks to be employed. Here, the paradigms of pragmatic and comprehensive sociology can help to analyse wartime controversies. Whether or not it is possible to practise this empirical-conceptual sociology is in itself a political indicator that can guide researchers. When debate and public controversy become impossible, because they are dangerous for researchers and respondents alike, a displacement is necessary, which can also provide heuristic support for thinking differently. Any move from Russia to former Soviet societies immediately confronts sociologists with the question of imperial domination, which is not just an academic and theoretical paradigm, it is also a tangible materiality which, in times of peace, can be overlooked, but which, in times of war, takes on its full meaning. Reflexivity can emerge from the ordinary, material and situated experience of the post-imperial condition and is widely shared by the citizens of contemporary societies affected by this war, from near or far. Ukrainian society is, of course, the most committed to experiencing and conceptualising the reality of the war, but all those concerned by the conflict must also make a reflexive effort to understand what is at stake in this situation. In this difficult yet worrying context, where new obstacles and frontiers are emerging, past academic landmarks are being called into question, but new scientific proposals are likely to appear. The challenge here is to document, through ethnographic fieldwork, the new critiques that are emerging in the societies concerned, their practical incarnations and the reflexive enquiry they cause. This is a decisive challenge for the social sciences, as it involves preparing the world after, a world of reconstruction and lessons learned from the difficulties of the present. To avoid imposing top-down political models, as in the democratic transition models of the 1990's, this reflection must be conducted in close consultation with the people who experience the dilemmas of war every day.

References

Aldrin, Philippe, et al. 2022. *L'enquête en danger: vers un nouveau régime de surveillance dans les sciences sociales*. Armand Colin.

Barthe, Yannick, et al. 2013. "Pragmatic sociology: a user's guide." *Politix* 103.3: 175–204.

Boltanski, Luc and Laurent Thévenot. 2006. *On justification: Economies of worth*. Princeton: Princeton University Press.

Boltanski, Luc. 2011. *On critique: A sociology of emancipation*. Polity.

Callon, Michel. 2013. "Pour une sociologie des controverses technologiques." *Sociologie de la traduction: Textes fondateurs* : 135–157.

Cefaï, Daniel. 2023. "Public inquiry in social sciences: a pragmatist outlook." *Research Handbook on Public Sociology*. Edward Elgar Publishing. 23–41.

Chakrabarty, Dipesh. 2000. *Provincializing Europe: Postcolonial Thought and Historical Difference*. Princeton: Princeton University Press.

Chateauraynaud, Francis. 2015. "L'emprise comme expérience. Enquêtes pragmatiques et théories du pouvoir." *SociologieS*.

Chateauraynaud, Francis. 2025. *L'empreneur et son double. Pragmatique du pouvoir et sociologie de l'emprise*. Editions du Croquant.

Clément, Karine. 2015. "Unlikely mobilisations: How ordinary Russian people become involved in collective action." *European Journal of Cultural and Political Sociology* 2.3–4: 211–240.

Clément, Karine, and Anna Zhelnina. 2020. "Beyond loyalty and dissent: Pragmatic everyday politics in contemporary Russia." *International Journal of Politics, Culture, and Society* 33: 143–162.

Colin Lebedev, Anna. 2013. *Le coeur politique des mères. Analyse du mouvement des meres de soldats en Russie*. Paris.

Daucé, Françoise. 2017. The civility of oppression. *Politika.*

https://www.politika.io/index.php/en/notice/the-civility-of-oppression

Daucé, Françoise, and Gilles Favarel-Garrigues. 2019. "La société civile au pluriel en Russie poutinienne." *L'Enjeu mondial. Populismes au pouvoir* : 237–247.

Daucé, Françoise, and Kathy Rousselet. 2023. "La recherche sur la Russie en France après le 24 février 2022. Le temps des tâtonnements." *Sociétés contemporaines* 100.3: 165–176.

Daucé, Françoise, Benjamin Loveluck, and Francesca Musiani. 2023. *Genèse d'un autoritarisme numérique. Répression et résistance sur Internet en Russie, 2012–2022*. Presses des Mines.

Daucé, Françoise. "Les "agents de l'étranger" dans la Russie en guerre: Genèse, expérience et recomposition d'un dispositif oppressif." *Revue française de science politique* 73.6 (2023): 861–881.

Diaz-Bone, Rainer, and Guillemette de Larquier. 2020. *Handbook of Economics and Sociology of Conventions*. Springer.

Dubina, Vera, and Alexandra Arkhipova. 2023. " "No Wobble": Silent Protest in Contemporary Russia." *Russian Analytical Digest* 291: 8–11.

Dufy, Caroline, et Céline Thiriot. 2013. "Les apories de la transitologie: quelques pistes de recherche à la lumière d'exemples africains et post-soviétiques", *Revue internationale de politique comparée*, vol. 20, no. 3, pp. 19–40.

Dunn, William N. 2022. "Improving Russian and East European studies by building on pragmatism and the lost legacy of the policy sciences." *Russian studies, political science, and the philosophy of technology. Lanham: Lexington books*: 255–274.

Eggard, Claudia. 2022. The End of the Post-Soviet. Zentrum for Osteuropa und internationale Studien (ZOIS). https://www.zois-berlin.de/en/publications/zois-spotlight/the-end-of-the-post-soviet

Erpyleva, Svetlan, Kappinen, Sasha (eds). "We need to carry on" Ethnography of Russian regions during wartime. *Academic Policy Papers Series*, n°5, December 2024.

Favarel-Garrigues, Gilles. *La verticale de la peur: ordre et allégeance en Russie poutinienne*. La Découverte, 2023.

Gabowitsch, Mischa. 2020. "Regimes of Engagement and Protest in Russia: A Reply to Arnold, Sidorkina, and Shevchenko." *Nationalities Papers* 48.2: 414–418.

Gautier, Claude, and Emmanuel Renault. 2020. "Pragmatism and Sociology: The French Debate." *Pragmatism and Social Philosophy*. Routledge. 71–85.

Gladarev, Boris, and Markku Lonkila. 2013. "Justifying civic activism in Russia and Finland." *Journal of Civil Society* 9.4: 375–390.

Gorodnichenko, Yuriy, Ilona Sologoub, Tetyana Deryugina. 2023. Why Russian studies in the West failed to provide a clue about Russia and Ukraine. *Berkeley News*, 21 June. https://news.berkeley.edu/2023/06/21/why-russian-studies-in-the-west-failed-to-provide-a-clue-about-russia-and-ukraine

Guriev, Sergei, and Daniel Treisman. 2022. *Spin dictators: The changing face of tyranny in the 21st century*. Princeton University Press.

Hansen, Magnus Paulsen. 2016. "Non-normative critique: Foucault and pragmatic sociology as tactical re-politicization." *European Journal of Social Theory* 19.1: 127–145.

Hendl, Tereza et al. 2024. '(En)Countering epistemic imperialism: A critique of "Westsplaining" and coloniality in dominant debates on Russia's invasion of Ukraine', *Contemporary Security Policy*, 45(2), pp. 171–209.

Holmqvist, Diana, and Kathryn Telling. "Pragmatic sociology and the economics of convention: Introducing a 'novel' approach to English-speaking education research." *European Educational Research Journal* (2024): 14749041241284004.

Kharkhordin, Oleg, and Risto Alapuro. 2011. *Political theory and community building in post-Soviet Russia*. Routledge.

Koleva, Svetla. 2020. "Doing Post-Western sociology in Central and Eastern Europe before and after the Great Change: some epistemological questions." *The Journal of Chinese Sociology* 7.1: 20.

Lamont, Michèle, and Laurent Thévenot. 2000. "Rethinking comparative cultural sociology." *Repertoires of Evaluation* 8.

Latour, Bruno, Peter Weibel (Ed). 2005. *Making Things Public: Atmospheres of Democracy*.

Latour, Bruno. 2005. "An introduction to actor-network-theory." *Reassembling the social*.

Lemieux, Cyril. 2018. *La sociologie pragmatique*. Paris, La Découverte, coll. Repères, 128 p.

Luhtakallio, Eeva, and Laurent Thévenot. "Politics of engagement in an age of differing voices." *European Journal of Cultural and Political Sociology* 5.1–2 (2018): 1–11.

Khromeychuk, Olesya. 2022. "Where is Ukraine?" *RSA Journal* 168.2 (5589): 26–31.

Lonkila, Markku 2012. Russian Protest On- and Offline: The role of social media in the Moscow opposition demonstrations in December 2011. *Briefing paper* 98.

Nash, Kate. 2014. "The promise of pragmatic sociology, human rights, and the state." *The spirit of Luc Boltanski: Essays on the Pragmatic Sociology of Critique*: 351–367.

Ogien, Albert, Matthews, Toby. 2014. Pragmatisms and Sociologies. *Revue française de sociologie*, n°3, pp. 414–428.

Quéré, Louis. 1999. "Action située et perception du sens." *Raisons pratiques (Paris)* 10: 301–338.

Quéré, Louis, and Cédric Terzi. 2014. "Did you say 'pragmatic'? Luc Boltanski's sociology from a pragmatist perspective." *The spirit of Luc Boltanski: Essays on the pragmatic sociology of critique*: 91–128.

White, Stephen and Ian McAllister. 2014. Did Russia (Nearly) have a Facebook Revolution in 2011? Social Media's Challenge to Authoritarianism. *Politics*, Volume 34, Issue 1, pages 72–84, February.

Oates, Sarah and Tetyana Lokot. 2013. Twilight of the Gods? How the Internet Challenged Russian Television News Frames in the Winter Protests of 2011–2012.

http://papers.ssrn.com/sol3/papers.cfm?abstract_id=2286727.

Méadel, Cécile. 2015. "Les controverses comme apprentissage." *Hermès* 3: 45–50.

Mjør, Kare Johan and Sanna Turoma (eds.). 2020, *Russia as Civilization: Ideological Discourses in Politics, Media and Academia*. Routledge.

Offe, Claus. "Capitalism by democratic design? Democratic theory facing the triple transition in Central and Eastern Europe." *The Political Economy of Transformation* (1994): 25–43.

Roulleau-Berger, Laurence. 2017. "L'invention de la sociologie post-occidentale: de l'Asie à l'Europe." *Le Portique. Revue de philosophie et de sciences humaines* 39–40.

Silber, Ilana F. 2016. "The cultural worth of 'Economies of Worth': French pragmatic sociology from a cultural sociological perspective." *The Sage handbook of cultural sociology*: 159–177.

Spivak, Gayatri Chakravorty. "Can the Subaltern Speak?" Marxism and the Interpretation of Culture, edited by Cary Nelson and Lawrence Grossberg, Macmillan Education, 1988, pp. 271–313.

Susen, Simon, and Bryan S. Turner, eds. *The spirit of Luc Boltanski: Essays on the 'pragmatic sociology of critique'*. Anthem Press, 2014.

Suslov, Mikhail. 2018. "Russian World" concept: Post-Soviet geopolitical ideology and the logic of "spheres of influence." *Geopolitics* 23.2: 330–353.

Thévenot, Laurent. 2017. "Des Liens du Proche aux Lieux du Public". Retour sur un Programme Franco-Russe Pionnier. *Revue d'études comparatives Est-Ouest* 3: 7–43.

Thévenot, Laurent, Kathy Rousselet, and Françoise Daucé. 2017. "Critiquer et agir en Russie." *Revue d'Etudes Comparatives Est-Ouest* 48.

Thévenot, Laurent. 2020. "How does politics take closeness into account? Returns from Russia." *International Journal of Politics, Culture, and Society* 33.2: 221–250.

Thévenot, Laurent. 2020. "How does politics take closeness into account? Returns from Russia." *International Journal of Politics, Culture, and Society* 33.2: 221–250.

Ylä-Anttila, Tuukka. "Familiarity as a tool of populism: Political appropriation of shared experiences and the case of Suvivirsi." *Acta Sociologica* 60.4 (2017): 342–357.

Zhuravlev, Oleg, Natalia Savelyeva, and Svetlana Erpyleva. 2020. "The cultural pragmatics of an event: The politicization of local activism in Russia." *International Journal of Politics, Culture, and Society* 33: 163–180.

CHAPTER 10

Chekhov's Gun

Security Services and Blind Spot of Russia Studies

Kirill Shamiev

1 Introduction*

The study of Russian politics is currently more relevant than ever. Yet before the full-scale invasion of Ukraine, some scholars of Russia often could neither correctly predict the imminent invasion nor explain the Russian performance in the war. Most researchers argued that the cost of the invasion would be enormous and thus deterring the Russian regime (Driedger and Polianskii 2023). They assumed the Russian regime had sufficient information to understand these deterring costs of the invasion (Borger 2021; Timofeev 2021; Trenin 2021; Yilmaz 2022; Ullman 2022; Gardner 2022). However, Russia decided to use military force. At the same time, many experts expected that Russia would overrun Ukrainian defences and quickly capture the Ukrainian capital. However, this did not happen; Russia did not live up to the expectations of its military performance, nor did it collapse after the first setbacks.

This chapter argues that the Russian decision to use military force and the military's underperformance could have been anticipated and explained by greater integration of research on Russian politics with studies of civil-military relations and technical military policy analysis. Despite a growing body of work focusing on Russian regime change, policy processes, reforms, and regional issues, research on Russian security, civil-military relations, and the military, while increasing, has been largely outside the mainstream of Russia studies, with limited interaction between the two strands. This undermined the quality of theories about Russian foreign policy and the understanding of Russia's defence capabilities. This ultimately limited to the predictive quality of analytical judgements.

Political predictions can refer to two different activities: pragmatic forecasting and scientific forecasting (Dowding and Miller 2019). Pragmatic prediction involves forecasting the quantitative probabilities of future events with suc-

* I acknowledge the invaluable assistance of Samuel Bendett, advisor at the Center for Naval Analysis (CNA). All errors are mine.

DOI:10.1163/9789004747364_011

cess measured by how accurately these probabilities match actual outcomes. In contrast, scientific prediction is concerned with the causal outcomes derived from theoretical models. These predictions are often generated by theoretical models and are integral to the explanatory power of political science. They imply that if a set of propositions explains a phenomenon under certain conditions, it should do so consistently in all relevant similar situations.

Successful predictions of Russia's behaviour benefited from explicitly using scientific forecasting by considering the dynamic multicausality of Russian behaviour and updating their theoretical assumptions about the Russian regime (Driedger and Polianskii 2023; Jonsson 2024; Gustafson et al. 2024). Just a few days before the full-scale invasion, Michael Kofman, an analyst of the Russian military who actively warned of the approaching invasion, made a sensible remark about the problem of making a valid prediction about Russia: "Folks who believe that something this dramatic is unlikely or improbable may not have observed that qualitative shift in the last two years" (Troianovski 2022). Dara Massicot, also a military-focused analyst of Russia, made specific policy suggestions about the weapons that could help Ukraine counter Russian military advantages before the invasion (Massicot 2022). The suggestions turned out to be strikingly similar to what politicians and military commanders would begin discussing and implementing several months within the invasion. A proper combination of theoretical political reasoning with attention to the technical and organisational military development could help successfully predict and explain Russia's foreign policy.

Two important caveats before going into the discussion. First, the problem with the exclusion of the security studies from the general regional studies is not neither unique nor new. As the editors of the handbook on civil-military relations claimed, mainstream studies in American or comparative politics often overlook the political role of the military too (Matei, Halladay, and Bruneau 2021). The classical theorists of civil-military relations warned about the broader issue discussed in this chapter and called it the "civil-military divide" (Huntington 1957; Janowitz 1960; Feaver 2005). In the academic realm, the civil-military divide fosters civilian researchers' biases and unsubstantiated assumptions about military topics and, conversely, military officers' inadequate assessment of civilian politics and social affairs.

Second, the analysts who made valid judgments about Russian intentions did not necessarily influence policymakers. The EU's foreign policy head Josep Borrel recalled how he did not believe U.S. Secretary of State Anthony Blinken that Russia would invade two days prior the actual invasion (Eckel 2023). In Germany, external experts, including journalists, generally provided timely and accurate warnings despite challenges in diagnosing and assessing Moscow's intentions. Yet the German government failed to heed these assessments (Michaels 2024). In November 2021, a retiring chief of the British Defence Staff expressed the belief that Russia would not launch a full-scale invasion. He emphasized that Russia would instead rely on a hybrid means, such as disinformation to avoid direct combat. (Mason and editor 2021). While the Biden administration was a leading voice about the Russian imminent full-scale invasion, the U.S. policy responses were far from clear with multiple officials disagreeing on China-Russia priorities, arms deliveries to Ukraine, and diplomacy with Vladimir Putin (Bertrand 2021; Borger 2021; Crowley and Barnes 2021; Dilanian et al. 2022).

This chapter addresses the gap in understanding Russian defence and security dynamics and argues that scholars can address it through a number of relatively simple analytical means. First, it urges narrowing the gap between scholars engaged in academic research and policy-oriented military analysts. The analysts' job is to develop evidence-informed, actionable analysis, often using fresh data from the field and their technical and/or operational expertise. Closer cooperation between theory-oriented scholars and practice-oriented analysts can positively reinforce the validity of both groups' arguments. The chapter also suggests that scholars of Russian regime change should incorporate civil-military relations and control over the security sector into their research on Russian regime dynamics. The Russian military and security services possess overwhelming powers to keep the Russian authoritarian regime stable, while the military is a strong political actor that limits the power of (un)elected politicians.

The chapter is structured as follows. First, I discuss the role of the military in Russian politics and how the misplaced debate on the military influence in Russian politics missed the rise of the dictatorial powers of the Russian president in the security domain. The section discusses the overwhelming civilian supremacy in Russian civil-military relations that undermined accountability and power sharing over the decision to use military force. Second, I review the literature on the Kremlin's struggle to make the armed forces more effective and how it contributed to the assessment of Russian military capabilities. Finally, I expand on the proposed ideas to improve the field of Russia studies and provide concluding remarks.

2 Owning the Gun: Civilian Control over Russian Military

The key enabler of Russia's aggressive and violet foreign policy behaviour is its authoritarian civilian control over the enormous armed forces and other security services. Weberian classical approach to the state claims that a political entity can only be called a state if it upholds the monopoly on the legitimate use of physical force within a given territory (Weber 1965). Therefore, the question of how politicians design their control over the monopoly to use violence is a paramount topic in political science. Theories of civil-military relations have historically focused on how civilian politicians can maintain power for the legitimate use of the military, the primary physical force available to the government (Huntington 1957). The tragic paradox of Russian civil-military relations is that the authoritarian civilian control over the institution designed to protect Russia consistently undermined its development and threatened the security of Russia's neighbours.

The first wave of scholarship on civil-military relations focused on identifying coup patterns, focusing on macro structural factors, economic crises, political instability, and military grievances (Brooks 2019). Russia was historically largely absent from this discussion because of its strong political control over the armed forces and security services. There has not been a successful military coup in Russia since the assassination of Tsar Pavel I in 1801 (Taylor 2003). In the Soviet Union, the security services were under strict party control. In the 1990s, the military did not attempt to organize a coup, although the leaders of the security services were involved in the unsuccessful Soviet coup of 1991 that expedited the fall of the USSR. Two years later, on the contrary, the military willingly carried out president Yeltsin's decision to shell the parliament. This decision is illustrative of the deficiencies in Russian control over the military. The Russian decision-making process for the use of force became unchecked and subject only to the will of the Russian president just two years after the fall of the USSR. 29 years later, the Russian military will be sent to Ukraine with virtually no consultation and validation with other political and governmental bodies.

The public debate somewhat ignored the dictatorial powers of the president in Russian civil-military relations. On the contrary, it was focused on the independent security agency's influence on the economy (Politkovskaya 2003; Bulavinov 2003; Yenikeev and Mekhdi 2013) and on Putin and his government (Ivanov and Vilgelm 2004; Krechetnikov 2013; Simoyanov 2013). Scholars contributed to this discussion by researching the composition of the Russian elite with security background and the potential domestic and foreign policy implications (Kryshtanovskaya and White 2003; 2005; Renz 2006; Bremmer and Charap 2007; Gel'man 2013; Hill 2016; Soldatov and Rochlitz 2018; Yakovlev 2021; Frye 2022; Hill and Gaddy 2012). The full-scale invasion showed again that there was little concern over the opinion of the Russian national security community, their opinions on the use of force and the lives of servicemembers (Zimmerman, Rivera, and Kalinin 2019; Roth and Sauer 2022). In analytical sense, the misguided first wave debate on seizing power by the military or other security services in Russia pulled the discussion away from more policy-relevant topics of how Russian political dynamics created the dictator's army.

Soviet and Russian domestic political dynamics enabled the centralization of political control over the military in Russia. A key feature of communist civil-military relations that Russia inherited was the relative autonomy of the military in defence policy. As a result, when communism collapsed, the Russian government faced serious civil-military challenges: appointing civilian defence ministers and staff to the military-dominated ministry; separating and subordinating the General Staff from the defence ministry; establishing institutions to manage defence policy, budgets and procurement; and establishing parliamentary committees and procedures for defence policy oversight (Cottey, Edmunds, and Forster 2005). However, Russia's first president Yeltsin inherited and accepted an institutional environment that allowed the military's politicization after the turmoil of Gorbachev's perestroika. The General Staff was poorly controlled by the Ministry of Defence and could report directly to the president. Active-duty senior generals and retired officers could argue about civilian political decisions in the media or oppose the orders from president Yeltsin.

In the first eight years in power, Putin enhanced tight control over the military (Barany 2008) by appointing the tradition of a civilian defence minister and gradually making the Russian Defence Ministry more civilian, as well as the institutional subordination of the General Staff to the Defence Ministry (instead of the parallel structures that existed before). Despite making a positive step towards better control over the armed forces, Putin did not develop legal and regular oversight mechanisms by legislative committees, courts, and audit institutions to ensure compliance with civilian directives (Matei, Halladay, and Bruneau 2021). Together with Russia's shift towards super-presidential authoritarianism, as result, civilian control over the armed forces in Russia effectively became the authoritarian presidential control.

The overwhelming allegiance of the military to the president is not surprising to the civil-military scholars. Soviet dictator Joseph Stalin reportedly claimed: "Cadres decide everything" (Grant 2022). In today's Russia, the president continues this tradition by enjoying overwhelming legal authority to decide who leads and commands Russia's military and security services. He forms and heads the Security Council of the Russian Federation, approves military doctrine, appoints and dismisses the command of the armed forces, including the appointment of senior officers without any consultation with other political bodies ("The Constitution of the Russian Federation" 1993). The president issues decrees on conscription, military training, and discharge from military service, and establishes the list of organizations operating in the interest of national defence and security, and seconds military personnel to these organizations. The president also decides on the creation of mobilization reserves, volunteer formations and the transfer of military units (Government of Russia 2021). Only the Russian president and his administration generally decide who and how will command Russian armed forces and security services. Similar to the research of Russian civilian elites, the analysis of Russian military leadership can shed light on both the Kremlin's priorities and potential military effectiveness.

The president also commands and oversees all other security services in Russia that can counterbalance any potential political moves from the military (Belkin and Schofer 2005; De Bruin 2018; Reiter 2020). The president approves the regulations of these bodies and appoints their heads and deputies (Government of Russia 2020). The president is the head of the National Guard, a large paramilitary force with heavy equipment, appoints its commander, defines its tasks, approves its structure and composition, and determines the number of its personnel, including the number of positions to be filled by senior officers, and development plans for the National Guard (Government of Russia 2016). The Russian president supervises the main counterintelligence service, the Federal Security Service (FSB) and appoints its head. He also directs the general priorities of foreign intelligence services, defines their objectives, controls and coordinates their activities and appoints the head of the Foreign Intelligence Service (SVR). In addition, the president commands and control his bodyguards, the Federal Guard Service (FSO), including the appointment of the FSO head.

Ironically, Prigozhin's unsuccessful mutiny in 2023 is a good example of the challenge with such a centralized control. After years of debate on the potential takeover of power by the state security services, the Russian president's inflated prerogatives in civil-military relations let Putin create a non-state PMC outside legal rules and institutional structure of the Russian security sector. This would have been unbelievable in a democratic country as creating an unofficial, formally illegal mercenary force would trigger several alarms in the national security community, the parliament, prosecutor's office and counterintelligence agencies. Ultimately, the experiment failed due to Prigozhin's attempt to organize a mutiny with his force. The Kremlin learned the lesson and incorporated his mercenaries into the official structures of the Russian Ministry of Defence and the National Guard. At the same time, during Putin's weakest moments in power, Russian official armed forces and security services remained apathetic or actively spoke out against Prigozhin's mutiny.

Apart from formal institutions, Russian informal norms ultimately regulate the individual behaviour of Russian servicemembers. Centuries of non-democratic civil-military relations made the Russian military culture conducive to the authoritarian control. It is known for its hierarchical organizational culture that nurtures the apolitical military and obedience to the decisions of the military commanders (P.K. Baev 2002; Shamiev 2023). Russian servicemembers tend to see themselves as the servants of Russia, its strong arm that protects the Russian people and the Russian state. This military culture instils a belief in civilian supremacy in Russian civil-military relations and reinforces civilian control in Russia. But it also inspires an exaggerated perception of the military's role outside a narrowly defined security sphere. It complicated Russian military development, which I discuss in the next section.

In general, Russian armed forces is a politically obedient military organization that has had a long tradition of centralized authoritarian control. Anyone replacing Putin in the presidential office would assume dictatorial powers towards the security institutions, including the military, that would give them a strong incentive to keep these powers unchecked. A combination of personalist authoritarian political institutions with an unrestrained and unchecked control over the "guns" prevent Russia from becoming more peaceful and development-oriented. Any potential democratic chance in Russia must support a reform of the Russian civil-military relations.

3 Learning to Shoot: Reforming the Russian Military

The military and other security services conceptually must be strong enough to deter or win wars. An insufficiently strong military could either invite aggression or lead to overconfidence and losses in real warfare (Feaver 1999). In the Russian case, any large-scale aggression threatening the existence of Russia (similar to the German invasion of the Soviet Union) is likely to be deterred by a large number of Russian nuclear weapons, deployed or in stockpile (Adamsky 2014; Ven Bruusgaard 2016; Anichkina, Péczeli, and Roth 2017). However, the nuclear weapons are useless in regional wars or for projecting Russia's power abroad. Despite enjoying an overwhelming control over the military, Moscow struggled with reforming its military.

Russia-focused analysts and researchers tended to have two opposite opinions on the state of the Russian military. Some scholars argued that Russian military development was largely a failure (P. Baev 2015; Valeriano and Maness 2015; Shiryaev 2015; Le Figaro 2016; Luzin 2019). After the fall of the USSR, the Russian army was in a dysfunctional state, threatening its service members, the Russian budget, not to mention struggling to fulfil its core duties. Russia inherited a decaying military machine with serious problems such as desertion, the mass exodus of officers, the collapse of the conscription system, rampant corruption, and the general unfitness of many military units for their tasks. The servicemembers were poorly paid, had inadequate housing, suffered from a shortage of qualified junior officers, and struggled to maintain a vast and technically obsolete infrastructure and equipment (Arbatov 1998; Douglas 2014; Golts 2017). Moscow initiated several reform attempts under each of the defence minsters since the 1990s. However, they did not produce the expected results, and cases of intramilitary violence, lack of training, and theft raised questions about the state of the Russian armed forces. Indeed, two years before a full-scale invasion private Ramil Shamsutdinov shot and killed eight fellow servicemembers for harassment and threats of sexual violence (Luxmoore 2020). Cases like that were never stopped from happening in Russia.

On the other hand, after a number of successful military operations and exercises, some researchers could exaggerate Russian military successes and military superiority (Campbell 2015; Trenin 2016; Radin et al. 2019; Dougherty 2022; O'Brien 2022). The 2008–2012 Serdyukov reform (also known as "The New Look" reform) reversed the negative trend of the Russian military development. Since 2012, Russia has achieved military successes in annexing Crimea, fighting in Eastern Ukraine, supporting the regime of Bashar Al'Assad and fighting Islamic terrorists in Syria as well as swiftly deploying peacekeeping forces to Nagorno-Karabakh and Kazakhstan. The speed and decisiveness of the deployments, the quality of the Russian equipment, and the tactical skills of Russian soldiers were generally seen in a positive light, compared to what was seen in the Russian-Georgian war of 2008 or in Chechnya in the 1990s.

A common problem with assessing Russian military successes was the lack of attention to the context in which the Russian military operated. In Crimea, Nagorno-Karabakh and Kazakhstan, Moscow conducted virtually swift interventions without organized armed resistance. In Syria, Russian forces, supported by the Syrian army, enjoyed clear skies against fragmented, ill-trained and ill-equipped rebel and terrorist groups. In the first year of the 2014 invasion of Ukraine, Russia allegedly lost up to 500 soldiers in a fight against a disorganized, poorly trained, underfunded Ukrainian armed forces and volunteers (Interfax-Ukraine 2015). Therefore, paying closer attention to the operational environment and socio-political context in which the Russian military operated would help produce a more realistic assessment of Russian military capabilities, which did improve after 2008 but were never as powerful as some were imagining.

The challenge of building a sufficiently strong military is perplexing and represents a key issue in Russian civil-military relations and politics. Similar to the political control, Russian president has had a lot of legal power to reform the military. The president of Russia is the commander-in-chief and the key decision-maker who determines the main directions of the domestic and foreign policy, including appointing the government ("The Constitution of the Russian Federation" 1993). He approves military statutes, regulations concerning the Ministry of Defence, the General Staff as well as concepts and development plans for the Armed Forces and other military formations, including the mobilization plan (Government of Russia 2021). Yet, despite these and other overwhelming powers, the Kremlin demonstrated underwhelming results in military development as the unfolding of the invasion of Ukraine demonstrated.

Gel'man's classical approach to policy change claims that Russian reforms are severely constrained by the regime's neopatrimonial nature. As a result, initially well-intentioned reforms are undermined and distorted by the private interests of the regime's actors and the survival of the autocratic centre (Gel'man 2016). Put simply, this approach argues that the Russian regime is based on corruption and that any well-intentioned reform is undermined by the interests of the ruling elite who want an illegal but informally acceptable piece of the reform pie in exchange for political loyalty.

Gel'man's approach became a leading way to understanding Russian reform process, including in public debate. It explains a lot of Russian reform failures and illuminates a tragic link between the regime political foundation and the wellbeing of the Russian people (Tarasenko 2015; Dekalchuk 2016; Starodubtsev 2016). Corruption also became an all-purpose tool in the public debate for explaining Russia's policy failures (Transparency International 2015; Romanowski 2018; Dixon 2024; Meijer 2024). However, this approach is less effective in explaining the failures of Russian defence reform.

First, the rent-seeking behaviour is not sufficient for explaining the poor outcomes of Russian defence reform. On the one hand, corruption, resource waste, and criminality have been part of the Russian security services since the fall of the USSR (Meakins 2018). On the other hand, since 2008, Russia has created a few effective and well-trained military units and formations under the Special Operations Forces (SSO), Airborne Troops (VDV), some GRU units, and some marine and motorized formations. Moreover, a bright counterfactual is the GRU-born PMC "Wagner," a relatively effective paramilitary force that was taken over by Putin's crony Prigozhin and his right-hand man Dmitry Utkin, a former military officer. Prigozhin personally was a highly corrupt magnate, a product of Putin's neopatrimonial system, and yet his paramilitary force became well-known for its performance as well as ruthlessness and brutality (Yapparova 2023).

Apart from corruption, the president's inability to overcome serious disagreements between civilian officials, the military and other security services were behind modest reform outcomes (Arbatov 1998; Gazukin 2001; Bartles 2011; Dvorkin and Arbatov 2013; Golts 2017; Shamiev 2019). These disagreements were largely driven by a set of divergent strategic (deep-core) and policy beliefs about what the Russian armed forces should be (Jenkins-Smith et al. 2014). The influence of complex belief systems are natural in all policy areas, including in the defence sector.

On the strategic level, the military elites were generally more activist (although not monolithic) than civilian elites to seeing military force as a legitimate and effective instrument for achieving policy outcomes, supporting offensive capabilities, and power projection during Putin's first two terms (Stewart and Zhukov 2009). The military had an expansive perception of its mission as a protector of the Russian people with domestic and foreign policy functions, such as an actor instilling the Russian patriotic identity (Barany 2008; Bruneau and Croissant 2019). Public military speeches also covered a wider range of security topics than the civilian ones, including those that were outside the institutional prerogatives of the armed forces (Stewart and Zhukov 2009). Coupled with the Russian anti-western strategic culture (Götz and Staun 2022; Kari 2019) and civilian support for the revisionist foreign policy (Radchenko 2021; Driedger 2023), it made the Russian military role harder to operationalize, hence, create relevant and coherent defence policies.

On the policy level, the debate over defence reform ideas has never stopped. Russian strategic defence documents contained unrealistic assessments, lacked the necessary internal cohesion, and failed to take account of financial constraints (Kofman et al. 2021). The size and composition of the armed forces caused persistent disputes, as Russian military officers reportedly saw future conflict as large-scale mechanized warfare, while civilian officials were inspired by NATO- and US-led missions with smaller, all-volunteer but more effective military contingents (Gazukin 2001; Bartles 2011; Mathers 2019; Shamiev 2019). The Russian military also had to catch up with the missed 20 years of international military thinking and technological development of the 1990s and 2000s over a much shorter time between the war with Georgia in 2008 and the full-scale invasion in 2022 (Kofman 2018). During the Serdyukov reform (2008–2012), despite the obvious need and support for a professional military, entrenched ideational biases and institutional legacies ultimately prevented the shift to an all-volunteer force in Russia. The resistance of military elites, coupled with divided opinions of civilian elite, hindered the reform process (Douglas 2014; Shamiev 2024). Russia consistently struggled to reconcile the conflicting policy beliefs on the military reform in the national security domain.

Additionally, the technical complexities of reforming Russian armed forces became a challenge. Russia invested a lot in the technological upgrade of its military, especially in the electronic warfare (EW), drones, and long-range non-strategic capabilities (Bendett 2019; Bendett et al. 2020). It did not achieve overwhelming technological superiority, but it did improve its weaponry in the competitively advantageous areas. However, understanding the nuances in the adoption of military technology and military concepts require some level of specific knowledge that civilian researchers usually lack. The Russia studies community would benefit from a strong intradisciplinary cooperation with specialists in military technology, strategy, the economy, and Russian military.

The Russian president can potentially appoint anyone he wants to the position of the ministry of defence and other security bodies. Putin opted for the selective appointments of individuals with similar ideas and compliant behaviour to the command and ministerial positions to expedite the defence reform. To make a military reform possible, Moscow had to appoint the chief of the General Staff and the minister of defence with similar reform ideas, effective and pragmatic relationship, and commitment to the policy process (Shamiev 2021). However, the list of candidates for the position of the chief of the General Staff is always limited to the most senior officers who were socialized and educated in the system the president wants to reform. Practically speaking, developing and validating innovative ideas for the military reform in Russia is hardly possible without becoming a servicemember, hence, committing to the same rules and procedures one wants to reform.

That is why both Yeltsin and Putin at times sought expertise from Western countries or even discreetly requested NATO assistance (in 2001) with the military reform (Blank 2002). The import of ideas was yet met with severe resistance and mockery from the conservative national security opinion makers (consisting of mostly retired officers and politicians with military background). The institutional complexity with appointing key military leaders and generating reform ideas distinguishes defence reform from other "civilian" policy processes, where both personal access and evidence generation are generally more open even in autocratic settings.

On top of that, Russian monitoring and accountability mechanisms prevented an effective oversight over the implementation of the reform. The president proposes to the Federation Council (controlled by the president) candidates for the post of Procurator General and Deputy Procurators and has the power to appoint and dismiss procurators of federal subjects and other prosecutors

("The Constitution of the Russian Federation" 1993). He also nominates candidates for judges of the Constitutional Court and the Supreme Court and appoints judges of other federal courts. This means that the whole system of military monitoring and accountability is in the hands of the Russian president. Considering the complexity and size of the Russian military, such a narrow approach to oversight was proven to be insufficient.

Corruption, the sheer complexity of the Russian defence sector, the lack of consensus between civilian and military leaders on military policy ideas, and centralized accountability mechanisms undermined Russian military development. Moreover, the armed forces under Putin also became more functionally politicized as demonstrated by the use of the military intelligence (GRU) against non-military targets, including Russian opposition (*BBC News* 2021). To break this system, Moscow has to fundamentally re-balance the military autonomy and bring the military under broader civilian control, including reforms such as increasing the civilian role in the Ministry of Defence, improving transparency and ensuring military accountability to political authorities and legislators. However, this would require a fundamental democratic change in Russian politics, hence, civil-military relations.

These ideas are not new but were ignored by the larger community of Russia analysts. For example, Stephen Blank observed in 2002 that the Russian military remained beyond effective political control, eliminating any chance for balanced military advice or genuine reform threatening the security, peace, prosperity and democracy of itself and its neighbours. Widespread brutality, corruption and criminality further alienated the military from democratic values and European integration. This persistent failure meant that Russia could not be a reliable partner for Europe or conduct a truly European security policy (2002). Moreover, examples from other non-democratic countries, such as China, show that even authoritarian civilian rulers can design their national civil-military relations in a way that would be more conducive to non-personalist authoritarian civilian control through institutional ties, historical legitimacy and political ideology (Escribà-Folch, Böhmelt, and Pilster 2020; Usman 2023). Neither democratic nor non-personalist civilian control has been achieved in Russia. Ultimately, the Chekhov's gun of Russian politics made its shot 22 years after Blank's original forecast.

4 Misfire: Russian Security Studies before February 2022

The roots of the problem with assessing Russian intentions and military capabilities are structural, institutional and normative. In Russia, while the civilian expertise on the country had gradually become more international, Russian military scholars remained in an encapsulated national military system of journals, conferences and reporting requirements, protected from access by the civilian academic community. Official Ministry of defence journals and scientific conferences could require a prior authorization from the responsible military organization. Civilian conferences usually did not have panels on the military or domestic security topics.

International Russia studies community has faced this issue too, mostly in Europe, as the United States achieved a stronger institutional integration between the national security community and civilian scholars (Pion-Berlin and Dudley 2020). Until recently, European security studies have tended to be more theoretical, focusing on post-structuralist approaches to explaining security issues rather than data-driven positivist studies of civil-military relations, armed forces, and various security topics. Ultimately, the parallel track system of civilian and military expertise negatively contributed to the theory-building on the role of the military in Russian politics, civilian control, and military development.

The Russian military and security services institutionally restricted their training exclusively to military personnel and blocked researchers from entering the field, which naturally reduced the civilian potential for understanding Russian military affairs. Few independent non-governmental organizations had stable access to the military in Russia. Organisations as Centre for Analysis of Strategies and Technology (CAST) and the Academy of Military Sciences (AVN) must perform a delicate balancing act by developing some military expertise without offending the RUMOD or the General Staff.

In the education sector, the ROTC programs at civilian universities (military departments) were reserved for male students only, were of poor educational quality, and were often used as a way to skip conscription rather than to learn anything new about the armed forces. Some Russian civilian higher education institutions adopted a conservative stance and limited the freedom of choice of dissertation topics for those Ph.D. students who wanted to study the military and other security services years before the full-scale invasion. Moreover, scientific research and publication of analyses on the Russian military became formally illegal in 2021 unless a researchers gets a prior permission from the FSB (Prokopieva 2021). Several independent military blogs disappeared after the passage of this law or switched to writing about the Second World War period.

Even pro-government reporters interviewed by the author for another study claimed that their access to the armed forces was severely restricted under Minister Shoigu. The Defence Ministry was very selective about the topics of their coverage and could censor certain moments if they did not like the footage. Naturally, the lack of verifiable evidence and independent research on the Russian military made the debate about Russian armed forces more biased.

This is in stark contrast to the experience of researchers in the 1990s and early 2000s, when scholars even from NATO countries could potentially gain access to military and security personnel, archives, and conferences to conduct research. This can potentially explain why a few critical scholarly contributions on Russian civil-military relations and national security appeared in the 1990s and early 2000s but then gradually faded away along with the growing censorship of security-related information in Russia. One of the last attempts to bridge military-related expertise in Russia and the West was a Carnegie Corporation-funded project at the CSIS that brought researchers from Russia, including from the aforementioned CAST, to Washington DC (Kofman 2017).

In addition, there was some normative divergence between more progressive/liberal civilian scholars and generally more conservative military analysts that prevented dialogue between the two. Russian liberal scholars could dismiss traditional realist concerns with the balance of power, deterrence, competition in the international system, and the use of force, even though these ideas were often the worldview of the military and security services internationally. They saw private interests (the infamous "Kremlin towers"), corruption, bureaucratic competition for power, and finances in Putin's autocratic system as key factors in explaining the dynamics of Russian politics.

At the same time, some realist analysts could be equally dismissive of the other side's issues. For them, the geopolitical vision took precedence over domestic concerns. Some of them saw domestic developments as either unimportant or beyond their reach because of the autocratic nature of the Russian regime. Great power competition and spheres of influence were the key factors in explaining Russia's behaviour toward Ukraine. Importantly, some of these people had to maintain these views in order to maintain access to the anti-Western national security and policymaking community. With the exception of some analysts (Valdai Club 2004; "About Mark Galeotti" 2008), critics of Russian domestic realities and the Putin regime gradually became less welcomed at major events such as the Valdai Forum and smaller, more routine conferences and seminars within the Russian government, the Ministry of Defence, and other state institutions.

However, realist positions of some military and civilian analysts in Russia do not necessarily explain their potential support for the decision to use force. Hawkish researchers are capable of making utilitarian judgements and understand the costs and benefits of a potential foreign policy decision. Remarkably, two Russian retired officers—three-star general Leonid Ivashov and colonel Mikhail Khodarenok—spoke out against the invasion from a realist perspective (Meduza 2022; Moore 2022). Both retired officers worked as military analysts/researchers held pro-government political views (Ivashov leaned towards far-right positions), opposed NATO enlargement, supported a strong Russian military, and yet provided a valid assessment of the enormous negative consequences of the invasion and the lack of Russian national interest in conducting it.

5 Way Forward

This chapter by no means offers a comprehensive solution to the challenges facing the Russia studies community. However, there are some tools that Russia analysts can try to integrate into existing efforts to understand Russian foreign and security policy dynamics.

First, as noted above, reversing the units of analysis can help to better understand Russian security. Scholars can examine how the country's domestic institutions and actors have influenced Russian civil-military relations, and thus the military capabilities and effectiveness of other security agencies. Although the Russian national security community can influence Russian political dynamics, the primacy of civilian supremacy in Russia is strong. The Russian military and security services have consistently shown that they are more concerned with protecting their institutional prerogatives from civilian interference than with overthrowing the government, even when Russian state capacity was weak, as it was in the 1990s. From president Yeltsin's decision to shoot the Russian parliament in 1993 to president Putin's order to begin the full-scale invasion of Ukraine in 2022, Russian armed forces and security services listened to their civilian leaders.

However, this does not mean that there are no problems with civilian control in Russia. A leading civil-military scholar, Risa Brooks, suggested several ideas to advance the civil-military field. Some of them propose scholars to integrate coup research into broader comparative studies, especially those of democratic backsliding (Brooks 2019). Russia is a strong case of failed democratization and can be used to illustrate how robust authoritarian institutions kept the military tightly under civilian control. A failed coup attempt by PMC Wagner leader Prigozhin is an interesting case of how both a political and military outsider with strong patrimonial ties to the president and sources of income tried to exploit his mercenary force for political purposes.

Moreover, the full-scale invasion of Ukraine also provided an opportunity to study other, more complex forms of Russian military insubordination, such as mutiny, fratricide, desertion, breakdown of discipline or cohesion, individual officer resignation, and failure to re-enlist. There are numerous independent media reports of hundreds of such cases, including of Russian servicemembers seeking international protection in Europe (France 24 2022; Barabanov 2023; Euronews 2024). Analysing these cases can facilitate the conceptualization of Russian civil-military relations by comparing them with other autocratic countries.

Second, the Russia studies community can potentially benefit from a stronger integration of academic experts with military and security analysts from policy-oriented think tanks and military organizations. Such organizations as the Center for Naval Analysis (CNA), the Conflict Intelligence Team (CIT), RAND, the Royal United Service Institute (RUSI), the Institute for Peace Research and Security Policy (IFSH), and others offer great technical and operational expertise that provide crucial insights into Russian military development. In addition, some militaries have conferences, workshops, and journals open to civilian scholars, including those from non-NATO countries, such as the Finnish Armed Forces Russia Seminar, the Annual West Point Social Sciences Seminar, the Modern War Institute at West Point, Ares & Athena journal of Centre for Historical Analysis and Conflict Research (CHACR) of the British Army, and the rest. This list is not exhaustive and can be expanded.

A prime example of civil-military knowledge exchange is the International Biennial Conference of the Inter-University Seminar on Armed Forces and Society (IUS). This conference has provided a venue for military, intelligence, private sector, and academic experts and professionals to present their research and exchange ideas on a wide range of topics related to national and international security. Russia topics have always been present at this conference and attracted diverse audience.

The European Initiative for Security Studies also organizes its annual conference on a diverse set of security topics with a more limited presence of Russia scholars. Yet, its venue provides an opportunity for European scholars to debate and exchange ideas on the future of war, military technology, defence cooperation, nuclear weapons and other topics. Narrowing the gap between academic expertise and policy-oriented military analysis of Russia can potentially strengthen the predictive qualities of theories of Russian politics and foreign affairs.

Conclusion

The full-scale invasion of Ukraine sparked a critical debate about the study of Russian politics. Prior to the invasion, many scholars and analysts could not accurately predict the war or foresee the subsequent underperformance of the Russian military. The goal of this chapter was to show how a hard look at Russian civil-military relations can help reveal policy-relevant and more valid arguments about the role of the Russian military and security services in Russian public affairs. Closer cooperation between scholars focused on academic research and policy-oriented military analysts could improve the validity of the findings for Russian policy. Incorporating civil-military relations into studies of Russian regime dynamics would potentially improve understanding of the role of the military and security services in maintaining the stability of Russia's authoritarian regime.

However, Russian scholars alone can hardly completely solve the problem of the civil-military divide in academic research and policy analysis. The problem is too complex, and it is also subject to national legal norms of civil-military engagement that scholars must follow. It is important to remember that most external analysis of Russia is ideally (far not always) used as a reference in the policy reports developed internally by dedicated government bodies and intelligence agencies. Most researchers possess neither full information about the host government's resources, defence and security programs and priorities nor access to it.

The Russian downshifting to a personalist dictatorship is further limiting access to traditional sources of empirical data, especially in the security field. Therefore, new innovative methods of conducting empirical research should be explored, such as the use of online sources (Telegram, Russian websites, including darknet), Soviet archives in Eastern Europe and Central Asia, publicly available leaked datasets, and Russian government and business data. Although some of them cause a plethora of ethical and analytical issues (especially novel online sources), there is a growing volume of work in this area (Khaund et al. 2021; Goonetilleke, Knorre, and Kuriksha 2023; Nizomutdinov and Filatova 2023).

Finally, it is important to reiterate that scholars are not to blame for the inadequate or flawed policy decisions made by politicians. There is enough evidence to confirm that some policymakers had access to alternative publicly available evidence and theories of Russian foreign policy behaviour but chose to ignore them before the full-scale invasion. As civil-military scholars would say, politicians have the right to be wrong because they are elected by the people and are accountable to them, including for the mistakes they have made. Scholars should provide the best possible evidence to decision-makers, academic peers, and the public, but it is ultimately up to politicians to decide what to do with that evidence. This issue is beyond the scope of this book, but it raises potential questions about political science and regional studies teaching and policy advisory systems in different countries. The author leaves this question for further study.

Bibliography

"About Mark Galeotti." 2008. *In Moscow's Shadows* (blog). August 24, 2008. https://inmoscowsshadows.wordpress.com/about/.

Adamsky, Dmitry (Dima). 2014. "Nuclear Incoherence: Deterrence Theory and Non-Strategic Nuclear Weapons in Russia." *Journal of Strategic Studies* 37 (1): 91–134. https://doi.org/10.1080/01402390.2013.798583.

Anichkina, Tatiana, Anna Péczeli, and Nickolas Roth. 2017. "The Future of US–Russian Nuclear Deterrence and Arms Control." *Bulletin of the Atomic Scientists* 73 (4): 271–278. https://doi.org/10.1080/00963402.2017.1338046.

Arbatov, Alexei G. 1998. "Military Reform in Russia: Dilemmas, Obstacles, and Prospects." *International Security* 22 (4): 83–134. https://doi.org/10.1162/isec.22.4.83.

Baev, Pavel. 2015. "Russia Is Not Strong. And Putin Is Even Weaker." Brookings. June 8, 2015. https://www.brookings.edu/articles/russia-is-not-strong-and-putin-is-even-weaker/.

Baev, Pavel K. 2002. "The Plight of the Russian Military: Shallow Identity and Self-Defeating Culture." *Armed Forces & Society* 29 (1): 129–146. https://doi.org/10.1177/0095327X0202900107.

Barabanov, Ilya. 2023. "Dmitry Mishov, Russian Airman Who Defected, Gives BBC Interview." *BBC*, June 11, 2023. https://www.bbc.com/news/world-europe-65867990.

Barany, Zoltan. 2008. "Civil-Military Relations and Institutional Decay: Explaining Russian Military Politics." *Europe-Asia Studies* 60 (4): 581–604.

Bartles, Charles K. 2011. "Defense Reforms of Russian Defense Minister Anatolii Serdyukov." *Journal of Slavic Military Studies* 24 (1): 55–80. https://doi.org/10.1080/13518046.2011.549038.

BBC News. 2021. "Why Russia's GRU Military Intelligence Service Is so Feared," April 19, 2021, sec. Europe. https://www.bbc.com/news/world-europe-56798001.

Belkin, Aaron, and Evan Schofer. 2005. "Coup Risk, Counterbalancing, and International Conflict." *Security Studies* 14 (1): 140–177. https://doi.org/10.1080/09636410591002527.

Bendett, Samuel. 2019. "The Rise of Russia's Hi-Tech Military." *Fletcher Sec. Rev.* 6:5.

Bendett, Samuel, Stephen Blank, Joe Cheravitch, Michael B. Petersen, Andreas Turunen, and Jeffrey Mankoff. 2020. *Improvisation and Adaptability in the Russian Military*. Center for Strategic and International Studies. http://csis-website-prod.s3.amazonaws.com/s3fs-public/publication/200430_Mankoff_Russian%20Military_web_v3_UPDATED%20FINAL.pdf.

Bertrand, Natasha. 2021. "Biden Keeping Ukraine at Arm's-Length." POLITICO. March 13, 2021. https://www.politico.com/news/2021/03/13/biden-ukraine-kyiv-putin-475546.

Blank, Stephen. 2002. "The Great Exception: Russian Civil-Military Relations." *World Affairs* 165 (2): 91–105.

Borger, Julian. 2021. "All Options Fraught with Risk as Biden Confronts Putin over Ukraine." *The Guardian*, November 25, 2021, sec. US news. https://www.theguardian.com/us-news/2021/nov/25/all-options-fraught-with-risk-as-biden-confronts-putin-over-ukraine.

Bremmer, Ian, and Samuel Charap. 2007. "The 'Siloviki' in Putin's Russia: Who They Are and What They Want." *The Washington Quarterly* 30 (1): 83–92. https://doi.org/10.1162/wash.2006-07.30.1.83.

Brooks, Risa A. 2019. "Integrating the Civil–Military Relations Subfield." *Annual Review of Political Science* 22 (1): 379–398. https://doi.org/10.1146/annurev-polisci-060518-025407.

Bruneau, Thomas C., and Aurel Croissant. 2019. *Civil-Military Relations: Control and Effectiveness Across Regimes*. Lynne Rienner Publishers, Incorporated.

Bulavinov, Ilja. 2003. "Koloda Rossiiskoj Federacii." Kommersant.Vlast. December 1, 2003. https://www.kommersant.ru/doc/432263.

Campbell, Garrett. 2015. "Russia's Military Is Proving Western Punditry Wrong." Brookings. October 23, 2015. https://www.brookings.edu/articles/russias-military-is-proving-western-punditry-wrong/.

Cottey, Andrew, Timothy Edmunds, and Anthony Forster. 2005. "Civil–Military Relations in Postcommunist Europe: Assessing the Transition1." *European Security* 14 (1): 1–16. https://doi.org/10.1080/09662830500042452.

Crowley, Michael, and Julian E. Barnes. 2021. "How Far Would Biden Go to Defend Ukraine Against Russia?" *The New York Times*, November 25, 2021, sec. U.S. https://www.nytimes.com/2021/11/25/us/politics/biden-putin-russia-ukraine.html.

De Bruin, Erica. 2018. "Preventing Coups d'état: How Counterbalancing Works." *Journal of Conflict Resolution* 62 (7): 1433–1458. https://doi.org/10.1177/0022002717692652.

Dekalchuk, Anna. 2016. "Choosing Between Bureaucracy and the Reforms." In *Authoritarian Modernization in Russia: Ideas, Institutions, and Policies*, 16. Routledge.

Dilanian, Ken, Carol Lee, Courtney Kube, and Dan De Luce. 2022. "Biden Admin Carefully Examining Legal Issues around Providing Arms to Ukraine." NBC News. February 25, 2022. https://www.nbcnews.com/politics/national-security/biden-admin-carefully-examining-legal-issues-providing-arms-ukraine-rcna17758.

Dixon, Robyn. 2024. "In Putin's Wartime Russia, Military Corruption Is Suddenly Taboo."

Washington Post, June 10, 2024. https://www.washingtonpost.com/world/2024/06/10/russia-corruption-defense-minstry-purge/.

Dougherty, Christopher. 2022. "Strange Debacle: Misadventures in Assessing Russian Military Power." War on the Rocks. June 16, 2022. https://warontherocks.com/2022/06/strange-debacle-misadventures-in-assessing-russian-military-power/.

Douglas, Nadja. 2014. "Civil–Military Relations in Russia: Conscript vs. Contract Army, or How Ideas Prevail Against Functional Demands." *The Journal of Slavic Military Studies* 27 (4): 511–532. https://doi.org/10.1080/13518046.2014.963397.

Dowding, Keith, and Charles Miller. 2019. "On Prediction in Political Science." *European Journal of Political Research* 58 (3): 1001–1018. https://doi.org/10.1111/1475-6765.12319.

Driedger, Jonas J. 2023. "Risk Acceptance and Offensive War: The Case of Russia under the Putin Regime." *Contemporary Security Policy* 44 (2): 199–225. https://doi.org/10.1080/13523260.2023.2164974.

Driedger, Jonas J., and Mikhail Polianskii. 2023. "Utility-Based Predictions of Military Escalation: Why Experts Forecasted Russia Would Not Invade Ukraine." *Contemporary Security Policy* 44 (4): 544–560. https://doi.org/10.1080/13523260.2023.2259153.

Dvorkin, Vladimir, and Alexei Arbatov. 2013. "Military Reform in Russia: Current State and Future Prospects." Moscow Carnegie Center. https://carnegie.ru/2013/07/11/ru-pub-52368.

Eckel, Mike. 2023. "How Did Everybody Get The Ukraine Invasion Predictions So Wrong?" *Radio Free Europe/Radio Liberty*, February 17, 2023, sec. Ukraine. https://www.rferl.org/a/russia-ukraine-invasion-predictions-wrong-intelligence/32275740.html.

Escribà-Folch, Abel, Tobias Böhmelt, and Ulrich Pilster. 2020. "Authoritarian Regimes and Civil–Military Relations: Explaining Counterbalancing in Autocracies." *Conflict Management and Peace Science* 37 (5): 559–579. https://doi.org/10.1177/0738894219836285.

Euronews. 2024. "Russians Who Desert the Military Face Criminal Charges." Euronews. April 12, 2024. https://www.euronews.com/2024/04/12/russian-deserters-in-limbo-facing-criminal-charges-and-unanswered-asylum-claims.

Feaver, Peter. 1999. "Civil-Military Relations." *Annual Review of Political Science* 2 (1): 211–241. https://doi.org/10.1146/annurev.polisci.2.1.211.

Feaver, Peter. 2005. *Armed Servants: Agency, Oversight, and Civil-Military Relations.* Revised edition. Cambridge, Mass.: Harvard University Press.

France 24. 2022. "Russian Soldier Seeks Asylum in France after Recounting 'Chaos' of Ukraine Invasion." France 24. August 30, 2022. https://www.france24.com/en/europe/20220830-soldier-seeks-asylum-in-france-after-recounting-russian-army-s-chaos-in-ukraine.

Frye, Timothy. 2022. *Weak Strongman: The Limits of Power in Putin's Russia*. Princeton University Press.

Gardner, Frank. 2022. "Ukraine Crisis: Five Reasons Why Putin Might Not Invade," February 21, 2022. https://www.bbc.com/news/world-europe-60468264.

Gazukin, Pavel. 2001. "Russian Armed Forces in the Post-Soviet Period. Problems of Reforming and Military Building: 1992–2001." *The Journal of Political Theory, Political Philosophy and Sociology of Politics Politeia* 20 (2): 44–71. https://doi.org/10.30570/2078-5089-2001-20-2-44-71.

Gel'man, Vladimir. 2013. *Iz Ognja Da v Polymja: Rossijskaja Politika Posle SSSR*. BKhV-Peterburg. https://books.google.com/books?hl=en&lr=&id=UUNj3XETVEkC&oi=fnd&pg= PA5&dq=%D0%B8%D0%B7+%D0%BE%D0%B3%D0%BD%D1%8F+%D0%B4%D0%B0+%D0%B2+%D0%BF%D0%BE%D0%BB%D1%8B%D0%BC%D1%8F&ots=6Gfq7x7Y4f&sig=cb9zNOriUlojPXfepKe8HN8imHo.

Gel'man, Vladimir. 2016. "The Vicious Circle of Post-Soviet Neopatrimonialism in Russia." *Post-Soviet Affairs* 32 (5): 455–473. https://doi.org/10.1080/1060586X.2015.1071014.

Golts, Aleksandr. 2017. *Military Reform and Militarism in Russia*. Acta Universitatis Upsaliensis.

Goonetilleke, Priyanka, Alex Knorre, and Artem Kuriksha. 2023. "Hydra: Lessons from the World's Largest Darknet Market." *Criminology & Public Policy* 22 (4): 735–777. https://doi.org/10.1111/1745-9133.12647.

Götz, Elias, and Jørgen Staun. 2022. "Why Russia Attacked Ukraine: Strategic Culture and Radicalized Narratives." *Contemporary Security Policy* 43 (3): 482–497. https://doi.org/10.1080/13523260.2022.2082633.

Government of Russia. 2016. "Federal'nyĭ Zakon Ot 03.07.2016 N 226-FZ 'O Voĭskakh Nat͡sional'noĭ Gvardii Rossiĭskoĭ Federat͡sii' [Federal Law of 03.07.2016 N 226-FZ 'On the National Guard Troops of the Russian Federation']." Garant. July 3, 2016. https://base.garant.ru/71433920/.

Government of Russia. 2020. "Federal Constitutional Law No. 4-FKZ 'On the Government of the Russian Federation.'" Rossiiskaya Gazeta. https://www.consultant.ru/document/cons_doc_LAW_366950/.

Government of Russia. 2021. *Federal'nyj Zakon "Ob Oborone" [Federal Law "On Defense"]*. http://pravo.gov.ru/proxy/ips/?docbody=&nd=102041583&rdk=5.

Grant, Susan. 2022. "5. Stalinist Care: Cadres Decide Everything." In *5. Stalinist Care: Cadres Decide Everything*, 94–116. Cornell University Press. https://doi.org/10.1515/9781501762604-009.

Gustafson, Kristian, Dan Lomas, Steven Wagner, Neveen Shaaban Abdalla, and Philip H.J. Davies. 2024. "Intelligence Warning in the Ukraine War, Autumn 2021—Summer 2022." *Intelligence and National Security* 39 (3): 400–419. https://doi.org/10.1080/02684527.2024.2322214.

Hill, Fiona. 2016. "Putin: The One-Man Show the West Doesn't Understand." *Bulletin of the Atomic Scientists* 72 (3): 140–144. https://doi.org/10.1080/00963402.2016.1170361.

Hill, Fiona, and Clifford G. Gaddy. 2012. *Mr. Putin: Operative in the Kremlin*. Rowman & Littlefield.

Huntington, Samuel P. 1957. *The Soldier and the State: The Theory and Politics of Civil-Military Relations*. Harvard University Press.

Interfax-Ukraine. 2015. "Nuland Says 400–500 Russian Soldiers Killed in Eastern Ukraine—Mar. 10, 2015." Kyiv Post. March 10, 2015. https://archive.kyivpost.com/article/content/war-against-ukraine/us-claims-400-500-russian-soldiers-killed-in-eastern-ukraine-383079.html.

Ivanov, Anatoliy, and Sergei Vilgelm, trans. 2004. "Predvybornyj Hod Putina: Chto Pishut Gazety." DW. February 25, 2004. https://www.dw.com/ru/%D0%BF%D1%80%D0%B5%D0%B4%D0%B2%D1%8B%D0%B1%D0%BE%D1%80%D0%BD%D1%8B%D0%B9-%D1%85%D0%BE%D0%B4-%D0%BF%D1%83%D1%82%D0%B8%D0%BD%D0%B0-%D1%87%D1%82%D0%BE-%D0%BF%D0%B8%D1%88%D1%83%D1%82-%D0%B3%D0%B0%D0%B7%D0%B5%D1%82%D1%8B/a-1122875.

Janowitz, Morris. 1960. *The Professional Soldier: A Social and Political Portrait*. Free Press.

Jenkins-Smith, Hank, Carol L. Silva, Kuhika Gupta, and Joseph T. Ripberger. 2014. "Belief System Continuity and Change in Policy Advocacy Coalitions: Using Cultural Theory to Specify Belief Systems, Coalitions, and Sources of Change." *Policy Studies Journal* 42 (4): 484–508. https://doi.org/10.1111/psj.12071.

Jonsson, Michael. 2024. "Swedish Intelligence, Russia and the War in Ukraine: Anticipations, Course, and Future Implications." *Intelligence and National Security* 39 (3): 443–457. https://doi.org/10.1080/02684527.2024.2325248.

Kari, Martti J. 2019. "Russian Strategic Culture in Cyberspace: Theory of Strategic Culture—a Tool to Explain Russia's Cyber Threat Perception and Response to Cyber Threats." Dissertation, Finland: University of Jyväskylä. https://jyx.jyu.fi/bitstream/handle/123456789/65402/978-951-39-7837-2_vaitos_2019_10_11_jyx.pdf.

Khaund, Tuja, Muhammad Nihal Hussain, Mainuddin Shaik, and Nitin Agarwal. 2021. "Telegram: Data Collection, Opportunities and Challenges." In *Information Management and Big Data*, edited by Juan Antonio Lossio-Ventura, Jorge Carlos Valverde-Rebaza, Eduardo Díaz, and Hugo Alatrista-Salas, 513–526. Cham: Springer International Publishing. https://doi.org/10.1007/978-3-030-76228-5_37.

Kofman, Michael. 2017. "Discussion at CSIS on Russian Combat Operations in Syria." *Russia Military Analysis* (blog). January 14, 2017. https://russianmilitaryanalysis.wordpress.com/2017/01/14/discussion-at-csis-on-russian-combat-operations-in-syria/.

Kofman, Michael. 2018. "Russian Performance in the Russo-Georgian War Revisited."

War on the Rocks. September 4, 2018. https://warontherocks.com/2018/09/russian-performance-in-the-russo-georgian-war-revisited/.

Kofman, Michael, Anya Fink, Dmitry Gorenburg, Mary Chesnut, Jeffrey Edmonds, and Julian Waller. 2021. *Russian Military Strategy: Core Tenets and Operational Concepts*. CNA.

Krechetnikov, Artem. 2013. "Zachem Putin menjaet svoego glavnogo telohranitelja?" BBC News Russian Service. August 16, 2013. https://www.bbc.com/russian/russia/2013/08/130816_zolotov_analysis.

Kryshtanovskaya, Olga, and Stephen White. 2003. "Putin's Militocracy." *Post-Soviet Affairs* 19 (4): 289–306. https://doi.org/10.2747/1060-586X.19.4.289.

Kryshtanovskaya, Olga, and Stephen White. 2005. "Inside the Putin Court: A Research Note." *Europe-Asia Studies* 57 (7): 1065–1075. https://doi.org/10.1080/09668130500302830.

Le Figaro. 2016. "Le Figaro: konkurirovat' s NATO rossiĭskoĭ armii pomeshaet korrupt͡sii͡a [Le Figaro: Corruption will prevent the Russian army from competing with NATO]." RFI. February 1, 2016. https://www.rfi.fr/ru/rossiya/20160201-figaro-konkurirovat-s-nato-rossiiskoi-armii-pomeshaet-korruptsiya.

Luxmoore, Matthew. 2020. "'I Could No Longer Endure The Humiliation': Russian Soldier Offers Explanation For Deadly Shooting Spree." *Radio Free Europe/Radio Liberty*, January 11, 2020, sec. Russia. https://www.rferl.org/a/i-could-no-longer-endure-the-humiliation-russian-soldier-offers-explanation-for-deadly-shooting-spree/30371947.html.

Luzin, Pavel. 2019. "How Russia's Military Makes Its Own Headlines." *RIDDLE* (blog). May 7, 2019. https://www.ridl.io/en/how-russia-s-military-makes-its-own-headlines/.

Mason, Rowena, and Rowena Mason Deputy political editor. 2021. "UK Must Be Ready for War with Russia, Says Armed Forces Chief." *The Guardian*, November 14, 2021, sec. UK news. https://www.theguardian.com/uk-news/2021/nov/14/uk-must-be-ready-for-war-with-russia-says-armed-forces-chief.

Massicot, Dara. 2022. "Ukraine Needs Help Surviving Airstrikes, Not Just Killing Tanks." Defense One. January 19, 2022. https://www.defenseone.com/ideas/2022/01/ukraine-needs-help-surviving-airstrikes-not-just-killing-tanks/360898/.

Matei, Florina Cristiana, Carolyn Halladay, and Thomas C. Bruneau, eds. 2021. *The Routledge Handbook of Civil-Military Relations*. 2nd edition. New York: Routledge. https://www.routledge.com/The-Routledge-Handbook-of-Civil-Military-Relations/Matei-Halladay-Bruneau/p/book/9780367540425.

Mathers, Jennifer. 2019. "The Rebuilding of Russian Military Capabilities." In *Routledge Handbook of Russian Security*, 1st ed., 10. Routledge. https://doi.org/10.4324/9781351181242.

Meakins, Joss I. 2018. "Squabbling Siloviki: Factionalism Within Russia's Security Services." *International Journal of Intelligence and CounterIntelligence* 31 (2): 235–270. https://doi.org/10.1080/08850607.2018.1417525.

Meduza. 2022. "All the world's leading media outlets are discussing how Russian TV doubted the success of the 'special operation.' This was done by military expert Mikhail Khodarenok, a regular guest on Skabeeva's show. Here's what he said [Vse glavnye mirovye SMI obsuzhdai͡ut, kak v ėfire rossiĭskogo TV usomnilis' v uspekhe "spet͡soperat͡sii" Ėto sdelal voennyĭ ėkspert Mikhail Khodarenok—postoi͡annyĭ gost' shou Skabeevoĭ. Vot chto on skazal]." Meduza. May 18, 2022. https://meduza.io/feature/2022/05/18/vse-glavnye-mirovye-smi-obsuzhdayut-kak-v-efire-rossiyskogo-tv-usomnilis-v-uspehe-spetsoperatsii.

Meijer, Nathalie Loiseau, Peter. 2024. "Russian Corruption Is an Urgent Security Threat." *Foreign Policy* (blog). August 15, 2024. https://foreignpolicy.com/2022/07/05/russia-corruption-security-threat/.

Michaels, Eva. 2024. "Caught off Guard? Evaluating How External Experts in Germany Warned about Russia's War on Ukraine." *Intelligence and National Security* 39 (3): 420–442. https://doi.org/10.1080/02684527.2024.2330133.

Moore, Mark. 2022. "Ex-Russia General Warns Putin against 'criminal' Ukraine Invasion." February 7, 2022. https://nypost.com/2022/02/07/ex-russia-general-warns-putin-against-criminal-ukraine-invasion/.

Nizomutdinov, Boris A., and Olga G. Filatova. 2023. "Testing Methods for Processing Comments from Telegram Channels and Public VKontakte to Analyze the Social Media." *International Journal of Open Information Technologies* 11 (5): 137–145.

O'Brien, Phillips Payson. 2022. "How the West Got Russia's Military So, So Wrong." *The Atlantic* (blog). March 31, 2022. https://www.theatlantic.com/ideas/archive/2022/03/russia-ukraine-invasion-military-predictions/629418/.

Pion-Berlin, David, and Danijela Dudley. 2020. "Civil-Military Relations: What Is the State of the Field." In *Handbook of Military Sciences*, edited by Anders Sookermany, 1–22. Cham: Springer International Publishing. https://doi.org/10.1007/978-3-030-02866-4_37-1.

Politkovskaya, Anna. 2003. "Siloviki Po Vyzovu." Novaya Gazeta. July 21, 2003. https://politkovskaya.novayagazeta.ru/pub/2003/2003-057.shtml.

Prokopieva, Svetlana. 2021. "'Pisat' nel'zi͡a voobshche ni o chëm'. Chto FSB skryvaet ot inostrant͡sev ['You can't write about anything at all.' What the FSB hides from foreigners]." Sever.Realii. July 22, 2021. https://www.severreal.org/a/perechen-svedenij-fsb-inoagent/31369983.html.

Radchenko, Sergey. 2021. "'Nothing but Humiliation for Russia': Moscow and NATO's Eastern Enlargement, 1993–1995." In *NATO in the Cold War and After*, 7–53. Routledge. https://www.taylorfrancis.com/chapters/edit/10.4324/9781003251637-2/nothing-humiliation-russia-moscow-nato-eastern-enlargement-1993%E2%80%931995-sergey-radchenko.

Radin, Andrew, Lynn E. Davis, Edward Geist, Eugeniu Han, Dara Massicot, Matthew Povlock, Clint Reach, et al. 2019. "The Future of the Russian Military: Russia's Ground Combat Capabilities and Implications for U.S.-Russia Competition." RAND Corporation. https://www.rand.org/pubs/research_reports/RR3099.html.

Reiter, Dan. 2020. "Avoiding the Coup-Proofing Dilemma: Consolidating Political Control While Maximizing Military Power." *Foreign Policy Analysis* 16 (3): 312–331. https://doi.org/10.1093/fpa/oraa001.

Renz, Bettina. 2006. "Putin's Militocracy? An Alternative Interpretation of Siloviki in Contemporary Russian Politics." *Europe-Asia Studies* 58 (6): 903–924. https://doi.org/10.1080/09668130600831134.

Romanowski, Michal. 2018. "Corruption, Terrorism, and Russia's Future | German Marshall Fund of the United States." April 10, 2018. https://www.gmfus.org/news/corruption-terrorism-and-russias-future.

Roth, Andrew, and Pjotr Sauer. 2022. "'I Don't See Justice in This War': Russian Soldier Exposes Rot at Core of Ukraine Invasion." *The Guardian*, August 17, 2022, sec. World news. https://www.theguardian.com/world/2022/aug/17/i-dont-see-justice-in-this-war-russian-soldier-exposes-rot-at-core-of-ukraine-invasion.

Shamiev, Kirill. 2019. "Against a Bitter Pill: The Role of Interest Groups in Armed Forces Reform in Russia:" *Armed Forces & Society* 47 (2). https://doi.org/10.1177/0095327X19852648.

Shamiev, Kirill. 2021. "Understanding Senior Leadership Dynamics within the Russian Military." Understanding Russian Military Today. CSIS. https://www.csis.org/analysis/understanding-senior-leadership-dynamics-within-russian-military.

Shamiev, Kirill. 2023. "Civil–Military Relations and Russia's Post-Soviet Military Culture: A Belief System Analysis." *Armed Forces & Society* 49 (2): 252–274. https://doi.org/10.1177/0095327X211062932.

Shamiev, Kirill. 2024. "Brass Tacks: Why Russia's Military Fails to Reform." ECFR. May 15, 2024. https://ecfr.eu/publication/brass-tacks-why-russias-military-fails-to-reform/.

Shiryaev, Valerii. 2015. "General nasheĭ armii [General of our army]." Novaya Gazeta. October 8, 2015. https://novayagazeta.ru/articles/2015/08/10/65185-general-nashey-armii.

Simoyanov, Aleksei. 2013. "Dialektika kremlevskih bashen." *Rabkor* (blog). November 11, 2013. https://rabkor.ru/columns/editorial-columns/2013/11/11/kremlin/.

Soldatov, Andrei, and Michael Rochlitz. 2018. "The Siloviki in Russian Politics." In *The New Autocracy: Information, Politics, and Policy in Putin's Russia*, edited by Daniel Treisman, 30. https://www.brookings.edu/book/the-new-autocracy/.

Starodubtsev, Andrei. 2016. "How Does the Government Implement Unpopular Reforms? Evidence from Education Policy in Russia." In *Authoritarian Modernization in Russia: Ideas, Institutions, and Policies*, by Vladimir Gel'man, 324. Routledge.

Stewart, Brandon M., and Yuri M. Zhukov. 2009. "Use of Force and Civil–Military Relations in Russia: An Automated Content Analysis." *Small Wars & Insurgencies* 20 (2): 319–343. https://doi.org/10.1080/09592310902975455.

Tarasenko, Anna. 2015. "Russian Welfare Reform and Social NGOs: Strategies for Claim-Making and Service Provision in the Case of Saint Petersburg." *East European Politics* 31 (3): 294–313. https://doi.org/10.1080/21599165.2015.1023895.

Taylor, Brian D. 2003. *Politics and the Russian Army: Civil-Military Relations, 1689–2000*. Cambridge University Press.

"The Constitution of the Russian Federation." 1993. Rossiiskaya Gazeta. http://www.constitution.ru/en/10003000-01.htm.

Timofeev, Ivan. 2021. "War Between Russia and Ukraine: A Basic Scenario?" 2021. https://russiancouncil.ru/en/analytics-and-comments/analytics/war-between-russia-and-ukraine-a-basic-scenario/.

Transparency International. 2015. "Russian Federation—Government Defence Anti-Corruption Index 2015." Government Defence Anti-Corruption Index. https://government.defenceindex.org/countries/russian-federation/.

Trenin, Dmitri. 2016. "The Revival of the Russian Military." *Foreign Affairs*, April 18, 2016. https://www.foreignaffairs.com/articles/russia-fsu/2016-04-18/revival-russian-military.

Trenin, Dmitri. 2021. "Russia's National Security Strategy: A Manifesto for a New Era." 2021. https://russiancouncil.ru/en/analytics-and-comments/comments/russia-s-national-security-strategy-a-manifesto-for-a-new-era/.

Troianovski, Anton. 2022. "Vladimir Putin: Crafty Strategist or Aggrieved and Reckless Leader?" *The New York Times*, February 18, 2022, sec. World. https://www.nytimes.com/2022/02/18/world/europe/putin-russia-ukraine.html.

Ullman, Harlan. 2022. "Why Putin Won't Invade Ukraine." *Atlantic Council* (blog). February 16, 2022. https://www.atlanticcouncil.org/blogs/new-atlanticist/why-putin-wont-invade-ukraine/.

Usman, Sahibzada Muhammad. 2023. "HISTORY AND FUTURE PERCEPTIVE OF CIVIL–MILITARY RELATIONS IN CHINA." *World Affairs* 186 (1): 190–209. https://doi.org/10.1177/00438200221136875.

Valdai Club. 2004. "First Conference of the Valdai Discussion Club 'Russia at the Turn of the Century: Hopes and Realities' [Pervai͡a Konferent͡sii͡a Diskussionnogo Kluba "Valdaĭ" "Rossii͡a Na Rubezhe Vekov: Nadezhdy i Realii"]." May 9, 2004. https://ru.valdaiclub.com/events/posts/articles/pervaya-konferentsiya-diskussionnogo-kluba-valday-rossiya-na-rubezhe-vekov-nadezhdy-i-realii/.

Valeriano, Brandon, and Ryan C. Maness. 2015. "Paper Tiger Putin." *Foreign Affairs*, April 30, 2015. https://www.foreignaffairs.com/russia-fsu/paper-tiger-putin.

Ven Bruusgaard, Kristin. 2016. "Russian Strategic Deterrence." *Survival* 58 (4): 7–26. https://doi.org/10.1080/00396338.2016.1207945.

Weber, Max. 1965. *Politics as a Vocation*. Fortress Press.

Yakovlev, Andrei. 2021. "Composition of the Ruling Elite, Incentives for Productive Usage of Rents, and Prospects for Russia's Limited Access Order." *Post-Soviet Affairs* 37 (5): 417–434. https://doi.org/10.1080/1060586X.2021.1966988.

Yapparova, Lilia. 2023. "'He Considered Himself Indestructible' Meduza Spoke to Wagner Mercenaries about the Plane Crash That Killed Yevgeny Prigozhin." Meduza. August 25, 2023. https://meduza.io/en/feature/2023/08/25/he-considered-himself-indestructible.

Yenikeev, Shamil, and Akhmed Mekhdi. 2013. "Gubernatory, Oligarhi i Silvoiki: Neft' i Vlast' v Rossii." Paris: Ifri. https://www.ifri.org/sites/default/files/migrated_files/documents/atoms/files/ifri_mehdi_yenikeyeff_russian_oil_industry_and_elites_rus_feb_2013.pdf.

Yilmaz, Harun. 2022. "No, Russia Will Not Invade Ukraine." Al Jazeera. September 2, 2022. https://www.aljazeera.com/opinions/2022/2/9/no-russia-will-not-invade-ukraine.

Zimmerman, William, Sharon Werning Rivera, and Kirill Kalinin. 2019. "Survey of Russian Elites, Moscow, Russia, 1993–2016." Inter-university Consortium for Political and Social Research [distributor]. https://doi.org/10.3886/ICPSR03724.v6.

CHAPTER 11

The Limits of 'Russian IR'

Challenges for a Particularistic International Relations Field

Ivan U. Kłyszcz

1 Introduction

Russia's full-scale aggression against Ukraine in 2022 brought a shock for researchers of Russia across all fields, including those concerned with the study of the country's foreign policy (see Renz 2024) This 'crisis of interpretation' (Kurnyshova & Makarychev, this volume) requires a look back to how we study how Russia engages in international relations.

The study of Moscow's foreign policy has its roots—with few exceptions—in the Cold War era. For example, in the US, by 1965 there were three thousand scholars with a primary interest in 'Slavic Studies' (Engerman 2004, 466–467). The Soviet collapse in 1991 did result in a drop in scholarly interest in the region and newly established Russia. Yet, the topic of Russian foreign policy was less affected by this gap, and documenting and scrutinising Moscow's external relations remained a viable—though no longer as central—topic of analytical and scholarly interest. The perception of Russia's return to major power aspirations after 2014 drove a new interest for expertise on Russia (La Lova 2023, 28–29). This drive was especially boosted by the increased access to archives and policymakers in Moscow, and the relative increase in academic freedom in Russia in the 1990s and early 2000s. In parallel, the Russian government itself sought to engage international—especially Western—scholars on topics of foreign relations, as evinced by the global list of attendees to the Valdai Club annual conference among other high-profile events. These expert interactions were interpreted as the more modern part of Russia's 'hybrid' status, between Western-style capitalism and democracy and its (recent) Soviet past (Zhuk 2024, vii).

The 2022 shock revealed several challenges for the field of research dedicated to scrutinising and theorising Russian foreign policy. Part of these challenges were institutional, as connections to Russia became impossible or unviable (see Gel'man, this volume). These challenges stem from many sources now well documented in the pre- and post-2022 literature. These include, among others, a commitment to the uniqueness myth in Russia (Kazharski, this

 | DOI:10.1163/9789004747364_012

volume), the Kremlin's 'great power' narrative (Gomza, this volume), Moscow-centrism, and unexamined imperialist assumptions (see Zhuk 2014). Other scholars have brought attention to the shortcomings to the discipline of Political Science specifically, with Lankina (2023) pointing to an over-reliance on elite-centred approaches and urban biases, as well as a reluctance to engage with historical and sociological approaches. As described below, several scholars have extended this reflection to the study of Russian foreign policy, its established notions and applications.

While the post-2022 scholarly reflection has engaged different fields and topics, the agenda is far from exhausted, including in the study of Russian international relations. The introspective look into the methods, methodologies and frameworks used to analyse Russia's external relations remains an imperative. In turn, this reassessment is part of the broader reflection on the conceptual instruments in the Social Sciences used to analyse the Russian case (e.g., La Lova 2023; Lankina 2023). In other words, the response to 2022 must include not just a look back at the causes of the war, but also a look back at our own methods and ways to look at Russian foreign policy (see Polianskii 2024).

Before proceeding, a distinction must be made; in the following, I use the label 'International Relations' (IR) to speak about the academic sub-discipline of Political Sciences dedicated to examining international relations. In turn, I speak about 'international relations' (not capitalised) to speak about the phenomenon of the international, the object of analysis (or interpretation) of the IR scholarship. In addition, as explained below, there is the study of IR in Russia, which consists of the scholarship drawing from the international IR literature. In parallel, there is what is termed 'Russian IR' (e.g., Tsygankov & Tsygankov 2014), the pluralistic study of international relations that rejects the universality of IR concepts applied to Russian foreign policy.[1] The discussion of these categories will pay attention to both Russia-based and non-Russia-based scholars and institutions, although the former play a larger role in the evolution of these fields.

This chapter aims to contribute to the growing body of the post-2022 reassessment of the literature on Russian foreign policy. Aliaksei Kazharski writes (2024, 88) about some applications of the Realist IR school to the Russo-Ukrainian war as 'operat[ing] in line with the Foucauldian description of the techniques of power', acting as a 'disciplining exercise' to build 'rational act-

1 To be clear, these categories do not refer to the origin of the scholars or the institutions they are affiliated with. One can be of a non-Russian background (however defined) and seek to theorise international relations relying only on 'Russian concepts' along the lines of Russian IR.

ors' out of small and large powers. This ultimately robs smaller actors—such as Ukraine—from their agency as independent states, in a rhetorical manoeuvre often identified as 'Westplaining' (Hendl et al 2024). Maria Mälksoo argues (2022, 9–10) for a 'less parochial' study of IR, namely, by decolonising the sub-discipline. Mälksoo stresses the centrality of memory politics in the formation of the categories deployed in IR, often reflective of unreformed colonial and imperial assumptions. In the case of Russia, this concerns, among other points, the country's relationship to NATO and Central and Eastern Europe, where the latter is not granted agency vis-à-vis Russian foreign policy.

The examples of criticism above address the IR in (or about) Russia. My interest in this chapter is to contribute to the broader reflection on the study of Russian foreign policy through a critique of the other field, that of pluralistic Russian IR. Continuing along the lines of Makarychev & Morozov (2013), this chapter will proceed in three steps. First, I will present the background of particularistic paradigms in IR, as the disciplinary context that Russian IR claims. Second, I will present a second context for Russian IR, that of different approaches to studying international relations inside Russia. I summarise this debate under the labels of 'IR in [or about] Russia', and 'Russian IR'. Third, I will zoom in to one framework within this Russian IR literature to highlight the enduring problematic assumptions and assessments in that field, namely, the framework of Russian 'foreign policy traditions'. Finally, I will contextualise these problematic assumptions in the broader policy-scholarship milieu where Russian IR operates.

2 Particularism in IR

For several years, the limits of IR scholarship have been highlighted as much as the sub-discipline is grounded on the works of Western and modernist thinkers (see Rengger 2015, 34), excluding the non-Western world (see Barnett & Zarakol 2023, 428–429). One of the critiques waged towards IR is that theories deployed in IR are universalistic and thus imposed on non-Western settings. For example, several scholars have pointed to the connection between the Western-centrism of IR as a consequence of Western hegemony, which, at worst, ends up justifying European imperialism.[2] Others also add the material disadvantages that higher education faces in the countries of the Global South, thus reducing

2 For a critical engagement of the question of IR theory and imperialism, see Hobson 2007; Fonseca 2019.

the opportunities for their scholars to impact the sub-discipline. Finally, many scholars criticise the standard narrative of the origins of international relations, overly focused on European history and notions of Western Enlightenment while neglecting (if not outright dismissing) global history and non-Western world orders (Zarakol 2022). As Colin Wight asked (2018, 64) rhetorically, 'What is the value of a subject that studies IR if it steadfastly refuses to be genuinely international?'

The turn towards the critique towards Western-centrism propelled a growing field broadly labelled as 'global IR', or non-Western approaches to IR, including in IR theory. The goal for these approaches, however, is not to start IR anew, but transforming rather than dismiss the existing theories (e.g. Realism, Liberalism) and concepts. Today, discussions on the state of IR often consider what to make of the *de facto* plurality of IR (e.g. Wight 2019), how to map and analyse the plurality (Ataka 2024), and whether plurality is valuable on its own right or as an analytical instrument. In turn, 'national schools' of IR have been recognised as belonging to the plurality of the sub-discipline, rather than being consigned to the role of local applications of universal IR precepts (see Vasilaki 2012, 7; Barnett & Zarakol 2023).

3 IR in Russia, Russian IR and the International Relations of Russia

I began engaging with the literature on Russian foreign policy in 2016, during the studies for my master's degree at an interdisciplinary Area Studies programme that drew its origins from Soviet Studies. As a student of Polish origin, I was motivated by my interest in Russian foreign policy and understanding Moscow's ability to project power despite its many post-Soviet weaknesses. Early on, the Constructivist interpretation of Russian foreign policy was presented in coursework, especially by the works of Andrey Makarychev, Viatcheslav Morozov and Andrey Tsygankov, among others. Unlike other mainstream approaches in IR, Constructivism offered me and other students of my cohort a way to integrate our aspiration to connect in-depth cultural and historical knowledge of Russia (typical of interdisciplinary programmes) with explaining Moscow's foreign policy. This made Russian IR an appealing area to explore in parallel to IR in Russia and other analyses of Russian foreign relations (Table 1).

The first category listed, IR in Russia, refers to the scholarship addressing Russian foreign policy from an uncommitted point of view towards the notion of pluralism in IR theory. Since 1991, IR in Russia drew from international conceptual frameworks common elsewhere, especially from the Realist

TABLE 11.1 Three understandings of 'international relations' in the Russian case

	IR in Russia	Russian IR	Analysis of int'l relations
Sources	Universal	Pluralistic	Uncommitted/heterodox
Modernity	Singular	Plural	Uncommitted
Values	Universal	Local	Uncommitted
Discipline	Singular	National	N/A
Contribution to ...	... the scholarship	... the scholarship	... policymaking, media commentary

SOURCE: OWN ELABORATION

school, which, as Shakleyina and Bogaturov argue (2004), was the mainstream approach to IR in Russia in the 1990s. Part of this proliferation was also the result of cooperation with European and US universities, which led to productive exchanges up until 2022 (see Gel'man, this volume). In addition, the topic of Russian foreign policy has remained a topic present in disciplinary IR journals, implementing concepts developed in non-Russian contexts. In turn, the application of IR frameworks to Russian foreign policy has been an essential feature of post-Soviet Russian scholarship. As reflected in the curricula of IR programmes at MGIMO—a leading international relations university in Moscow—, applied rational choice and history of international relations were the approaches taught the most often (Hagmann & Biersteker 2018). The former reflects the enduring presence of IR in Russia among the country's leading institutions.

In this context, Russian IR operates under two assumptions: the existence of a 'national school' of Russian IR, and that this school is 'obliged to assert and realize' non-Western values (Lebedeva 2010; Lebedeva 2021, 109). In that sense, Russian IR articulates a narrative that resonates with the Global IR concern with 'subsuming' the existing categories and frameworks of IR, albeit with an emphasis on the *Russian* character of this subsumption. Whether there is indeed a distinct Russian 'school' of IR has been the object of debate as the ideological character of the interpretations of the field (see below) have been an obstacle for its formation (Omelicheva & Zubytska 2016, 41–42). Some articulate Russian IR as reflecting not only a pluralism in values and knowledge production, but also in terms of plurality in modernity, where different cultures have different paths to modernity (Voskressenski 2017). In that sense, Russian IR has been conceptualised as a theorising of international relations that addresses Russia's 'semi-peripheral' status in the world hierarchy, with its unique values and sources of modernity (Tsygankov & Tsygankov 2010). From this point of view, Russian IR adopts some of the abovementioned critiques of the Western-centric IR sub-discipline, especially the centrality of the US. This critique of US-centrality in IR echoes—if not directly borrows from—the critiques of the asymmetry between the 'Global North' and the 'Global South' in knowledge production (see Voskressenski 2017). In this narrative, the Russian government is sometimes a patron of the Russian IR but ultimately a passive actor regarding the values and ideas involved in Russian IR theorising, as it ultimately 'lacks the power to establish them' (Tsygankov & Tsygankov 2021, 12).

The third category refers to the analytical—rather than scholarly—study of the international phenomenon. This category points to the analysis of international relations as seen outside of the university milieu and housed in think tanks, state institutions, private sector, and other 'third sector' organisations. Like elsewhere, many analysts operating in this category may draw from either the IR in Russia or Russian IR, but their contribution is oriented to policymaking and not to the scholarship. That said, in a Russian context, these analysts may be scholars themselves, primarily employed at universities and engaging policymakers in an *ad hoc* manner (see Kaczmarska 2020). While a thorough analysis of this category goes beyond the scope of this chapter, it is part of the picture as it plays an essential role in the context of knowledge production in and about Russia (see below).

These three categories can be helpful as analytical categories, though they do not preclude the existence of grey areas. As addressed below, some IR Russian scholars and scholars of Russian IR are part of the policy milieu that demands conformity and in exchange contributes to the legitimacy of the Kremlin (see Bakir 2023).[3] As Omelicheva argues (2022), a portion of Russian IR developed in government-approved echo-chambers and has been driven by Kremlin-aligned academic influencers.

4 The Pitfalls of Russian IR: The Case of the 'Foreign Policy Traditions' Framework

In the Russian IR literature, one specific framework of analysis stands out: the foreign policy traditions framework. Grounded on a Constructivist interpretation of Russian foreign policy, what I call the 'traditions framework' is an attempt to define the parameters of continuity and change in Russian foreign

3 In the context of Russia's highly centralised and personalistic political system and the long-standing intelligence services management of academic relations, and of Russia's full-scale aggression against Ukraine, this dual career track deserves special scrutiny.

policy across several centuries (see below). In the following, I present this traditions framework as an exemplar of the Russian IR field, one that exposes its enduring problematic foundations.

Is it adequate to discuss the foreign policy traditions as a representative of Russian IR as described above? Indeed, the topic of examining Russian IR in the context of Russia's full-scale invasion of Ukraine deserves a more thorough treatment, going beyond one framework within this pluralistic field of global IR. Yet, the foreign policy traditions framework is influential in the study of Russian IR and of Russian foreign policy in general, and it conspicuously presents itself as a conceptual tool attuned to Russia's particular culture and history (Tsygankov & Tsygankov 2014). Going even further, Tsygankov and Tsygankov (2023) present this framework as encompassing the field of Russian IR itself.

The traditions framework is influential in the study of Russian international relations broadly understood. For example, Andrey Tsygankov's textbook *Russia's Foreign Policy. Change and Continuity in National Identity* is a touchstone for many of the post-2000 cohorts in Russian Studies—including my own—, and a frequent reference for the Constructivist interpretation of Moscow's external relations. The book consists of an in-depth examination of Russian diplomatic history along the lines of the traditions framework. First published in 2006, *Russia's Foreign Policy* has been re-edited six times, with the latest edition published in 2022. According to *Open Syllabus* (n.d.), 212 syllabi in English-language courses across the world assign Tsygankov's book as part of their reading list. According to *Google Scholar* (n.d.), the 2016 edition of the book was cited 1196 times. To put these figures into perspective, the Russian foreign policy monographs by Bobo Lo (2015), Robert H. Donaldson and Vidya Nadkarni (2018), and Jeffrey Mankoff (2009) were in 51, 171 and 123 syllabi, and were cited 492, 644, 1090 times, respectively. Inasmuch as these numbers matter, they demonstrate the relevance of the foreign policy traditions framework in the broader study of Russian international relations.

4.1 *Russia's Foreign Policy Traditions*

Based on an unconformity with Realism's focus on power over ideas and with Liberalism's focus on modernisation, the foreign policy traditions framework assumes that the West is Russia's meaningful 'Other', defining—in opposition or in proximity—Russia's sense of 'Self'. The formulation of the sense of Self is based on both the Russia's situation domestically (the state of its economy and politics) and internationally (diplomacy, alliances, conflict). The upshot is a causal mechanism, where change in domestic and international circumstances leads to change in the ideas on national identity as espoused by new power coalitions, in turn leading to change in foreign policy (Tsygankov 2016).

TABLE 11.2 The Russian foreign policy traditions framework

	Westernisers	Statists	Civilisationists
Russian identity	European	Major power	Own civilisation
Focus	Modernisation	Security, *'Derzhava'*	Distinct values
Relationship with the West	Positive, filial	Pragmatic	Antagonistic

SOURCE: OWN ELABORATION BASED ON TSYGANKOV 2016

In sum, the framework is Constructivist and empiricist, drawing from sociology of knowledge and intellectual history. It assumes a causal framework rather than an interpretivist ontology. While this approach reflects the pluralistic assumptions of Russian IR, the foreign policy traditions framework is inspired by the study of US foreign policy traditions (Mead 2002). (The analysis in this chapter does not extend to this US inspiration.)

To explain change and continuity in Russian foreign policy, the traditions framework considers three broad categories that describe the parameters of distinct Russian approaches to foreign relations. The differentiating factor of these categories is the way Russian identity is construed, particularly in relation with the West (Table 11.2). Each of these categories encompasses a different foreign policy as well as a different set of individuals advocating for these directions: the Westernisers, the Statists (also rendered as the 'Realists') and the Civilisationists (Tsygankov 2016, 2021, 2022, 2024). The Westernisers tend to identify Russia with the West, casting Russia as a member of European civilisation. In turn, they seek convergence and coexistence with the West. The Statists tend to identify Russia as a major power, one that does not require to draw from foreign sources to engage in its own path of modernisation. For the Statists, coexistence with the West is possible on pragmatic terms, but competition tends to dominate. Finally, the Civilisationists define Russia in ethno-cultural terms as a distinct civilisation, whose values and identity are incommensurable with those of the West. The result is an inevitably antagonistic relationship between the two. In these three categories there are several sub-groups, such as liberal Statists and Westerniser supporters of a social state (Tsygankov 2024, 149), but the three main categories remain.

According to this narrative, these three broad directions have lineages deep into Russian diplomatic and intellectual histories. The Westernisers share a heritage with emperors Peter the First and Catherine the Second, as each of them pursued westwards expansion, modernisation projects and conspicuously and consciously sought their empire to converge with the European

Enlightenment. Other highpoints—according to this narrative—include the Holy Alliance, the Entente and the Grand Alliance, key moments of alignment and alliance between Russia and the European powers. In more recent times, Russian President Boris Yeltsin, his first foreign minister Andrey Kozyrev, and Yegor Gaidar are cast as the modern proponents of the Westerniser tradition (Tsygankov 2024). The Statists descend from a lineage of *Realpolitik* thought exemplified by diplomat Alexander Gorchakov and, more recently, former KGB director and Prime Minister Yevgeny Primakov. The Statists generally focus on the idea of great power status, national sovereignty and security, and the foundations thereof, which, according to this narrative, are different from those of the West. Namely, Russia's 'semi-peripheral' status compels Russian Realist thinkers to incorporate notions of 'national values' to their assumptions, thus deviating from the parsimonious tendencies of Western Realist IR. Here, the notion of '*Derzhava*'—'great power'—allegedly captures a more complex notion than its literal translation into English (Tsygankov 2022, 1–8; see also Reshetnikov 2024, 23). The Civilisationists, in turn, draw from the messianic traditions in Russian political thought, embracing the notion of Russia having a particular mission in the world granted to it by culture (frequently articulated in religious terms), history and geography. This exceptionalist tradition often points to referent figures such as Ivan the Fourth, Vladimir Solovyev, and the Slavophile movement of the Romanov empire (Tsygankov 2023, 4–9).

In short, the framework manages to contextualise Russia's changing foreign policy within broader and historical debates in Russian political thought, notably between Slavophiles and Westernisers. As described above, this framework is popular. It encourages a reflection on the role of ideas and ideology in shaping foreign policy, which is a notion I continue to embrace. To this day, scholars of Russia often categorise foreign policy debates in Russia along the categories of Westerniser, Statist and Civilisationist, regardless of their commitment to universal or pluralistic IR (e.g., Kuchins & Zevelev 2012; Cannady & Kubicek 2014; Wilhelmsen 2019).

4.2 *The Limits of the 'Foreign Policy Traditions' Framework: Eurocentrism*

In the past two decades, a rich literature has developed demonstrating the role of interactions with non-Europeans in shaping the identities of the Romanov Empire, the Soviet Union and of the Russian Federation. Presenting an overview of how and when these interactions happened goes beyond the scope of this chapter. Indeed, here I will briefly point to three dimensions of Russian identity and its role in international relations that are lost due to the framework's Eurocentrism.

First, there is an important colonial heritage that shapes the interplay between Russian identity and foreign policy. Indeed, the Romanov colonial enterprise across Asia and North America contributed to the formation of the 'European' identity of the Romanov empire and its claims to belonging to the European Enlightenment and other trends in the continent. For example, by the XIX century, the colonial administration of the Romanov Caucasus was articulated in terms like those of other European empires of the time. Namely, as an enlightened *mission civilisatrice,* with the accompanying sense of prestige, valued by the other European empires, albeit with the difference of the Romanov empire being continental rather than an overseas one (Schorkowitz 2019). These experiences left lasting legacies in the Soviet Union and Russia, contributing to Russia's present-day sense of imperial entitlement towards the Caucasus and Central Asia (Koplatadze 2019; Kassymbekova & Marat 2022).

Second, the encounters with non-Western empires—among others, the Ottoman, Ming and Qing China, and the Persian empire—also played a significant role in the Westernisation of the Romanov Empire. In the case of the Ottoman Empire, as Victor Taki recounts (2016), the rulers of Muscovy first sought to portray themselves and their realm as equals to the Ottomans, a position that went against the custom of the *Porte* in the XI and XVI centuries. By the XVIII century however, Romanov writers began to adopt Orientalist framings to portray the Ottomans, following the trends prevalent in Europe, including, among others, the cliche of 'Oriental immobilism' and the discussions on the causes of the Ottoman decline. While this trend rendered evident the empire's Westernisation, it also featured an essentialising portrayal of the West in what may be termed 'Occidentalism' and a critique of Western society (Taki 2011, 350–351). In short, the encounters with non-Western empires did serve just to affirm the European identity of the Romanov empire, but also to highlight its difference from the West, thus establishing a trichotomy of the West, the East and Russia (see Khalid 2000, 699).

Third, the Eurocentrism of the framework tends to underplay—especially after 1991—the relevance of Belarus and Ukraine as referents for Russian identity. Indeed, Eurocentrism can sometimes be construed rather as Westerncentrism, occluding the role of Belarus and Ukraine in Russia's discourse of its own identity. While located in Europe, these two countries are erased by the discourse of a tripartite Russian identity, namely, the XIX century definition of the Russian nation as consisting of White Russians, Little Russians and Great Russians, respectively. Since the late 1990s, this notion of Russia's identity was recaptured under the label of the 'Russian world', an intellectual construct made by Kremlin-linked 'political technologists' (Wilson 2024) meant to undermine the agency of Ukrainians and other peoples and countries formerly and currently under Moscow's control. As Kassymbekova and Marat argue (2022), Russia's contemporary entitlement over Belarus and Ukraine also draws from the same sense of victimhood of loss of empire as with the losses of the Caucasus and Central Asia. Indeed, Russia's aggression against Ukraine—since 2014 and especially since 2022—has been portrayed by the Kremlin as the fight against Western aggression, namely, against the Russian nation as constituted by Greater and Little Russians (Pyoneimeni 2024).

The result is that, in attempting to account for lineages of Russian foreign policy, the traditions' framework tends towards the 'essentialist trap' (Barnett & Zarakol 2023). Namely, it replicates a great power narrative that addresses the ontological bases of Russian identity only in relation to other great powers (see Gomza, this volume), remaining blind to other relations and other relevant scales of analysis.

4.3 *The Limits Of The 'Foreign Policy Traditions' Framework: A Flattening Of Intellectual Legacies*

The second limitation of the foreign policy traditions framework is its insistence on intellectual lineages. Here, the limitation is more straightforward. The imposition of supposed lineages simplifies Russia's philosophical and political debates, removing them from their intellectual and sociological contexts. This results in—as Andrey Makarychev remarked (2014, 36)—flat narrations of Russian foreign policy, of the ideas that animate it, and of the individuals that promote it. The upshot is that the framework struggles to provide persuasive interpretations of concrete shifts in Russian foreign policy, in turn drawing attention away from Russian agency. To illustrate this latter point, I will focus on a central episode in the framework's narrative: the 1996–1998 appointment of Yevgeny Primakov to the role of Yeltsin's second foreign minister.

The standard narrative of the foreign policy traditions framework presents Primakov's appointment as the rise of a new power coalition in the Kremlin—that of the military, industrialists, and security services—, and as a shift to Russia embracing an assertive foreign policy, open to diverge from the West. The rise of Primakov is often interpreted in light of Moscow's alleged proximity to the West during the early 1990s, namely, as the result of a Russian disappointment with the West for the latter's lack of consideration for Russian great power claims (e.g., Kuchins & Zevelev 2012, 150; Tsygankov 2016; see also Omelicheva & Zubytska 2016, 34). The new external relations would be oriented towards defending Russia's role as an 'independent great power' based on the balance of power and a vision of a multipolar world. In the terminology of the framework, the rise of Primakov signified the shift from a Westernising to a Statist foreign policy.

This interpretation has become the conventional wisdom on the mid-1990s Russian foreign policy (see Lannon 2018, 14–17). Yet, this narrative is not unassailable nor self-evident. In the following, I raise four objections to it, each one pointing to the limitations of the foreign policy traditions framework, especially when applied to specific shifts in Russian foreign policy.

First, the narrative overemphasises change over continuity in foreign policy. Crucially, the narrative hinges on the idea that, in pursuing a Western course before, Russia had abandoned its great power aspirations. Moscow's foreign policy in the early 1990s was already driven by the desire to restore Russia's status as a major power. This is captured in elite and general polling done in 1992, where preferences were overwhelming over Moscow preserving superpower status, even if in a different shape as during the Cold War (Clunan 2009). However, the pursuit of this status was informed by *perestroika*-era assumptions, namely, that Moscow will benefit from becoming Washington's indispensable partner in solving global issues. In other words, by partnering with the West—inasmuch as it did—, Moscow was already pursuing a great power strategy.

Second, this narrative overestimates the unwillingness of Moscow to assert an autonomous foreign policy. Indeed, the early 1990s featured a combination of cooperation and confrontation with the West, with Russia defending China's human rights record in 1992, re-establishing a defence partnership with India in 1993, and abandoning its pro-Kuwait policy in 1993 as well. More significant still, the 1994 launch of the first Chechen war demonstrated that Yeltsin was not moved by Western or international concerns over human rights or peaceful conflict resolution. Western objections of human rights' violations in Chechnya (e.g., Forsberg & Herd 2005, 459–460) did not impede the continuation of the campaign, ending only in 1996 and only after military defeat.

Third, the interpretation over-estimates the prevalence of political pluralism in Russia after 1991. Indeed, the sociological approach adopted misses the fact that Soviet-era elites largely remained in power after the Soviet collapse, as no lustration or truth and reconciliation processes were pursued (Snegovaya 2023). Without a doubt, there was more political pluralism in Russia in 1996 when compared to the Soviet era. But in the mid-1990s the Russian elite was still largely made up of former members of the Communist Party and of the *nomenklatura*. The so-called 'New Russians' would only gain power and influence by becoming integrated into the Kremlin's networks, rather than by creating their own networks and alternative power poles (Baturo & Elkink 2022). In short, when applied to contemporary Russia, the foreign policy traditions framework tends to overestimate the degree of pluralism within Moscow's foreign policy establishment.

Fourth, the interpretation of Primakov's rise advanced by this narrative ironically undermines the relevance of national identity in Russia's foreign policy. Namely, what ideas Kozyrev and Primakov might have about Russia become secondary to their ideas about the West. Indeed, from this point of view, and taken to the extreme, Russia's national identity becomes a billiards ball, always reactive to the political and intellectual life of the West. This results in an overly shallow interpretation of Russian politics (see Gel'man, this volume). As mentioned in the previous section, contemporary Russian identity draws from many sources and heritages, including its encounters with non-Western peoples and empires, its sense of victimhood and of imperial entitlement, and the legacies of the Romanov empire and the Soviet Union. The West plays a role in each of these discourses, but its overemphasis precludes other sources and obscures Russia's agency.

There are alternatives to the interpretation of Primakov's 1996 appointment. For example, a Constructivist approach to grand strategy (e.g., Balzacq & Krebs 2021) can account for the 1996 Primakov appointment, namely, as a shift in strategy rather than a comprehensive reappraisal of what it means to be 'Russian'. Having realised that the great power strategy of partnering with Washington was not progressing, Yeltsin decided on Primakov's foreign policy based on a 'multipolar' worldview, emphasising the autonomy of Russian foreign policy and enhancing relations with China and India. Drawing from his background in the Middle East and experience as Gorbachev's ally during *perestroika*, Primakov favoured a foreign policy narrative oriented towards the UN and great power bargaining. However, this foreign policy also contemplated maintaining dialogue with the West, thus maintaining a degree of continuity with the supposedly pro-Western Kozyrev line (see Lannon 2018, 14–25).

5 The Paradox of Russian IR: A Closed Space with Reach

As Omelicheva argued, after a period of intensive exchange with international scholarship, Russian IR became a bubble of its own, where *sui generis* Realist interpretations became dominant (2022, 2). Yet, as Omelicheva points to, Russian IR has produced conceptual innovations in the past, before entering an epistemologically closed loop. In this chapter, I argued that we must also consider the importance of Russian IR—illustrated by the foreign policy traditions framework—in shaping our understanding of Russian foreign policy in general. The reach of the three categories of 'Westernisers', 'Statists' and 'Civilisationists' demonstrate the relevance of Russian IR beyond the boundaries of its epistemic bubble. However, this is a paradox. Russian IR is, on the one hand, a closed circle, operating on faulty assumptions and a conspicuous rejection of 'Western' frameworks. On the other hand, many of its key ideas have influenced the way the literature and generations of students understand Russian foreign policy. What explains the popularity and reach of the Russian foreign policy traditions framework?[4]

This paradox can be interpreted in light of what Fisher-Onar describes (2022) as the puzzle of emerging, post-imperial powers leveraging postcolonial rhetoric for their own imperial projects (see also Callahan 2008). While this rhetorical manoeuvre principally concerns the articulation of state narratives for diplomacy and strategic communication, it also has a presence in IR theorising and academic engagements. Namely, emerging powers cast their aggressive foreign policies into a narrative of victimhood at the hands of the West, ultimately producing narratives of exceptionalism that justify foreign aggression. In turn, these narratives of victimhood incentivise academics in the emerging powers to operate within a scepticism towards universal concepts, in turn lending their academic credibility to legitimating the authoritarian system and its foreign policy (Fisher-Onar 2022, 9; see Owen et al 2018; Bakir 2023). In the case of Russia, as exemplified by the case analysed here, the Russian IR field continues to conflate a pluralistic approach to IR theorising with an exceptionalist narrative about Russia, in turn used to launder Russia's aggressive foreign policy. In that sense, Russian IR often reflects the broader problems of the 'non-Western IR', global IR research agenda, such as ethno-cultural exceptionalism and operating in a false dichotomy between 'Western' and 'non-Western' IR (Owen et al 2018; Barnett & Zarakol 2023). As Omelicheva & Zubytska explained (2016, 42):

> Russian IR perspectives seem able to account for Russian foreign policy and views of the world because the academic and foreign policy discourses have been co-evolving. The emphasis on Russia's great power identity, concerns with the US domination of global affairs, and the search for models of a "just" and "progressive" world order have been pivotal for both academic and policy discourses in Russia.

The upshot is that the Kremlin is not a passive participant but an active promoter of Russian IR. As Omelicheva argues (2022, 4), elite educational institutions specialised in Political Science are part of a broader state-funded milieu

4 I thank a participant of the workshop 'Russia and its War(s): A scholarly audit and political implications' (Tallinn, 7 June 2024) for presenting this question in such clear terms.

of think tanks, research institutions, and media platforms. This network is sustained by state funding, interpersonal relations and patronage networks, and is often overtly run from state institutions. MGIMO—mentioned above—is the main higher education institution run by the Ministry of Foreign Affairs. The Valdai Club is closely tied to the Russian Presidential Administration. The Russian Institute for Strategic Studies—a major think tank—is known to be run by the Foreign Intelligence Service (SVR in Russian). In addition, Russian security services are active in monitoring the curricula and activities of Russian higher education, with implications also for the contacts with Western academia (see Bertelsen 2023). While none of this implies that Russian IR in whole or in part is a Kremlin project, it points to a section of Russian IR being part of an epistemic bubble that exists in 'symbiosis' with Russian politics (Omelicheva 2022, 5–6). Russian IR is not the only framework part of this network. For example, Gomza (this volume) points to the persistent narrative of Russia belonging to the great power club, a trope prevalent both in the Kremlin's communications and propaganda, and in the academic literature on Russian foreign policy.

6 A Way Forward

There is value in addressing Russian notions of IR. Indeed, they can serve as a bridge between elite-focused, 'shallow' analysis and deterministic, 'deep' appraisals of Russian politics (Gel'man, this volume). In that sense, there is interest in understanding the 'Russian perspective'—if there is one—according to its own categories. This interest has become even more urgent since the launch of the full-scale war of aggression against Ukraine, the question being, namely, what decisions, ideas, and trends in Russia led the country to engage in war, and to do so in the way it did (i.e., illegally and without regard to human life).

Therefore, there are two necessary steps that a pluralistic view of IR in Russia would require for the way ahead. First, a separation from the foreign policy discourse of the Kremlin is necessary. If one is to take pluralism in IR as valuable on its own, the aforementioned 'co-evolution' of Russian IR and the Kremlin's foreign policy discourse should give pause before accepting this framework as valuable on its own terms.[5] Approaching Russian IR categories must be done with a critical attitude towards their underlying narratives, including

5 For example, the German school of geopolitics was co-opted by the Nazi regime in the interwar period, leading to its abandonment for years after the Second World War.

their empirical and normative bases. This chapter has highlighted the role of Eurocentrism, but other dimensions deserve scrutiny just as urgently, such as the Kremlin's 'great power' ambitions (see Gomza, this volume).

Second, Russia's decision to wage war against a peaceful neighbour needs to be examined from the point of view of the particular categories to IR thought in Russia. Here, more productive than the foreign policy traditions framework can be an examination of how normative perspectives on international relations legitimised the use of force in the case of Russia in 2022 (and other times). A critical approach to those notions that are common currency in Russian discussions of foreign affairs—for example, in society more broadly—could lead to a productive engagement while avoiding the essentialist trap.

Conclusion

Russia's full-scale invasion of Ukraine has prompted a reconsideration about how to study Russia, including how we approach the country's foreign relations. As this volume has shown in its various chapters, the 'old Russia studies' (see introduction) came to accept certain Kremlin narratives, unwittingly eroding its foundations as a field of scholarly pursuit. In that vein, since 2022, several scholars have highlighted how the IR literature contained problematic assumptions and assessments about Moscow's foreign policy, especially towards its former imperial subjects. Contributing to this broader reflection on the literature, this chapter proposes to extend this intellectual project to the pluralistic study of Russian IR, a field that has deep influence in the ways we understand Russian foreign policy both inasmuch it shapes the broader IR literature and its role in higher education curricula. While this project goes beyond the scope of this chapter, here I have attempted to critically engage with the foreign policy traditions framework. Inasmuch as the traditions framework operates within the pluralistic assumptions of Russian IR and constitutes a major touchstone for the study of Russian international relations, its scrutiny in light of the global IR project and the post-2022 Russia Studies points to broader pitfalls in the Russian IR scholarship. As explored in this chapter, these are, namely, its Eurocentrism and its insistence on lineages.

The popularity of the foreign policy traditions framework is understandable. It combines in-depth knowledge of Russian history, Russian intellectual history, a sociological framework, and a Constructivist approach to International Relations. These are all merits. However, the foreign policy traditions framework risks castling the study of Russian foreign policy in its own, idiosyncratic quasi-field. Adopting this perspective risks isolating the study of Russia, making it dependent on potentially flawed assumptions, and losing the connection to disciplinary peers. A pluralistic approach to Russian international relations can benefit from critically engaging the origins of Russian concepts of the international rather than the skewed image created by Eurocentrism—and the implicated erasure of colonial legacies—and lineage-focus of the field.

References

Bakir, Caner. 2023. "The Vicious Circle of Policy Advisory Systems and Knowledge Regimes in Consolidated Authoritarian Regimes." *Policy and Society* 42 (3): 419–439.

Balzacq, Thierry, and Ronald R. Krebs. 2021. "The Enduring Appeal of Grand Strategy." In *The Oxford Handbook of Grand Strategy*, 1–21.

Barnett, Michael, and Ayşe Zarakol. 2023. "Global International Relations and the Essentialism Trap." *International Theory* 15 (3): 428–444.

Baturo, Alexander, and Johan A. Elkink. 2021. *The New Kremlinology: Understanding Regime Personalization in Russia*. Oxford University Press.

Bertelsen, Olga. 2023. "Russian Front Organizations and Western Academia." *International Journal of Intelligence and CounterIntelligence* 36 (4): 1184–1209.

Callahan, William A. 2008. "Chinese Visions of World Order: Post-Hegemonic or a New Hegemony?" *International Studies Review* 10 (4): 749–761.

Cannady, Sarah, and Paul Kubicek. 2014. "Nationalism and Legitimation for Authoritarianism: A Comparison of Nicholas I and Vladimir Putin." *Journal of Eurasian Studies* 5 (1): 1–9.

Clunan, Anne L. 2009. *The Social Construction of Russia's Resurgence: Aspirations, Identity, and Security Interests*. JHU Press.

Donaldson, Robert H., and Vidya Nadkarni. 2018. *The Foreign Policy of Russia: Changing Systems, Enduring Interests*. Routledge.

Engerman, David C. 2004. "The Ironies of the Iron Curtain: The Cold War and the Rise of Russian Studies in the United States." *Cahiers du monde russe* 45 (3–4): 465–495.

Fisher-Onar, Nora. 2022. "The Capitulations Syndrome: Why Revisionist Powers Leverage Post-Colonial Sensibilities Toward Post-Imperial Projects." *Global Studies Quarterly* 2 (4).

Forsberg, Tuomas, and Graeme P. Herd. 2005. "The EU, Human Rights, and the Russo-Chechen Conflict." *Political Science Quarterly* 120 (3): 455–478.

Hendl, Tereza, Olga Burlyuk, Mila O'Sullivan, and Aizada Arystanbek. 2024. "(En)Countering Epistemic Imperialism: A Critique of 'Westsplaining' and Coloniality in Dominant Debates on Russia's Invasion of Ukraine." *Contemporary Security Policy* 45 (2): 171–209.

Hobson, John M. 2007. "Is Critical Theory Always for the White West and for Western Imperialism? Beyond Westphilian towards a Post-Racist Critical IR." *Review of International Studies* 33 (S1): 91–116.

Kaczmarska, Katarzyna. 2020. *Making Global Knowledge in Local Contexts: The Politics of International Relations and Policy Advice in Russia*. Routledge.

Khalid, Adeeb. 2000. "Russian History and the Debate Over Orientalism." *Kritika: Explorations in Russian and Eurasian History* 1 (4): 691–699.

Kassymbekova, Botakoz, and Erica Marat. 2022. "Time to Question Russia's Imperial Innocence." *PONARS Eurasia* Policy Memo No. 771.

Kazharski, Aliaksei. 2024. "On 'Westsplaining,' Realism, and Technologies of the Self: A Foucauldian Reading of the Realist Commentary on Ukraine." *Journal of Regional Security* 19 (1): 77–96.

Koplatadze, Tamar. 2019. "Theorising Russian Postcolonial Studies." *Postcolonial Studies* 22 (4): 469–489.

Kuchins, Andrew C., and Igor A. Zevelev. 2012. "Russian Foreign Policy: Continuity in Change." *The Washington Quarterly* 35 (1): 147–161.

La Lova, Lanabi. 2023. "Methods in Russian Studies: Overview of Top Political Science, Economics, and Area Studies Journals." *Post-Soviet Affairs* 39 (1–2): 27–37.

Lankina, Tomila. 2023. "Branching Out or Inwards? The Logic of Fractals in Russian Studies." *Post-Soviet Affairs* 39 (1–2): 70–85.

Lannon, Gregory P. 2018. "Testing the Limits of Realism and Liberalism: Andrei Kozyrev and Russian Foreign Policy, 1990–1996." PhD diss., University of Virginia.

Lebedeva, Marina M. 2004. "International Relations Studies in the USSR/Russia: Is There a Russian National School of IR Studies?" *Global Society* 18 (3): 263–278.

Lo, Bobo. 2015. *Russia and the New World Disorder*. Brookings Institution Press.

Makarychev, Andrey, and Viatcheslav Morozov. 2013. "Is 'Non-Western Theory' Possible? The Idea of Multipolarity and the Trap of Epistemological Relativism in Russian IR." *International Studies Review* 15 (3): 328–350.

Makarychev, Andrey. 2014. *Russia and the EU in a Multipolar World: Discourses, Identities, Norms*. Vol. 127. Ibidem-Verlag/Ibidem Press.

Mankoff, Jeffrey. 2009. *Russian Foreign Policy: The Return of Great Power Politics*. Rowman & Littlefield Publishers.

Mead, Walter R. 2002. *Special Providence: American Foreign Policy and How It Changed the World*. Routledge.

Omelicheva, Mariya. 2022. "Influencers, Echo Chambers, and Epistemic Bubbles." *PONARS Eurasia* Policy Memo no. 777.

Omelicheva, Mariya Y., and Lidiya Zubytska. 2016. "An Unending Quest for Russia's Place in the World: The Discursive Co-Evolution of the Study and Practice of International Relations in Russia." *New Perspectives* 24 (1): 19–51.

Owen, Catherine, John Heathershaw, and Igor Savin. 2018. "How Postcolonial is Post-

Western IR? Mimicry and Mētis in the International Politics of Russia and Central Asia." *Review of International Studies* 44 (2): 279–300.

Polianskii, Mikhail. 2024. "Russian Foreign Policy Research and War in Ukraine: Old Answers to New Questions?" *Communist and Post-Communist Studies*, 57(2), 156–172.

Rengger, Nicholas. 2015. "Pluralism in International Relations Theory: Three Questions." *International Studies Perspectives* 16 (1): 32–39.

Renz, Bettina. 2024. "Western Estimates of Russian Military Capabilities and the Invasion of Ukraine." *Problems of Post-Communism* 71 (3): 219–231.

Reshetnikov, Anatoly. 2024. *Chasing Greatness: On Russia's Discursive Interaction with the West Over the Past Millennium*. University of Michigan Press.

Schorkowitz, Dittmar. 2019. "Was Russia a Colonial Empire?" In *Shifting Forms of Continental Colonialism: Unfinished Struggles and Tensions*, edited by Dittmar Schorkowitz, John R. Chávez, Ingo W. Schröder, *Palgrave MacMillan*, 117–147.

Shakleyina, Tatyana A., and Aleksei D. Bogaturov. 2004. "The Russian Realist School of International Relations." *Communist and Post-Communist Studies* 37 (1): 37–51.

Silove, Nina. 2018. "Beyond the Buzzword: The Three Meanings of 'Grand Strategy'." *Security Studies* 27 (1): 27–57.

Smyth, Regina. 2023. "Plus Ça Change: Getting Real About the Evolution of Russian Studies After 1991." *Post-Soviet Affairs* 39 (1–2): 10–26.

Snegovaya, Maria. 2023. "Why Russia's Democracy Never Began." *Journal of Democracy* 34 (3): 105–118.

Taki, Victor. 2011. "Orientalism on the Margins: The Ottoman Empire Under Russian Eyes." *Kritika: Explorations in Russian and Eurasian History* 12 (2): 321–351.

Tsygankov, Andrei P. 2016. *Russia's Foreign Policy: Change and Continuity in National Identity*. Rowman & Littlefield.

Tsygankov, Andrei P., and Pavel A. Tsygankov. 2021. "Constructing National Values: The Nationally Distinctive Turn in Russian IR Theory and Foreign Policy." *Foreign Policy Analysis* 17 (4).

Tsygankov, Andrei P. 2024. *Russian Westernizers and Change in International Relations: The Promised West*. Taylor & Francis.

Tsygankov, Andrei P., and Pavel A. Tsygankov. 2023. "Three Traditions in Russian IR Theory: Westernism, Statism, and Eurasianism." In *The Routledge Handbook of Russian International Relations Studies*, edited by Maria Lagutina, Natalia Tsvetkova, Alexander Sergunin. *Routledge*.

Vasilaki, Rosa. 2012. "Provincialising IR? Deadlocks and Prospects in Post-Western IR Theory." *Millennium* 41 (1): 3–22.

Voskressenski, Alexei D. 2016. *Non-Western Theories of International Relations: Conceptualizing World Regional Studies*. Springer.

Wight, Colin. 2019. "Bringing the Outside In: The Limits of Theoretical Fragmentation and Pluralism in IR Theory." *Politics* 39 (1): 64–81.

Wilhelmsen, Julie. 2019. "Putin's Power Revisited: How Identity Positions and Great Power Interaction Condition Strategic Cooperation on Syria." *Europe-Asia Studies* 71 (7): 1091–1121.

Wilson, Andrew. 2023. "The Methodology of the 'Russian World' as the Technological Foundation for Ukrainophobia." *Centre for Democratic Integrity*. https://democratic-integrity.eu/the-methodology-of-the-russian-world/

Zarakol, Ayşe. 2022. *Before the West: The Rise and Fall of Eastern World Orders*. Cambridge University Press.

Zhuk, Sergei I. 2014. "Ukrainian Maidan as the Last Anti-Soviet Revolution, or the Methodological Dangers of Soviet Nostalgia (Notes of an American Ukrainian Historian from Inside the Field of Russian Studies in the United States)." *Ab Imperio* 2014 (3): 195–208.

CHAPTER 12

Beyond the Ivory Tower

Countering Simplistic Narratives about the Russo-Ukrainian War with Open-Source Intelligence (OSINT)

Anselm Schmidt and Sanshiro Hosaka

1 Introduction

The past decades have seen significant shortcomings in analyses of Russia and Ukraine in worldwide academia. From an overestimation of Russia's military strength to an understatement of its willingness to wage genocidal wars against its neighbours, scholars of Eastern Europe in general, and of Russia in particular, have too often ignored or overlooked signs of Russian aggression. At the same time, they have often downplayed the agency of actors in-between themselves and the Russian Federation, including that of Ukrainians, the Ukrainian state, and Ukrainian society. Russia's full-scale invasion of February 2022 has surprised many who thought to possess intimate knowledge about Russia and its foreign policy. It has also shed new light on a war that had been downplayed, ignored, or forgotten since Russia invaded Ukraine in 2014.

In this chapter, we trace flawed analyses of Russia's invasion of Ukraine by focusing on academic narratives about a "civil war" in Ukraine, denying or downplaying Russia's role in the armed conflict since 2014. These narratives were not only widely disseminated by Russia's overt propaganda and covert influence operations but also by unsuspecting academics among scholars worldwide. While many of these narratives are rooted in weak ontological, epistemological, and ideological perspectives on Russia and its neighbours, they can be countered by utilising novel methodologies from beyond the ivory tower of classical academia. Unfortunately, some scholars overemphasise their professional delineation from investigative journalism and demonstrate little willingness to learn from past mistakes. Others, however, are starting to realise the importance and embrace the potential of novel digital sources and methodologies.

Thus, we take inspiration from the field of investigative journalism and introduce two sets of new sources and methods for collecting and analysing empirical data. On the one hand, data provided by open-source intelligence (OSINT),

 | DOI:10.1163/9789004747364_013

vetted by investigative journalists and various online communities, has demonstrated Russia's responsibility for the war early on—including for shooting down flight MH17. They have been used in quests for truth and accountability for Russian war crimes worldwide—from Syria to Ukraine. However, their validity and reliability largely depend on the quality and transparency of journalistic work and volunteer communities. Authors who single-handedly pick data points to their liking engage with OSINT in name only.

On the other hand, large-scale data leaks and their analysis give rare glimpses into the Kremlin's inner workings and demonstrate that not Ukrainian division, but Russian political technology informed so-called separatism in Ukraine, negating widespread narratives about a supposed "civil war" at the core of the Russo-Ukrainian War since 2014. However, there has been limited discussion over the specific methodological and ethical challenges in utilising digital communications for research purposes that were unlawfully obtained and subsequently made publicly available by third parties, such as anonymous hackers. We will touch upon the added value but also some potential pitfalls scholars need to be aware of before utilising such novel sources and methodologies.

The chapter demonstrates how, if scholars overcome their reservations against sources and methodologies of neighbouring professions such as investigative journalism, they can become more aware of biassed narratives in academia which ultimately only benefit the aggressor and its war. If not, OSINT will nevertheless continue to give others a more reliable picture of Russian aggression worldwide, including in Ukraine.

2 "Civil War" in Ukraine: Simplistic Narratives and Flawed Methodologies

The academic debate over whether the armed conflict in Ukraine constitutes a "civil war," or a "Russian invasion" is well documented (for a history of these debates, see Hauter 2021). Some scholars argued that the conflict is driven by domestic dynamics within the Donbas, while others emphasised Moscow's involvement, framing it as an externally orchestrated event. Adding further weight to the latter argument, a recent report by the Joint Investigation Team (JIT), established by police and criminal justice authorities from the Netherlands, Australia, Belgium, Malaysia, and Ukraine to investigate the downing of MH17, demonstrates that the "Donetsk People's Republic" (DPR) was under the "overall control" of the Russian Federation. Notably, Alexander Borodai, a Russian citizen and then self-declared prime minister of the DPR, referred to a Russian government official as his "boss" during a phone conversation and acted under direct instructions from the Kremlin (JIT 2023, 10).

The civil war hypothesis is often backed by simplistic narratives and recurring stereotypes about Ukraine's historical, cultural, linguistic and political cleavages, or the concept of "two Ukraines" (for critiques, see Riabchuk 2002, 2015; Zhurzhenko 2014), which portrays the country as divided between a pro-European, Ukrainian-speaking West and a pro-Russian and Russian-speaking East. As Tymofii Brik (2021, 201–202) argues, scholars often propose theoretical models to explain empirical findings in Ukraine without assessing their actual applicability to the data. However, beyond this overly simplistic understanding by non-Ukrainian scholars, there are also methodological shortcomings in the way some Western scholars have approached the armed conflict in Ukraine.

One key issue is that scholars who failed to seriously engage with high-quality OSINT reports from major investigative groups such as Bellingcat, InformNapalm, the Conflict Intelligence Team (Toler 2018a), as well as Kremlin-related leaked materials like the Surkov leaks (Digital Forensic Research Lab 2016) and the Glazyev tapes (Reggae Mortis 2016), risk underestimating Moscow's leading role in the subversions and armed conflict in Eastern Ukraine. Such oversight can lead to an incomplete understanding of the conflict's external factors.

Jakob Hauter (2022) categorises 63 academic articles into five groups: 1) explicitly supports the civil war hypothesis; 2) implicitly supports the civil war hypothesis; 3) remains neutral; 4) implicitly supports the invasion hypothesis; and 5) explicitly supports the invasion hypothesis. We further examined whether these articles conducted independent OSINT analysis or cited dedicated OSINT reports, and whether they considered Kremlin-related leaked materials, or referenced reports using such materials (see Figure 12.1).

According to Hauter's classification, 13 peer-reviewed academic articles explicitly support the Ukraine civil war narrative. All of these (Tsygankov 2015; Davies 2016; Katchanovski 2016; Kudelia 2016; Loshkariov and Sushentsov 2016; Matveeva 2016; Robinson 2016; Sotiriou 2016; Strasheim 2016; Matsuzato 2017; Sakwa 2017; Giuliano 2018; Nitsova 2021) either disregard or underestimate reputable OSINT findings. Paul Robinson (2016, 12) cites Bellingcat's report on the origin of artillery attacks on Ukrainian military positions in Eastern Ukraine in the summer of 2014, but comments that "its methodology is controversial and far from universally accepted." Silviya Nitsova (2021, 1835–1836) references leaked materials such as the Surkov leaks and Glazyev tapes but argues that these leaks are "insufficient to determine the scale or effectiveness of Moscow's involvement," instead emphasising "the grassroots actions in

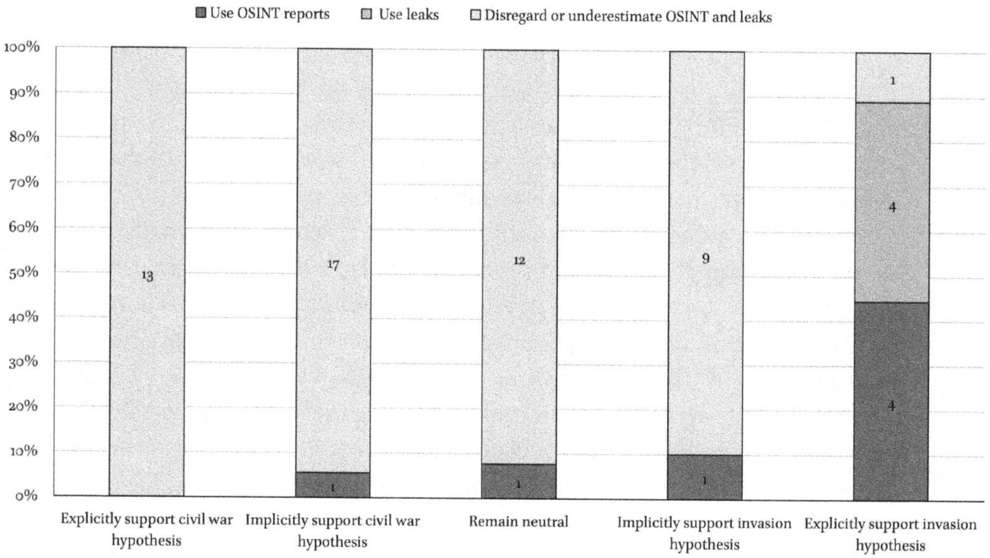

FIGURE 12.1 Assessments of the armed conflict in Ukraine and the use of OSINT and leaks
SOURCE: AUTHOR'S OWN COMPILATION BASED ON HAUTER 2022

Donbas" by citing Anna Matveeva (2018), who gives voice to so-called Russian Spring "rebels." Ivan Katchanovski (2016) references publicly available videos, photos, and radio intercepts, but closely aligns with Russia's messaging by disseminating a "Maidan massacre" conspiracy theory.[1]

In contrast, out of nine peer-reviewed articles that explicitly support the Russian invasion hypothesis, eight (Galeotti 2016; Wilson 2016; Bowen 2019; Hosaka 2019; Kuromiya 2019; Kuzio 2019; Melnyk 2019; Mykhnenko 2020) reference OSINT analyses from sources such as Bellingcat, InformNapalm, and the Conflict Intelligence Team, or secondary analyses of leaked materials, including the Surkov leaks or the Glazyev tapes. One article goes further by conducting its own in-depth analysis of the Surkov leaks (Hosaka 2019).

There are multiple reasons for scholars' reluctant engagement with such new sources of data and novel methodologies. On the one hand, there are limits to the scope of data and methodologies scholars can master which need to be taken into account. On the other hand, however, there are some self-imposed barriers which need to be challenged: (1) an overemphasis on academic delin-

1 Using publicly available data, such as videos and photos, Katchanovski argues that the Maidan massacre was a false-flag operation by the Ukrainian far-right. He misrepresents, however, the conclusions of ballistics experts' examinations and Ukraine's medical reports. See, for example, the critique by Rons (2022).

eation to neighbouring professions such as investigative journalism, (2) a tendency to value source material based on personal contacts, "access" to power and government officials, including overt propagandists and covert agents of influence, (3) over-emphasis on simplistic theories of world politics compared to empirical data on the ground, and (4) ignorance toward long-established findings among investigative journalists and OSINT communities.

While some influential scholars of Russian studies dismiss journalism per se as "popular writing" rather than serious analytical work (Frye 2021, 6), others begin to realise the importance of digital methods and sources used in investigative journalism due to the deficiencies of data sources in Russian studies, especially Russia's full-scale invasion of Ukraine (Gel'man 2023, 6–7).[2] Hauter (2022, 2023), in particular, has started developing methodology closely mirroring investigative journalism. Especially in authoritarian settings, the findings of these methods can be more reliable than focusing on officials and taking their statements at face value. "Access" to the powerful in fora such as the Valdai Discussion Club are active measures carefully scripted by Russian propagandists and "polit-technologists" to influence academia towards ignoring their acts of aggression (Bentzen 2018).

Simplistic theories of world politics, often grounded in theories of bipolarity after WW2, tend to ignore the agency of anyone in-between (for example, Mearsheimer 2014). However, without considering findings from OSINT and Kremlin-related leaks, scholars from both positivist and interpretivist traditions underestimated Russia's systematic interference in Ukraine's domestic policies and its responsibility for the "conflict" in Ukraine. Positivist epistemology failed to explicitly recognize the difficulty of accessing the data necessary to gauge the causal weight of external factors of the war (for example, Giuliano 2018), thus underestimating Russia's covert participation. In contrast, interpretivist traditions, analysing identity narratives of seemingly local actors, including their pronouncements and overt symbolic actions (for example, Laruelle 2016; Suslov 2017; Malyarenko and Wolff 2019), had difficulties to detect covert relationships between the local actors they were studying and the external actor who micromanaged the "people's republics". Russian warfare's "plausible deniability" (Jonsson and Seely 2015, 4) disoriented many scholars.

3 Attack on MH17: Ignoring Agency and Responsibility

One stark example of simplistic narratives leading to flawed academic contributions about the Russo-Ukrainian War before February 2022 evolves around

2 See also Vladimir Gel'man's contribution to this volume.

the attack on flight MH17 by Russian mercenaries, including "former" intelligence assets, using weaponry directly provided by the 53rd Anti-Aircraft Missile Brigade of the Russian Armed Forces. All 298 people aboard died, in addition to the already hundreds of Ukrainian casualties of Russia's war in Donbas at the time. Investigative journalists quickly identified many of the perpetrators and the weapon system responsible (Allen 2014, 2016; Huis 2019). Nearly all their findings would later be collaborated by the JIT (MH17 Animation 2016; JIT Press Meeting 2018; JIT 2023). Meanwhile, Russian propaganda went into overdrive—producing one conspiracy theory after another, often contradicting itself (Higgins 2020).

Academia struggled to keep up. Across disciplines, many authors chose to exclude questions of Russia's agency and responsibility for the attack. Legal scholars found various ways to avoid the issue of Russian aggression, invasion, and state-sponsored terrorism. For example, Mark Gibney (2015, 169) writes that MH17 was shot down only "allegedly" and ignores the overarching question of responsibility by framing "insurgents" and "rebels" in Donbas as independent from Russia. Sarah Williams (2016, 214) admits that "it has been suggested that the weapon was provided by Russian forces" but fails to provide sources or context beyond Russia's denial. Framing the attack on MH17 as a "tragedy", Marieke de Hoon (2017) removes all agency from the perpetrator. Tragedies happen—they are not committed, such as a crime, an attack, or an act of aggression.

Some jurists did incorporate the findings of various journalists into their analyses, consequently accusing Russia of becoming a state-sponsor of terrorism. Despite using the euphemism of "pro-Russian separatist forces", Sascha-Dominik Bachmann (2014, 14) mitigates it by citing evidence of "incriminating communications between the rebels and their Russian handlers." Others use similar euphemisms but discuss legitimate legal questions, such as how to circumvent Russia blocking any accountability in its role as a permanent member of the UN Security Council (for example, Ramsden 2017).

In political geography, Gerard Toal and John O'Loughlin (2018), despite acknowledging the findings of the JIT, contextualise the attack as part of "broader struggles over Ukraine," not a Ukrainian struggle against Russian aggression. Without adopting the "civil war" narrative per se, the authors fall for one of the primary goals of Russian propaganda adopting such narratives—denying subjectivity and agency to Ukraine. Continued ignorance about Ukrainian agency demonstrates the limits to novel data and methodologies when analyses are grounded in simplistic assumptions. This is further exemplified by Toal (2024)'s recent support for territorial concessions to appease Russia even after its war crimes in Ukraine, which has been heavily criticised by his colleagues (for example, Anghel 2024; Oksamytna 2024).

In media studies, Matt Sienkiewicz (2015, 221) uses the work of volunteers investigating MH17 online to attest to Ukraine "a valuable strategic dividend from online labourers" as opposed to "Russia's vast media and public relations resources". However, his analysis equates Russian propaganda with the work of global OSINT communities and journalists, playing into the hands of Russian disinformation by discarding the empirical data underlying open-source investigations of actual evidence. Sara Oates (2016), on the other hand, demonstrates how Russia's media coverage of the attack can be theorised without relativizing and falsely equating Russian propaganda with democratic practices in (and beyond) Ukraine. Other scholars in the field tried to avoid the issue of responsibility by focusing exclusively on the discursive level, disregarding vital lessons of contextualization from (critical) discourse analysis (see, for example, Mölder and Sazonov 2019; cf. Fairclough 2015).

In 2018, Manchester University Press published a whole volume of Russian propaganda narratives. Wrongly claiming that "whether the actual downing of Flight MH17 [...] involved a jet attack, an anti-aircraft missile, or both, ultimately cannot be established with certainty" years after multiple investigations established the opposite, Kees van der Pijl (2018, 159) spreads conspiracy theories connecting George Soros, Nord Stream, and other Gazprom pipelines to the attack on MH17. Ultimately, he blames a supposed "Kiev [*sic*!] war party", "US imperialist positions", and NATO for the "tragedy" of Russian mercenaries shooting down the plane (Pijl 2018, 8, 99, 160). Despite the scarcity of such brazen cases of propaganda in academic publications outside of Russia, even some security scholars acknowledging "Russian interference" are quick to blame Ukrainians. Reputable academic publishers, such as Brill and Wiley, accept articles blaming Kyiv's weakness for the attack on MH17 (Haar 2014) or claiming that "after years of stable relations with Russia, the new Crimean war came suddenly very close when an airplane full of tourists was suddenly shot from the sky in Europe's backyard [*sic*!]" (Kuipers, Verolme, and Muller 2020, 377).

In the end, Russian propaganda successfully muddied the waters and let large parts of academia ignore that Russian aggression was to blame for the killing of MH17's passengers and crew. Scholars rarely connected the attack on MH17 and the war in Donbas with Russia's invasion of Crimea, despite both being part of its wider war of aggression against Ukraine—its population, sovereignty, and territorial integrity. It was again up to journalists to track Russian spins and lies about the war in general and the attack on MH17 in particular (Nimmo 2016; Toler 2018b; Higgins 2020). Nevertheless, their tireless efforts started to pay off, with some scholars explicitly referencing journalists and their investigative work (see, for example, Rietjens 2019).

4 Towards Reliable OSINT: Community Standards and Verification

As the case of MH17 demonstrates, the vast amount of content available online has turned OSINT into a powerful tool for research, journalism, and international investigations (Fiorella 2021). It typically involves publicly available information that anyone can access without requiring special permissions (Berkeley Human Rights Center and UN High Commissioner for Human Rights 2022, 6). Nevertheless, its analysis might also involve data that is only commercially available—such as high-resolution satellite imagery (Vincent 2023). War crimes by Russia and its allies worldwide, including in Syria, have been met with increasingly advanced and sophisticated methods to analyse such content in search of truth and accountability (Ahmad 2024).

Just as Assad's war against his own population, Russian war crimes in Ukraine have demonstrated the dire need for facts and evidence countering the propaganda of dictatorships. Investigative journalists have made extensive use of OSINT in their reporting on Russian atrocities, such as investigations into the massacre of more than 500 civilians, including children, by Russian soldiers in the Kyiv suburb of Bucha (see, for example, Froliak, Al-Hlou, and Willis, Haley 2022). Acknowledging that Russia is committing similar atrocities all over Ukraine, various authors chose to highlight Bucha as an example of the genocidal nature of Russia's war against Ukraine, utilising a combination of privately gathered and openly available data (Vincent 2023).

One of the pitfalls for scholars in successfully engaging with OSINT is unfamiliarity with the specific methods of verification applied in different OSINT communities. Although community standards fulfil a similar function as peer review, they work very differently. With an unknown number of people involved, transparency about the underlying data and the methods of its analysis are key. Lacking knowledge about this process, authors might fall for OSINT in name only—as in the case of Katchanovski discussed earlier. However, cherry-picking data points to fit a predetermined narrative does not withstand public scrutiny, especially within communities gathering specific methodological and topical expertise. Nevertheless, even among experts, the same data can lead to differing conclusions. When satellite pictures between February and March 2022 showed thousands of Russian military vehicles on the road to Kyiv, some experts speculated about an upcoming siege (see, for example, Beaumont 2022; Specia 2022). Others took a closer look at the data and determined that the convoy, just as the whole invasion in the north, had simply stalled (see, for example, Mendick 2022; cf. Press and Libet 2023).[3]

3 For different interpretations of the same data by policymakers, see the conclusion of Kirill Shamiev's contribution to this volume.

One of the advantages of large OSINT communities is that they require transparency and accountability by design. A handful of people can build such communities only if they share their data, its collection, their methods, and their analyses. To be taken seriously by experts in their field, they need to publish their results with accompanying "proofs," including links to their sources and a reliable methodology, such as details used for geolocating footage (Toler 2022). One of the most widely cited OSINT analyses of Russia's war against Ukraine, Stijn Mitzer and Joost Oliemans' list of equipment losses during its full-scale invasion (Oryx 2022) would not have been possible without the ability for anyone to validate its findings. Linking one or multiple pictures of each loss allows for a vast community of online volunteers to submit new losses or identify duplicates on the list. It also allowed for Jakub Janovsky and other contributors to take over once the original authors did not have the resources to maintain it (Allison 2023; Mitzer and Oliemans 2023). It stands to argue that—even in the digital age—counting only visually confirmed losses will underreport actual losses despite some duplicates likely remaining in the data (York 2024). Nevertheless, Oryx has provided a sound basis for approximating Russian (and Ukrainian) losses. With both sides' officials providing unreliable data, it is often considered the gold standard for quantifying military losses in Ukraine.

Even though widely cited, Oryx's list can still be considered rudimentary compared to other OSINT communities. These, in turn, are complemented by individual "war mappers", geographically mapping the frontlines of Russia's war in Ukraine.[4] Geolocating losses makes the identification of duplicates more reliable by reducing the need to compare them to only those that appeared physically close to each other. In the end, the transparent approach followed by Oryx and other OSINT community projects allows for mutual control and complementation of each approach, as the classification of a picture on one site will necessarily lead to (re-)evaluation on the others. Mapping locations alone does not address the question of timing, as vehicles can be moved or stem from before the full-scale invasion. However, even here, more and more sophisticated methods are used, such as analysing weather patterns, foliage, and war-related changes in terrain assisted by publicly available or commercial satellite imagery (Baker 2020).

4 Large OSINT communities focused on military losses in Russia's full-scale invasion of Ukraine include WarSpotting (https://ukr.warspotting.net/) and lost_warinua (https://bit.ly/map_lostwarinua). Prominent "war mappers" of the invasion include Andrew Perpetua in the US (https://map.ukrdailyupdate.com/), DeepState in Ukraine (https://deepstatemap.live/), and @DefMon3 in Sweden (https://www.scribblemaps.com/maps/view/Operational_Map_Ukraine/nBT8ffpeGH).

OSINT communities vary in their political, ideological, and professional background. Many take a clear stance against the invasion, often avoiding the spread of information that would seriously impede the work of the Armed Forces of Ukraine, such as geolocating leaked Ukrainian troop movements as they happen. Many OSINT communities support fundraisers by Ukrainians and their military. However, even OSINT accounts with an "anti-Ukrainian bias" can provide valuable source material, although they often have to backtrack pro-Russian claims in the face of public scrutiny possible by adopting a relatively transparent methodology (@WarVehicle 2024). These are not to be confused with the ecosystem of Russian "milbloggers", such as FighterBomberPilot, Rybar, and WarGonzo. They openly support many Russian war crimes, do not share their methods and sources, are generally unreliable, sometimes on the Kremlin's payroll, and nevertheless constantly risk jail due to authoritarian "propaganda laws".

Of course, even community-driven verification is far from perfect, especially given selective publication strategies and increasingly sophisticated, including AI-generated, fake material. However, scholars' mapping of relevant communities, as well as the communities' own transparency, guidelines, expertise, and scrutiny, have the potential for a much closer approximation of reality than most officials are willing to provide. This is particularly true in a state of war, and especially in an authoritarian setting such as in contemporary Russia. Scholars can draw from their own expertise in theory, data collection, and traditional methodologies to enrich the work of OSINT communities by incorporating OSINT into their work.

Depending on their resources and capabilities, they are presented with multiple possibilities. Specialising in the field, Hauter (2022, 80–104, 2023) combines process tracing with novel methodologies for analysing what he calls "digital open source information (DOSI)". But scholars can also use more traditional approaches, such as interviews, to map OSINT communities and their reliability without learning the specifics of collecting or analysing digital data. They can rely on trustworthy journalists and volunteers, if they acquire sound judgement of whom to trust in their sources and methods. In the end, scholars' intensity of engagement can vary from merely evaluating OSINT outputs to participating in OSINT communities themselves. However, choosing ignorance towards OSINT might lead scholars to miss important data points in their analysis of Russian aggression and Ukrainian resistance, as well as the inner workings of Russia itself. Instead, scholars can learn from neighbouring professions, such as journalism, how to cooperate on equal footing in the "common effort" of collecting reliable data and conducting insightful analysis (Charlton, Mayer, and Ohme 2024).

Journalists by profession and OSINT communities by nature are driven by the news of the day. Constantly, there is an abundance of new information to be analysed as quickly as possible. Scholars, on the other hand, theoretically avoid chasing after the latest news cycle in their work. Instead, they should rely on investigating the past to analyse the present. In turn, they can avoid premature conclusions and provide deeper analysis. As discussed before, this has unfortunately not been the case in many analyses of Russia's war against Ukraine since 2014. Instead, their difference in approach often resulted in a wrong sense of entitlement to a deeper insight per se, ignoring important data points unearthed by mutual cooperation between investigative journalists and OSINT volunteers.

OSINT analysts primarily focus on the day-to day events by examining photos, videos, audio, and other digital materials. Russia's strategic and tactical plans in its political warfare against Ukraine, in contrast, were early on revealed through leaked emails from Kremlin officials.

5 Large-Scale Email Leaks: Rare Glimpses into the Kremlin's Inner Workings

Leaked emails, alongside audiovisual materials, have emerged as pivotal sources for examining the origins of Russia's war against Ukraine in 2014. They are particularly effective for analysing non-democratic regimes, where secrecy often impedes scholarly understanding. Leaks can expose the hidden workings of these regimes, jeopardising their efforts to maintain secrecy. Despite its potential, many scholars of Russian studies have yet to embrace leaked email data as a primary source (Hosaka 2025). Thus, the reluctance to engage with leaked data has led to a gap in understanding Russia's inner workings between scholars and practitioners.

The so-called Surkov leaks—data dumps from email accounts belonging to Vladislav Surkov, a key Russian presidential aide on Ukrainian affairs—revealed critical details about Russia's covert operations in Ukraine since 2014. The first (2,347 emails from the mailbox of prm_surkova@gov.ru) and second tranche (435 emails from pochta_mg@mail.ru) were published in October and November 2016. While many of the emails are routine reports, a meticulous comparison with open-source information revealed Surkov's orchestration of propaganda and disinformation efforts against Ukraine and the international community. These included manipulating public opinion through pseudo-experts, fake organisations, and various media techniques. Moscow promoted the "Novorossiya" myth, later rebranded as "Donbas," to portray recruited thugs as independent political actors resisting the Ukrainian government. This illusion aimed to legitimise Russian involvement while creating a narrative of the local identity of "rebels." The so-called "Donetsk People's Republic" and "Lugansk People's Republic," viewed by some scholars as non-state actors, were, in fact, artificial entities controlled by the Kremlin, with their creators comfortably issuing instructions from Moscow (Hosaka 2019).

Scholars who analyse overt speech acts of allegedly local political actors without considering Moscow's covert actions tend to overrate their political agency. For example, in November 2015, Alexander Kazakov, an adviser to "DPR" leader Alexander Zakharchenko, informed Surkov about US scholar Alexander J. Motyl (2015)'s article in *Foreign Affairs*, "When Ukraine Lost Donetsk: The World According to Alexander Zakharchenko". The article took the "DPR" leader's public statements at face value, portraying him as an independent political actor. Motyl argued that peace in eastern Ukraine depended on both Putin and Zakharchenko, highlighting Zakharchenko's claim that the fate of the Donbas was decided locally, not by external powers. Kazakov saw this as a triumph of Surkov's manipulation, as the article—based on a crafted public image of Zakharchenko—successfully reached Western policymakers through a respected international publication (Hosaka 2019, 764).[5]

The third tranche of the Surkov leaks, which contained communications from Surkov's deputy, Inal Ardzinba (1,046 emails from viktor_vinogr@mail.ru), was released in November 2017. This leak demonstrated how Russia continued subversive actions in 2015 and beyond to disinform Ukrainian and international audiences and discredit the Ukrainian government. The Russian Presidential Administration orchestrated and financed separatist projects not only in Donbas but also in Odesa, Kharkiv, Dnipro, Zaporizhzhia, and Mykolaiv under such innocuous banners as "Porto Franco," "special economic zone," "special ecological status," "all-Ukrainian constitutional forum", and "federalization" of Ukraine (Shandra and Seely 2019). Other valuable leaks include the Frolov leaks (Hosaka 2018) and the Rapoport leaks (Zolotukhin 2020), which provided further insights into Russian covert operations.

However, large data dumps from leaked emails have brought not only opportunities but also possible pitfalls for researchers. Major challenges include 1) the verification of the authenticity of the leaked emails, 2) the overestimation and underestimation of the significance of emails, 3) the problem of tainted leaks, and 4) the ethical concerns in using leaked data.

---

5 For a visual presentation of this OSINT investigation, see Hosaka 2020.

First, validating leaked accounts typically involves a comprehensive forensic examination. The 2016 hacking of the Democratic National Committee (DNI 2017) and the 2017 "Macron Leaks" (Schultheis 2017) are examples where the ownership of hacked accounts was confirmed by the account holders themselves. However, when alleged owners deny being hacked, as in the case of the Surkov leaks, the process becomes more complex. The Russian government rejected the very fact of these leaks altogether, as the leaked data implicated Russian interventions in Ukraine's sovereignty.

Nevertheless, gigabytes of data dumps, containing not only emails, but also numerous attached files, such as daily media monitoring reports and other briefs, confidential reports, CVs and valid metadata, are unfeasible to fabricate. Experts have validated the authenticity of the Surkov leaks, noting that trivial details, such as the identities of Surkov's secretaries, align with other available data. Techniques like online IP location finders and DKIM (DomainKeys Identified Mail) signature verification are essential tools for confirming the authenticity of emails, although DKIM verification leaves some types of emails unverifiable.

Along with identifying a possible account holder, it is important to determine the account's purpose and usage. For example, Ukrainian hacktivists published a dump of an email account attributed to Boris Rapoport, a former Kremlin official (007 2020). This account, containing around 27,500 messages for the period from May 2008 to March 2020, was assumed to have been primarily used as a subaccount for receiving and storing documents, given the fact that 99% of emails are incoming. The sheer volume of emails highlights the challenge researchers face in analysing such data, and experts suggest focusing on specific periods of interest based on the account owner's track record (Zolotukhin 2020).

Cross-referencing from multiple leaks can provide further insights and correct previous assumptions. For example, the Surkov leaks suggested that Surkov was responsible for appointing Alexander Zakharchenko as the leader of the "DPR," while the Rapoport leaks, which were published three years later, provided additional evidence of this arrangement by containing a biography of Zakharchenko sent to Surkov before his appointment became public knowledge (see Figure 12.2).

Second, the interpretation of digital evidence is not value-neutral and can be influenced by researchers' biases, potentially leading to either overestimation or underestimation of the data's significance. Analysts inclined towards sensationalism may misinterpret leaked confidential communications in ways that amplify their speculative claims. Conversely, researchers may underestimate the significance of relationships between covert political actors if they rely

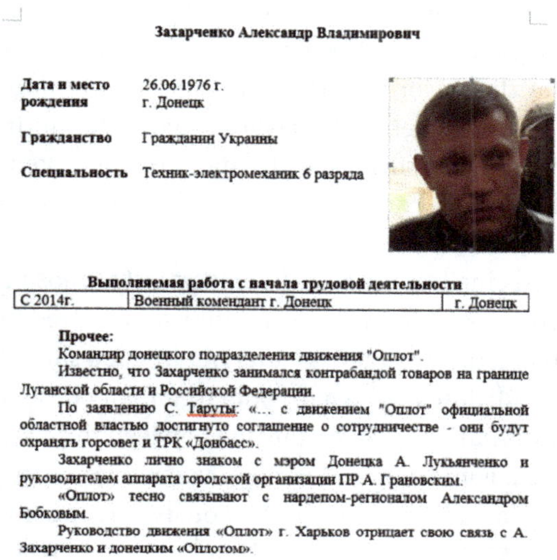

Захарченко Александр Владимирович

Дата и место рождения	26.06.1976 г. г. Донецк
Гражданство	Гражданин Украины
Специальность	Техник-электромеханик 6 разряда

Выполняемая работа с начала трудовой деятельности

С 2014г.	Военный комендант г. Донецк	г. Донецк

Прочее:

Командир донецкого подразделения движения "Оплот".

Известно, что Захарченко занимался контрабандой товаров на границе Луганской области и Российской Федерации.

По заявлению С. Таруты: «… с движением "Оплот" официальной областной властью достигнуто соглашение о сотрудничестве - они будут охранять горсовет и ТРК «Донбасс».

Захарченко лично знаком с мэром Донецка А. Лукьянченко и руководителем аппарата городской организации ПР А. Грановским.

«Оплот» тесно связывают с нардепом-регионалом Александром Бобковым.

Руководство движения «Оплот» г. Харьков отрицает свою связь с А. Захарченко и донецким «Оплотом».

FIGURE 12.2
Zakharchenko's biography as sent to Surkov before his appointment became public knowledge
SOURCE: SCREENSHOT OF LEAK FROM HTTPS://ORDILO.ORG/DAMP-E-MAIL-BORIS-RAMAIL-RU-YAKYI-NALEZHYT-B-RAPOPORTU-P/

too heavily on leaked emails, overlooking the possibility that highly sensitive decisions in foreign policy are likely to occur in private settings, bypassing official email accounts, which are frequently managed by secretaries.

The Surkov leaks, which revealed two email accounts—one official and one unofficial—administered by secretaries, did not include direct communications with key figures like Alexander Borodai, the self-proclaimed "DPR" prime minister in spring 2014. Researchers hypothesised that Surkov controlled Borodai indirectly through Konstantin Malofeev (see, for example, Hosaka 2019, 753). However, intercepted phone calls by the JIT (2023) on the MH17 downing confirmed that Surkov and Borodai had direct communications, demonstrating the limitations of relying solely on leaked emails to trace political relationships.

Third, stolen authentic digital documents can be selectively modified and published to discredit targeted individuals or organisations, a tactic experts refer to as "tainted leaks." Such leaks involve the manipulation of genuine data, often to distort facts or create misleading narratives. For instance, emails stolen from a prominent U.S. journalist and vocal critic of Putin were tampered with before being publicly "leaked" in an attempt to undermine the credibility of Russian opposition activists and journalists (Hulcoop et al. 2017). This form of information manipulation complicates efforts to verify the authenticity of leaked materials, posing significant challenges for researchers, although such manipulation may leave some detectable traces.

Fourth, using leaked data raises legal and ethical concerns, as the data is obtained without the owners' consent. Nonetheless, researchers can conduct studies ethically by thoroughly considering the data generation process, potential biases, and associated risks (Boustead and Herr 2020). However, Serge Egelman et al. (2012, 129) argue that even if data is available in the public domain, it remains private and raises ethical questions about its use in research. Thus, balancing the ethics of using hacked data with the potential benefits of disseminating knowledge for the greater public good is crucial.

Another issue with using leaked emails is the risk of inadvertently exposing the names and private email addresses of individuals not directly involved in the subject matter. On the one hand, the pursuit of greater public good should not come at the cost of harming individuals who have not given their consent for research. On the other, in authoritarian regimes, the line between the state and private sectors is often blurred. For instance, Russia's covert activities and disinformation campaigns engage a wide range of professionals who are not overtly linked to the state. Putin's regime utilises not only politicians, bureaucrats, and military personnel but also scholars, journalists, media specialists, religious figures, philanthropists, bankers, lawyers, artists, athletes, and other available societal resources (Hosaka 2019, 765).

These pitfalls have been addressed by investigative journalists to a certain extent. They often have direct access to confidential sources who share digital data and implement rigorous security measures to protect these sources. The publication of raw data risks exposing the identity of the sources and other individuals. Journalists, therefore, often verify the authenticity of leaked materials collectively, without disclosing the original data. For example, Delfi, an Estonian news website, obtained documents leaked from the Russian presidential administration—huge Excel files with dozens of hidden sheets, records of work meetings, as well as presentations and reports created in the Kremlin's Directorate of Domestic Policy. An analysis demonstrates secret presidential decrees, projects to ensure Putin's electoral victory, patriotic education, and blocking access to Western information sources not only in Russia but also in the occupied territories in Ukraine. These documents were shared and verified by journalists from other countries and explicated by external experts (Roonemaa 2024). In another example, leaked internal documents from the Fund for Support and Protection of the Rights of Compatriots Living Abroad (*Pravfond*) revealed that it spent millions of euros to finance propaganda and influence operations in 48 countries worldwide. The documents obtained by the Danish public broadcaster DR from a European intelligence source were shared within a consortium of European journalists (Roth 2024).

As with OSINT and its communities, scholars have the choice to directly engage with the leaked materials after incorporating their potential and pitfalls into their methodologies. More feasibly, they can rely on the work of investigative journalists to validate large-scale data leaks through mutual corroboration.

Conclusion

As evident throughout this volume, Russia's full-scale invasion of Ukraine has revealed serious flaws in many analyses of Russia and Ukraine before February 2022. Nowadays, instead of trying to construct a "civil war" with actors covertly controlled by the Kremlin, the Russian Armed Forces are openly operating on Ukrainian territory, and Russia continues its illegal annexations that started with the Ukrainian territories of Crimea and Sevastopol in 2014. Meaningfully analysing the current invasion and addressing academic shortcomings of the past is hardly possible without considering the entirety of the war between 2014 and today.

A review of academic articles from before the full-scale invasion reveals missed opportunities to utilise already available data which clearly demonstrated Russia's responsibility not only for its invasion and subsequent annexation of Crimea but also for the war in Donbas. Investigative journalists and OSINT communities used novel digital sources and methodologies to prove Russia's involvement in the attack on flight MH17. Leaked emails demonstrate Moscow's tight control over the so-called separatists in southeast Ukraine. Documented losses and analyses of satellite material continue to provide valuable insight into the diminishing strength of the Russian Armed Forces and counteract propaganda about their continuous success.

Digital investigations and datasets can also expand potential research questions, especially when scholars face methodological challenges and limitations in conducting fieldwork, interviews, data collection, and accessing informants in Russia and Ukraine—challenges that have been exacerbated since Russia's full-scale invasion (see, for example, Libman 2023; Smyth 2023).[6] While this topic is beyond the scope of the current chapter, it is worth noting that many investigative journalists also address Russia's domestic affairs. Numerous email dumps, such as the Rapoport leaks, reveal the Kremlin's manipulation of the information agenda within Russia, yet these resources have not yet been fully utilised.

We have discussed how scholars can benefit from OSINT and leaked materials, particularly those researching Russia's war against Ukraine. However, we do

6 See also Francoise Dauce's contribution to this volume.

not suggest that all scholars must master OSINT methodologies (though they certainly can). Rather, our argument is that scholars studying Russia should have a foundational understanding of the processes by which this new type of knowledge is generated. Novel sources and methodologies also come with new challenges, and they need to consider the limitations in verifying digitally collected and leaked materials. Being able to assess their credibility, scholars can integrate findings from credible OSINT communities and investigative journalists that meet specific criteria of validation and serve the purposes of their research questions into their empirical data.

Although academia and journalism serve distinct purposes—academia seeks to generate knowledge, while journalism aims to establish facts—Nicolas Kayser-Bril (2018, 39) argues that investigative journalism, through the development of structured methodologies, can produce falsifiable knowledge. Digital investigative journalism can enhance the quality of academic research by providing factual insights, a task that academia has sometimes struggled to achieve, as in the context of Russia's undeclared war against Ukraine. Digital sources independent from governments serve as a counterbalance to official data and narratives, especially in times of war and particularly in the context of autocracies.

References

007. 2020. 'Damp e-mail boris.ra@mail.ru, yakyy nalezhyt' B. Rapoportu—pomichnyku Surkova-Putina [Mail Dump boris.ra@mail.ru, Belonging to B. Rapoport—an Assistant of Surkov-Putin]'. *Ordilo*. https://ordilo.org/damp-e-mail-boris-ramail-ru-yakyi-nalezhyt-b-rapoportu-p/.

Ahmad, Muhammad Idrees. 2024. 'Open Source Journalism, Misinformation and the War for Truth in Syria'. In *Open Source Investigations in the Age of Google*, Security Science and Technology, eds. Henrietta Wilson, Olamide Samuel, and Dan Plesch. New Jersey: World Scientific, 133–151. https://doi.org/10.1142/9781800614079_0007.

Allen, Timmi. 2014. *MH17: Source of the Separatists' Buk*. Amsterdam: Bellingcat. Bellingcat Investigation. https://www.bellingcat.com/app/uploads/2014/11/Origin-of-the-Separatists-Buk-A-Bellingcat-Investigation1.pdf.

Allen, Timmi. 2016. *The Lost Digit: Buk 3x2*. Amsterdam: Bellingcat. Bellingcat Investigation. https://www.bellingcat.com/wp-content/uploads/2016/05/The-lost-digit-BUK-3x2_EN_final.pdf.

Allison, George. 2023. 'Analyst Sheds Light on Ukraine and Challenges in OSINT'. *UK Defence Journal*. https://ukdefencejournal.org.uk/analyst-sheds-light-on-ukraine-and-challenges-in-osint/.

Anghel, Veronica. 2024. 'How Wars Don't End: A Response to Gerard Toal's Analysis of Ceasefire Negotiations in Ukraine'. *Environment and Planning C: Politics and Space*: 23996544241276298. https://doi.org/10.1177/23996544241276298.

Bachmann, Sascha-Dominik. 2014. 'Malaysia Airlines Flight MH17: The Day Russia Became a Sponsor of Terrorism'. *Amicus Curiae: Journal of the Society for Advanced Legal Scholars* (97): 14–16.

Baker, Rae. 2020. 'How Nature May Reveal Secrets in Geolocation'. *Medium.* https://wondersmithrae.medium.com/how-nature-may-reveal-secrets-in-geolocation-e34dc9f42ed.

Beaumont, Peter. 2022. 'Vast Russian Military Convoy May Be Harbinger of a Siege of Kyiv'. *The Guardian.* https://www.theguardian.com/world/2022/mar/01/vast-russian-military-convoy-kyiv-siege-ukraine.

Bentzen, Naja. 2018. *Foreign Influence Operations in the EU.* Brussels: European Parliament Research Service. Briefing. http://www.europarl.europa.eu/RegData/etudes/BRIE/2018/625123/EPRS_BRI(2018)625123_EN.pdf.

Berkeley Human Rights Center and UN High Commissioner for Human Rights. 2022. *Berkeley Protocol on Digital Open Source Investigations.* New York. Protocol. https://humanrights.berkeley.edu/wp-content/uploads/2024/02/Berkeley-Protocol.pdf.

Boustead, Anne E., and Trey Herr. 2020. 'Analyzing the Ethical Implications of Research Using Leaked Data'. *PS: Political Science & Politics* 53(3): 505–509. https://doi.org/10.1017/S1049096520000323.

Bowen, Andrew S. 2019. 'Coercive Diplomacy and the Donbas: Explaining Russian Strategy in Eastern Ukraine'. *Journal of Strategic Studies* 42(3–4): 312–343. https://doi.org/10.1080/01402390.2017.1413550.

Brik, Tymofii. 2021. 'The Donbas and Social Science: Terra Incognita?' In *Civil War? Interstate War? Hybrid War? Dimensions and Interpretations of the Donbas Conflict in 2014–2020*, Soviet and Post-Soviet Politics and Society, ed. Jakob Hauter. Stuttgart: Ibidem, 191–214.

Charlton, Timothy, Anna-Theresa Mayer, and Jakob Ohme. 2024. 'A Common Effort: New Divisions of Labor Between Journalism and OSINT Communities on Digital Platforms'. *The International Journal of Press/Politics*: 1–22. https://doi.org/10.1177/19401612241271230.

Davies, Lance. 2016. 'Russia's "Governance" Approach: Intervention and the Conflict in the Donbas'. *Europe-Asia Studies* 68(4): 726–749. https://doi.org/10.1080/09668136.2016.1173652.

Digital Forensic Research Lab. 2016. 'Breaking Down the Surkov Leaks'. *Medium.* https://medium.com/dfrlab/breaking-down-the-surkov-leaks-b2feec1423cb.

DNI. 2017. *Background to 'Assessing Russian Activities and Intetions in Recent US Elections': The Analytic Process and Cyber Incident Attribution.* Washington, DC: Director of National Intelligence. https://www.dni.gov/files/documents/ICA_2017_01.pdf.

Egelman, Serge, Joseph Bonneau, Sonia Chiasson, David Dittrich, and Stuart Schechter. 2012. 'It's Not Stealing If You Need It: A Panel on the Ethics of Performing Research Using Public Data of Illicit Origin'. In *Financial Cryptography and Data Security*, Lecture Notes in Computer Science, eds. Jim Blyth, Sven Dietrich, and L. Jean Camp. Berlin: Springer, 124–132. https://doi.org/10.1007/978-3-642-34638-5_11.

Fairclough, Norman. 2015. *Language and Power*. 3rd ed. London: Routledge.

Fiorella, Giancarlo. 2021. 'First Steps to Getting Started in Open Source Research'. *Bellingcat*. https://www.bellingcat.com/resources/2021/11/09/first-steps-to-getting-started-in-open-source-research/.

Froliak, Masha, Yousur Al-Hlou, and Alistair Willis, Haley. 2022. 'Their Final Moments: Victims of a Russian Atrocity in Bucha'. *The New York Times*. https://www.nytimes.com/interactive/2022/12/21/world/europe/bucha-ukraine-massacre-victims.html.

Frye, Timothy. 2021. *Weak Strongman: The Limits of Power in Putin's Russia*. Princeton: Princeton University Press.

Galeotti, Mark. 2016. 'Hybrid, Ambiguous, and Non-Linear? How New Is Russia's "New Way of War"?' *Small Wars & Insurgencies* 27(2): 282–301. https://doi.org/10.1080/09592318.2015.1129170.

Gel'man, Vladimir. 2023. 'Exogenous Shock and Russian Studies'. *Post-Soviet Affairs* 39(1–2): 1–9. https://doi.org/10.1080/1060586X.2022.2148814.

Gibney, M. 2015. 'The Downing of MH17: Russian Responsibility?' *Human Rights Law Review* 15(1): 169–178. https://doi.org/10.1093/hrlr/ngu036.

Giuliano, Elise. 2018. 'Who Supported Separatism in Donbas? Ethnicity and Popular Opinion at the Start of the Ukraine Crisis'. *Post-Soviet Affairs* 34(2–3): 158–178. https://doi.org/10.1080/1060586X.2018.1447769.

Haar, Barend ter. 2014. 'Lessons from the MH17 Disaster'. *Security and Human Rights* 25(4): 454–461. https://doi.org/10.1163/18750230-02504007.

Hauter, Jakob, ed. 2021. *Civil War? Interstate War? Hybrid War? Dimensions and Interpretations of the Donbas Conflict in 2014–2020*. Stuttgart: Ibidem.

Hauter, Jakob. 2022. 'A Digital Open Source Investigation of How War Begins: Ukraine's Donbas in 2014'. PhD Thesis. University College London. https://discovery.ucl.ac.uk/id/eprint/10145765/2/Jakob-Hauter-PhD-Thesis-Final.pdf.

Hauter, Jakob. 2023. 'Forensic Conflict Studies: Making Sense of War in the Social Media Age'. *Media, War & Conflict* 16(2): 153–172. https://doi.org/10.1177/17506352211037325.

Higgins, Eliot. 2020. 'The Impossible Conspiracy Fueling Russia's MH17 Defence'. *Bellingcat*. https://www.bellingcat.com/news/uk-and-europe/2020/03/09/the-impossible-conspiracy-fuelling-russias-mh17-defence/.

Hoon, Marieke de. 2017. 'Navigating the Legal Horizon: Lawyering the MH17 Disaster'. *Utrecht Journal of International and European Law* 33(84): 90–119. https://doi.org/10.5334/ujiel.368.

Hosaka, Sanshiro. 2018. 'The Kremlin's Active Measures Failed in 2013: That's When Russia Remembered Its Last Resort—Crimea'. *Demokratizatsiya* 26(3): 321–364.

Hosaka, Sanshiro. 2019. 'Welcome to Surkov's Theater: Russian Political Technology in the Donbas War'. *Nationalities Papers* 47(5): 750–773. https://doi.org/10.1017/nps.2019.70.

Hosaka, Sanshiro. 2020. 'Political Technology in International Conflicts'. https://rgdoi.net/10.13140/RG.2.2.34577.34409.

Hosaka, Sanshiro. 2025. 'Leaked Email Data: A New Source for the Study of Authoritarian Regimes'. *Digital War* 6(1): 1. https://doi.org/10.1057/s42984-024-00097-w.

Huis, Pieter van. 2019. *'A Birdie Is Flying towards You': Identifying the Separatists Linked to the Downing of MH17*. Amsterdam: Bellingcat. Investigation. https://www.bellingcat.com/app/uploads/2019/06/a-birdie-is-flying-towards-you.pdf.

Hulcoop, Adam, John Scott-Railton, Peter Tanchak, Matt Brooks, and Ron Deibert. 2017. *Tainted Leaks: Disinformation and Phishing with a Russian Nexus*. Toronto: The Citizen Lab. https://tspace.library.utoronto.ca/bitstream/1807/97197/3/Report%2392--Taintedleaks.pdf.

JIT. 2023. *Findings of the JIT MH17 Investigation into the Crew Members of the Buk TELAR and Those Responsible in the Chain of Command*. The Hague: Openbaar Ministerie. Report. https://www.prosecutionservice.nl/topics/mh17-plane-crash/documents/publications/mh17/map/2023/report-mh17.

JIT MH17 Press Meeting MH17, May 24, 2018. 2018. Bunnik. https://www.youtube.com/watch?v=fqdW3_opPr4.

Jonsson, Oscar, and Robert Seely. 2015. 'Russian Full-Spectrum Conflict: An Appraisal after Ukraine'. *The Journal of Slavic Military Studies* 28(1): 1–22. https://doi.org/10.1080/13518046.2015.998118.

Katchanovski, Ivan. 2016. 'The Separatist War in Donbas: A Violent Break-up of Ukraine?' *European Politics and Society* 17(4): 473–489. https://doi.org/10.1080/23745118.2016.1154131.

Kayser-Bril, Nicolas. 2018. 'Measuring the Unmeasured with Data'. In *Digital Investigative Journalism*, eds. Oliver Hahn and Florian Stalph. Cham: Springer, 31–41. https://doi.org/10.1007/978-3-319-97283-1_4.

Kudelia, Serhiy. 2016. 'The Donbas Rift'. *Russian Politics & Law* 54(1): 5–27. https://doi.org/10.1080/10611940.2015.1160707.

Kuipers, Sanneke, Ellen Verolme, and Erwin Muller. 2020. 'Lessons from the MH-17 Transboundary Disaster Investigation'. *Journal of Contingencies and Crisis Management* 28(4): 376–385. https://doi.org/10.1111/1468-5973.12287.

Kuromiya, Hiroaki. 2019. 'The War in the Donbas in Historical Perspective'. *The Soviet and Post-Soviet Review* 46(3): 245–262. https://doi.org/10.1163/18763324-04603003.

Kuzio, Taras. 2019. 'Russian Stereotypes and Myths of Ukraine and Ukrainians and Why Novorossiya Failed'. *Communist and Post-Communist Studies* 52(4): 297–309. https://doi.org/10.1016/j.postcomstud.2019.10.007.

Laruelle, Marlène. 2016. 'The Three Colors of Novorossiya, or the Russian Nationalist Mythmaking of the Ukrainian Crisis'. *Post-Soviet Affairs* 32(1): 55–74. https://doi.org/10.1080/1060586X.2015.1023004.

Libman, Alexander. 2023. 'Credibility Revolution and the Future of Russian Studies'. *Post-Soviet Affairs* 39(1–2): 60–69. https://doi.org/10.1080/1060586X.2022.2148446.

Loshkariov, Ivan D., and Andrey A. Sushentsov. 2016. 'Radicalization of Russians in Ukraine: From "Accidental" Diaspora to Rebel Movement'. *Southeast European and Black Sea Studies* 16(1): 71–90. https://doi.org/10.1080/14683857.2016.1149349.

Malyarenko, Tetyana, and Stefan Wolff. 2019. *The Dynamics of Emerging De-Facto States: Eastern Ukraine in the Post-Soviet Space*. London: Routledge.

Matsuzato, Kimitaka. 2017. 'The Donbass War: Outbreak and Deadlock'. *Demokratizatsiya* 25(2): 175–201.

Matveeva, Anna. 2016. 'No Moscow Stooges: Identity Polarization and Guerrilla Movements in Donbass'. *Southeast European and Black Sea Studies* 16(1): 25–50. https://doi.org/10.1080/14683857.2016.1148415.

Matveeva, Anna. 2018. *Through Times of Trouble: Conflict in Southeastern Ukraine Explained from Within*. Lanham: Lexington Books.

Mearsheimer, John J. 2014. 'Why the Ukraine Crisis Is the West's Fault: The Liberal Delusions That Provoked Putin'. *Foreign Affairs* 93(5): 77–89.

Melnyk, Oleksandr. 2019. 'From the "Russian Spring" to the Armed Insurrection: Russia, Ukraine and Political Communities in the Donbas and Southern Ukraine'. *The Soviet and Post-Soviet Review* 47(1): 3–38. https://doi.org/10.1163/18763324-04603009.

Mendick, Robert. 2022. 'How Russia's Vast Military Convoy Was Ambushed on the Road to Kyiv'. *The Telegraph*. https://www.telegraph.co.uk/world-news/2022/02/28/russias-17-mile-convoy-thwarted-road-kyiv/.

MH17 Animation Regarding the Transport Route and the Launch Site. 2016. https://www.youtube.com/watch?v=Sf6gJ8NDhYA.

Mitzer, Stijn, and Joost Oliemans. 2023. 'Putting Down the Pen: Reflecting on Oryx's Journey'. *Oryx*. https://www.oryxspioenkop.com/2023/08/putting-down-pen-reflecting-on-oryxs.html.

Mölder, Holger, and Vladimir Sazonov. 2019. 'The Impact of Russian Anti-Western Conspiracy Theories on the Status-Related Conflict in Ukraine: The Case of Flight MH17'. *Baltic Journal of European Studies* 9(3): 96–115. https://doi.org/10.1515/bjes-2019-0024.

Motyl, Alexander J. 2015. 'When Ukraine Lost Donetsk: The World According to Alexander Zakharchenko'. *Foreign Affairs*. https://www.foreignaffairs.com/articles/ukraine/2015-11-22/when-ukraine-lost-donetsk.

Mykhnenko, Vlad. 2020. 'Causes and Consequences of the War in Eastern Ukraine: An Economic Geography Perspective'. *Europe-Asia Studies* 72(3): 528–560. https://doi.org/10.1080/09668136.2019.1684447.

Nimmo, Ben. 2016. 'How MH17 Gave Birth to the Modern Russian Spin Machine'. *Foreign Policy*. https://foreignpolicy.com/2016/09/29/how-mh17-gave-birth-to-the-modern-russian-spin-machine-putin-ukraine/.

Nitsova, Silviya. 2021. 'Why the Difference? Donbas, Kharkiv and Dnipropetrovsk after Ukraine's Euromaidan Revolution'. *Europe-Asia Studies* 73(10): 1832–1856. https://doi.org/10.1080/09668136.2021.1912297.

Oates, Sarah. 2016. 'Russian Media in the Digital Age: Propaganda Rewired'. *Russian Politics* 1(4): 398–417. https://doi.org/10.1163/2451-8921-00104004.

Oksamytna, Kseniya. 2024. 'The Moral and Strategic Clarity of Supporting Ukraine's Self-Defense: Why Accepting Russian Colonialism Should Remain a Taboo'. *Environment and Planning C: Politics and Space*: 23996544241277319. https://doi.org/10.1177/23996544241277319.

Oryx. 2022. 'Attack on Europe: Documenting Russian Equipment Losses during the Russian Invasion of Ukraine'. *Oryx*. https://www.oryxspioenkop.com/2022/02/attack-on-europe-documenting-equipment.html.

Pijl, Kees van der. 2018. *Flight MH17, Ukraine and the New Cold War: Prism of Disaster*. Manchester: Manchester University Press.

Press, Claire, and Svitlana Libet. 2023. 'How Russia's 35-Mile Armoured Convoy Ended in Failure'. *BBC*. https://www.bbc.com/news/world-europe-64664944.

Ramsden, Michael. 2017. 'Uniting for MH17'. *Asian Journal of International Law* 7(2): 337–360. https://doi.org/10.1017/S2044251316000217.

Reggae Mortis. 2016. 'Kto est' kto v proslushke peregovorov Sergeia Glaz'eva [Who is Who in the Wiretapping of the Negotiations of Sergei Glazyev]'. *Conflict Intelligence Team*. https://citeam.org/glazyev-whoiswho/.

Riabchuk, Mykola. 2002. 'Ukraine: One State, Two Countries? With Comments'. *Tr@nsit online* 23. https://www.iwm.at/transit-online/ukraine-one-state-two-countries-with-comments.

Riabchuk, Mykola. 2015. '"Two Ukraines" Reconsidered: The End of Ukrainian Ambivalence?' *Studies in Ethnicity and Nationalism* 15(1): 138–156. https://doi.org/10.1111/sena.12120.

Rietjens, Sebastiaan. 2019. 'Unraveling Disinformation: The Case of Malaysia Airlines Flight MH17'. *The International Journal of Intelligence, Security, and Public Affairs* 21(3): 195–218. https://doi.org/10.1080/23800992.2019.1695666.

Robinson, Paul. 2016. 'Russia's Role in the War in Donbass, and the Threat to European Security'. *European Politics and Society* 17(4): 506–521. https://doi.org/10.1080/23745118.2016.1154229.

Rons, Ian. 2022. 'A Load of Ballistics'. *The Daily Sceptic*. https://dailysceptic.org/2022/07/17/a-load-of-ballistics/.

Roonemaa, Holger. 2024. 'Kremlin Leaks: Secret Files Reveal How Putin Pre-Rigged His Reelection'. *VSquare.org*. https://vsquare.org/kremlin-leaks-putin-elections-russia-propaganda-ukraine/.

Roth, Andrew. 2024. 'Revealed: Russian Legal Foundation Linked to Kremlin Activities in Europe'. *The Guardian*. https://www.theguardian.com/world/article/2024/jun/02/revealed-russian-legal-defence-foundation-pravfond-europe.

Sakwa, Richard. 2017. 'The Ukraine Syndrome and Europe: Between Norms and Space'. *The Soviet and Post-Soviet Review* 44(1): 9–31. https://doi.org/10.1163/18763324-04401003.

Schultheis, Emily. 2017. 'The Macron Leaks Probably Came Too Late to Change the French Election'. *The Atlantic*. https://www.theatlantic.com/international/archive/2017/05/france-macron-leak-hack/525738/.

Shandra, Alya, and Robert Seely. 2019. *The Surkov Leaks: The Inner Workings of Russia's Hybrid War in Ukraine*. London: Royal United Services Institute for Defence and Security Studies. RUSI Occasional Paper. https://static.rusi.org/201907_op_surkov_leaks_web_final.pdf.

Sienkiewicz, Matt. 2015. 'Open BUK: Digital Labor, Media Investigation and the Downing of MH17'. *Critical Studies in Media Communication* 32(3): 208–223. https://doi.org/10.1080/15295036.2015.1050427.

Smyth, Regina. 2023. 'Plus Ça Change: Getting Real about the Evolution of Russian Studies after 1991'. *Post-Soviet Affairs* 39(1–2): 10–26. https://doi.org/10.1080/1060586X.2022.2161232.

Sotiriou, Stylianos A. 2016. 'The Irreversibility of History: The Case of the Ukrainian Crisis (2013–2015)'. *Southeast European and Black Sea Studies* 16(1): 51–70. https://doi.org/10.1080/14683857.2016.1150700.

Specia, Megan. 2022. 'Analysts Say Russian Convoy Likely Signals a Shift to Invasion Strategy.' *The New York Times*. https://www.nytimes.com/2022/03/01/world/europe/russian-convoy-ukraine.html.

Strasheim, Julia. 2016. 'Power-Sharing, Commitment Problems, and Armed Conflict in Ukraine'. *Civil Wars* 18(1): 25–44. https://doi.org/10.1080/13698249.2016.1144494.

Suslov, Mikhail. 2017. 'The Production of "Novorossiya": A Territorial Brand in Public Debates'. *Europe-Asia Studies* 69(2): 202–221. https://doi.org/10.1080/09668136.2017.1285009.

Toal, Gerard. 2024. 'The Territorial Taboo: Explaining the Public Aversion to Negotiations in the Ukraine War Support Coalition'. *Environment and Planning C: Politics and Space*: 23996544241268335. https://doi.org/10.1177/23996544241268335.

Toal, Gerard, and John O'Loughlin. 2018. '"Why Did MH17 Crash?": Blame Attribution, Television News and Public Opinion in Southeastern Ukraine, Crimea and the De Facto States of Abkhazia, South Ossetia and Transnistria'. *Geopolitics* 23(4): 882–916. https://doi.org/10.1080/14650045.2017.1364238.

Toler, Aric. 2018a. 'Crowdsourced and Patriotic Digital Forensics in the Ukrainian Conflict'. In *Digital Investigative Journalism*, eds. Oliver Hahn and Florian Stalph. Cham: Springer, 203–215. https://doi.org/10.1007/978-3-319-97283-1_19.

Toler, Aric. 2018b. 'The Kremlin's Shifting, Self-Contradicting Narratives on MH17'. *Bellingcat*. https://www.bellingcat.com/news/uk-and-europe/2018/01/05/kremlins-shifting-self-contradicting-narratives-mh17/.

Toler, Aric. 2022. 'How We Geolocated a Photo of a Russian Missile Programming Team'. *Bellingcat*. https://www.bellingcat.com/resources/2022/10/28/how-we-geolocated-a-photo-of-a-russian-missile-programming-team/.

Tsygankov, Andrei. 2015. 'Vladimir Putin's Last Stand: The Sources of Russia's Ukraine Policy'. *Post-Soviet Affairs* 31(4): 279–303. https://doi.org/10.1080/1060586X.2015.1005903.

Vincent, Brandi. 2023. '"Have the Kids Come Home?": How Open Source Technologies Are Reshaping Conflict and Humanitarian Operations'. *DefenseScoop*. https://defensescoop.com/2023/08/11/have-the-kids-come-home-how-open-source-technologies-are-reshaping-conflict-and-humanitarian-operations/.

@WarVehicle. 2024. 'Tweet'. *Twitter*. https://x.com/WarVehicle/status/1825163596176752890.

Williams, Sarah. 2016. 'MH17 and the International Criminal Court: A Suitable Venue?' *Melbourne Journal of International Law* 17(1): 210–237.

Wilson, Gary. 2016. 'Secession and Intervention in the Former Soviet Space: The Crimean Incident and Russian Interference in Its "Near Abroad"'. *Liverpool Law Review* 37(3): 153–175. https://doi.org/10.1007/s10991-016-9187-x.

York, Chris. 2024. 'Russian Equipment Losses in Ukraine Surpass 15,000, Says Oryx OSINT Group'. *The Kyiv Independent*. https://kyivindependent.com/russian-equipment-losses-in-ukraine-surpass-15-000-says-oryx-osint-group/.

Zhurzhenko, Tatiana. 2014. 'A Divided Nation? Reconsidering the Role of Identity Politics in the Ukraine Crisis'. *Die Friends-Warte* 89(1–2): 249–267.

Zolotukhin, Dmytro. 2020. 'Vzlom Borisa Rapoporta, Chast' 1: Ch'i ushi torchat? [Hacking Boris Rapoport, Part 1: Whose Ears Stick Out?]'. *Detektor Media*. https://detector.media/infospace/article/177879/2020-06-11-vzlom-borysa-rapoporta-chast-1-chy-ushy-torchat/.

Conclusion

Ivan Gomza and Andrey Makarychev

The volume is a collaborative effort of authors who, in one or another respect, embody the peripheries of the erstwhile imperial Russian state. Some contributors are affiliated with institutions situated on the "imperial fringes", such as the University of Helsinki, the International Centre of Defence and Security in Tallinn, and the University of Tartu, while others represent the periphery through their ethnic backgrounds, with many hailing from Belarusian, Kazakh, or Ukrainian descent. In the 21st century, the independent states of Lithuania, Estonia, Ukraine, Belarus, and Kazakhstan became for the Kremlin first the laboratory to try out the alchemy of the "Russian world" ideology. Eventually, they morphed into the "lost Russian bastions" to be reclaimed. Taken collectively, therefore, the authors might be considered the "privileged outsiders" scrutinising Russia's destructive neo-imperial agenda. The peculiarity of such a dual positionality is that although being intimately linked to Russia—by mastery of the Russian language, upbringings, and social ties—the authors nevertheless are estranged from the typical Russian vantage point, which enables them to suggest some novel ways to studying Russia anew.

There are both methodological and normative reasons for this endeavour, each amply explored in the present volume. The task of this concluding chapter is to highlight for the readers the significant intersections that emerge from the plethora of arguments advanced by the individual contributors. These overlaps are likely valuable for Political Science, International Relations, and History students whose interests situate outside—or on the outer limits of—the traditional scope of "Russian Studies". It might also interest the general readership looking for a fresh take on the already-known questions.

When considered holistically, this volume treats five essential topics and their nuanced applications: (1) a plea to recalibrate the study of Eastern Europe, (2) an urgent need to account for distorted knowledge-generation processes within the discipline, (3) new vistas emerging within the field, (4) possibilities for comparative research, and (5) politicisation and depoliticisation of Russian Studies. The eruption of the Russo-Ukrainian war in 2022, the scope of physical destruction and human misery it caused, and the geopolitical shockwaves it provoked intensified and made these four questions even more pressing and meaningful.

 | DOI:10.1163/9789004747364_014

1 Getting Eastern Europe Outside of Russia's Shadow

When on February 24, 2022, Russia's troops invaded Ukraine, and Putin branded Ukraine an artificially created hostile "anti-Russia", the broader context of Russia's war aims became indisputable: the Kremlin was intent on restoring its supremacy over the biggest East European country. Historically, East Europe played the role of a landed bridge, which Moscow used to project its power into Europe, hence Russia's fixation on territorial control over this region. However, viewing East Europe solely as a borderland / "bloodland" between Europe and Russia deprives regional nations of both identity and agency. Yet historically, the broad (and utterly inaccurate) catchwords of Orthodoxy, Slavic ethnicity, shared cultural heritage, or "post-Communism" camouflaged the mosaic-like variety of Eastern Europe incommensurable with Russia. Far too often, the general public and academia prefer to treat nations eastwards of Vienna as "variations on a Russian theme" and thus open to the transposition of tropes, concepts, and visions applicable to Russia. One of the unforeseeable outcomes of the 2022 invasion was a set of initiatives to understand the conflict, its participants, its historical and regional dimensions, and its challenges that eventually overarched as a growing scholarly consensus to decentre Russia. Although sometimes erroneously rendered as a quest to decentralise (even dismember) Russia as a political entity, the endeavour focuses on epistemological de-privilegisation and de-exceptionalisation of Russia, namely a move away from default adoption of Russian interpretation of historical events, naming, values, and preferences when dealing with history, politics, and culture of Eastern Europe.

Most authors of the present volume concur that it is imperative to decentre Russia. Some do it explicitly and with gusto (Chapters 2, 6, 7 and 9), others prefer more cautious approach being aware of limitations of the initiative (Chapters 3, 4, and 8); nevertheless, it is a consensus that Lithuanian, Hungarian, Polish, Ukrainian, or Crimean Tatar vantage points on the historical encounters and current political developments at the genuine frontier line where three religions and numerous ethnicities meet cannot and must not be abridged to Russian studies. Naturally, a pivot away from Russia to the multiplicity of Eastern Europe will require scholars to master languages and local contexts, combined with a general eagerness to shed old interpretations. Still, the most essential geopolitical outcome of the Russo-Ukrainian war is the growing sense of agency and uniqueness proper to Eastern Europe, which suggests that the typical approach when knowing Russian was enough to do business and research in Riga and Praga, in Kharkiv and Khiva is bound to fade away.

2 Taking Stock of the Distorted Knowledge-Generation Processes

The general impetus to add new (or, rather, rediscover previously suppressed) voices to the study of the region becomes more articulated as scholars grow more sensitive to the issue of distorted knowledge production long in effect within the discipline. The individual chapters of this volume highlighted two aspects of the phenomenon.

First, there are typical and rather well-known power-related distortions when the political centre imposes its agendas on both peripheries and external actors. Moscow and St. Petersburg practiced this strategy extensively through bureaucratic rule, asymmetric warfare (see Chapter 2), ethnoreligious suppression and homogenising of the national space (see Chapter 7), condemnation of barbarity ostensibly proper to Others (see Chapter 6), and post-imperial manipulation of relations with new states (see Chapter 11). A not-so-surprising reaction by the targeted peoples was a surge in nationalist sentiments and religious revival from Makhachkala to Kazan to Krakow, combined with the external balancing, the most prominent example being the flocking of East European countries to the EU and NATO. A recent rendition of this trope is the academic realignment in Finland and Poland with the EU (see Chapter 8).

Second, since access to knowledge about a given society constitutes the ultimate manifestation of power relations, the current regime in Russia practices a not-so-subtle limitation of academic freedoms and manipulations with relevant data about Russian society and politics, as well as its adversaries. Whereas the latter strategy exemplified by the barrage of mis- and disinformation about the "Kyiv junta," "US bio laboratories," or "British-provoked tanker crashes" remains beyond the scope of this volume, many authors addressed no less pressing methodological issues of how to study Russia today. In particular, can one believe the opinion polls collected after Putin's regime took a full turn towards authoritarianism? Can the official statistics provide reliable information on the Russian economy? Chapters 3, 4, and 9 flag these issues and stipulate how improper confidence in unreliable data distorted the scholarly understanding of the Kremlin's ability to conduct war, maintain economic stability, and impose elite cohesion. An additional challenge further limiting the purview of academic research is that it mostly remains rooted in models and theories advanced to understand post-Communist regime transition, economic liberalisation, neopatrimonial corruption, and other models which proved to be unable to predict (let alone avert) the anti-liberal and anti-democratic drift espoused by Putin and his cronies. The Russo-Ukrainian war, which has already brought numerous alterations in areas ranging from military capabilities to civil-military relations, from the covert introduction of a war economy to seismic shifts in regional geopolitics, made the need for new optics on and new knowledge of East Europe and Russia only more urgent.

3 Charting New Horizons

The present volume charts several promising ways to study East Europe and Russia. One inspiring technique suggests how to penetrate behind the veil of Putinesque secrecy and explore the riddle wrapped in a mystery inside an enigma. Chapter 4 discusses the possibilities tangentially, while Chapter 12 offers a passionate plea to use open-source intelligence (OSINT) and digital open-source information (DOSI) to excavate, weigh, and verify data relative to Russia's policies. Admittedly, not all scholars will happily substitute field research with investigative journalism. However, given the limitations on academic exchange imposed by Putin, the continuation of the anti-West propaganda, and the propensity of the regime to use Westerners (regardless of their actual occupation) as hostages to be exchanged for real Russian spies, DOSI is likely to be a preferable option. Moreover, OSINT has already proved to be valuable for insights on a plethora of topics ranging from Russia's military losses (InfoNapalm 2025) to the identification of Russian military personnel responsible for targeting civilians in Ukraine (The Insider 2022) to local commanders' opinions on the situation on the ground (Volya 2025). Given the growing thrust to integrate AI with intelligence acquisition and analysis, exemplified by platforms such as *Logically* and *Pimeyes*, and the availability of OSINT training, and further enhancement of tools via GeoINT and ImgINT strategies, OSINT are likely to offer a viable supplement to traditional academic approaches.

A second innovation traced in this volume is an appeal to account for necropolitics, namely the conflation of a politics of care and control with those of death as the hallmark of illiberal sovereignty (see Chapter 5). There are at least two promising research avenues relevant to necropolitics. To begin with, there is a practical dimension of necropolitics exemplified by deliberate attacks on hospitals, destruction of civilian infrastructure, and the recent trend of "second strikes" when Russian UAVs and rockets hit the same target twice, with a 30–40-minute pause, in order to eliminate Ukrainian first-respondents like firefighters and medics. In its practical dimension, necropolitics illustrate how control over death produces control over life; however, the exact mechanisms that enable such control require further investigation. Next, the genealogy of Russian necropolitics waits for an explanation: is it rooted in Russian military doctrine and strategic culture, Putin's worldview, or the neo-colonial nature of the Russo-

Ukrainian war? Or should some other factors be accounted for? According to the author of Chapter 6, Moscow's voluntary break with the liberal international society and disdain for its normative basis can be seen not as proof of Russia's civilizational peculiarity, but rather as self-barbarization manifested in the evidence of multiple atrocities and war crimes committed by the invading army in the occupied territories.

A third research avenue worth exploring is Russia's great power ideology (see Chapter 2). Although it offers hardly anything substantial for study (since tropes of Russia's greatness, uniqueness, messianic role, and international prestige are likely to be repeated *ad nauseum* and without significant changes throughout history), two GPI functions are of scholarly interest. First is the domestic consumption of GPI. Does it contribute to regime stabilisation? Should GPI be considered a missed element of Putin's rule? Taking longitudinally, how to measure the GPI impact during imperial, Soviet, and contemporary periods in order to compare its role in domestic politics. Second is the GPI international dimension. Clearly, the need to resort to GPI is related to status dissatisfaction. However, is there a kind of positive feedback loop when Russian elites, constantly reminding their international counterparts about Russia's greatness, accept their deception and start acting more brazenly? For policymakers, it might be interesting to weigh pros and cons of the "symbolic substitution policy", namely to grant Russia's elites tokens of prestige without conceding anything substantial to leverage the acute status dissatisfaction. A possible argument against this policy is that *l'appetit vient en mangeant*, and thus, it might encourage further demands. An enticing way to test either possibility is to conduct a randomised controlled trial, an approach gaining momentum in IR studies.

4 Promoting Comparative Dimensions

This volume's most important—though admittedly the least articulated—conclusion is the necessity to bring Russia to comparative studies. Far too long, Russia's geographical vastness, convoluted path to modernity, and Messianic stance promoted compartmentalisation and exclusion. The whole discipline of Russian studies—just like its counterpart, the American studies—articulated, rationalised, and, eventually, reified Russia's uniqueness. This is not to say that a better understanding of Russia does not require specially trained analysts—the same is true for China, Chile, South Korea, or Spain. Rather, it is to reject the centuries-old misperception alleging that "one cannot fathom Russia with a common yardstick". Social science, by definition, must strike a careful balance between ideographic language and fleshless abstraction. To follow Tyutchev's old lie is precisely to opt for ideographic language instead of making science.

As Richard Rose has long argued,

> the norm is to conduct studies relying upon explicit concepts defined with sufficient clarity to permit systematic comparison with other countries. The concepts employed focus upon particular segments of society, such as class and voting, public expenditure as a percentage of GNP, and so on. The subject is not countries per se but analytic attributes common to many countries. (Rose 1991, 454)

Applying this principle to the intersections of this volume's individual chapters, four promising domains of comparative analysis seem to emerge. All are linked by the idea, explicitly expressed in Chapter 4, that mid-range theories are required to move the discipline further. First is the question of institutional dysfunctionality and its effects. This question bridges together the need to consistently apply lessons from the historical sociology of state-making to Russia's trajectory with studies on neopatrimonial bad governance and authoritarian regime types. The urgency for bridging gained prominence due to the central role of war-making in state-building. The surprising strength of the corrupt Russian state (see Chapter 3) might shed its initial surprises when one considers war-making efforts exhibited by other patrimonial and neopatrimonial states—be it absolutist France in the 18th century, imperial Russia in the 19th century, or the Saddam's Iraq in the 20th century. Despite expectations that corrupt and top-heavy bureaucracies would blunt the war effort, each of these states practiced long and expansive warfare. Moreover, contrary to standard Tillyan account, war-making was not responsible for the emergence of effective merit-based bureaucracy and enhanced state capacity in none of these cases. Rather, (neo)patrimonial polities were waging war and inflicting further and further damage on themselves until war unmade them (Teschke 2017). Arguably, Putin has adopted this path: maximise the war effort while simultaneously rewarding his cronies and ignoring the larger outcomes and long-term consequences. Circumstances of the postponement effect, as well as exact mechanisms through which warfare is related to neopatrimonial decomposition, must be refined in a broader set of cases.

The second important domain, logically related to the first one, is civil-military relations. Chapter 10 addresses this particular issue, but the scope should be even larger. The nexus between coup-proofing strategies, combat-effectiveness, and regime type is to be further explored, especially within the comparative perspective. Another noteworthy point of interest, which does not necessarily have to focus exclusively on Russia and should account for the Ukrainian case, is that of the home front during 21st-century warfare. Does the availability of long-range missiles and UAVs effectively erode the differences between front and rear? How does this affect war-weariness? How about the gap between veterans coming home and civilians who never left for the front: is it growing deeper or shallower because of 21st-century warfare?

Next, civil-military relations get additional rendition when one considers what soldiers of one belligerent party do with the civilians representing the other. Here, one might spot a fruitful intersection of research on necropolitics and occupation policies. Does the ethno-religious composition of occupying troops relate to the magnitude of war crimes (e.g., a platoon recruited in non-Slavic or non-Orthodox regions of Russia display a more dehumanised attitude toward civilians)? Are there individual officers and rank-and-file trying to minimise civilian casualties, and, if positive, what might be the causes? Another important aspect of necropolitics is the effect of wreaked destruction upon an individual's eagerness to support the war effort: do people who suffered material damage want vengeance or closure? These are heartbreaking questions that scholars today are especially well-positioned to explore. Lastly, one might find it helpful to study Russian necropolitics from a comparative perspective: was the strategic bombing of Japan or the spoliation of Palatinate different from Russia's leverage of necropolitics in 2022–2025? Are there differences between Russian, Western, and non-Western necropolitics? Have necropolitics evolved with technical innovations in weaponry throughout history? Answering these questions will contribute to proper triangulations of narratives on civilisation and barbarity treated in Chapter 6.

One might also apply a comparative perspective to explore great power Ideologies in other contexts. After all, Russia is not unique in claiming its exceptional status and mission. There are national variations of GPI in France, the US, the UK, and China (to name just a few). A plethora of research questions emerge: when do the elite produce GPIs? Is GPI tied to foreign policy blunders (Suez adventure for France and the UK, misguided Vietnam policy for the US, counter-productive "Wolf warrior diplomacy" for China), or can it contribute to finding opportunities? Is there a link between GPI and militarism? Or regime type? IR and political science will benefit when tackling these questions.

5 Political (Dis)engagement of the Academic Community

Finally, some authors of this edited volume discussed Russia studies from the viewpoint of politicization and depoliticization. Two interrelated arguments proffered in chapters 4 and 8 add some polemical undertones to this topic. One is a critique of the European academia for the tendency of growing over-politicisation and polarisation of research on Russia and Ukraine which to some extent might allegedly mirror authoritarian practices of black-and-white type of thinking in non-democratic countries. What may logically stem from this reasoning is that to preserve its ability to seek the truth, the academia has to remain normatively neutral and stay beyond ideological and axiological divides. Another argument suggests that the Russian academic community perceived the invasion of Ukraine as an "exogenous shock", which implies that this community preferred either to dissociate itself from Putin's regime or to disregard the multiple academic contributions to legitimation and maturing the Kremlin's imperial policies towards Ukraine.

Both claims share a presumption that academia may or could be detached from the domain of the political. However, this dissociation can often be problematic. In the case of the West, universities as social institutions have reacted to Russia's violation of basic international laws by cutting connections with counterparts located in Russia and funded by the government that started the violence. However, this normative position did not lead to an alleged uniformity within Western epistemic communities that generated a variety of intellectual products that in one way or another reflected the diversity of characteristics of the event of February 24, 2022, as Russo-Ukraine war, Russia's full-scale invasion, Russia's aggression against Ukraine, or as a 'conflict'. There is also a broad range of conceptual approaches applied by Western scholars belonging to different disciplines and schools, including—as this book illustrates—political sociology, media and communication studies, political theory, and many others. As chapter 11 suggests, some new analytical methods may compensate for the deficit of more traditional data that otherwise could have come from field research in Russia or from reliable statistics.

As for the Russian academia, its distancing from politics is equally dubious, since numerous Russian scholars and policy analysts have for years been contributing to discursive legitimation of such consequential components of Moscow's foreign policy as the Russian world doctrine, securitization of the West and Ukraine, criticism of liberal values as detrimental for Russian identity, and glorification of Russian imperial past. These narratives produced by Russian academics leave no ground for considering universities, research institutions and think tanks as staying aloof from the Kremlin in the preparation of the war against Ukraine and its normalisation. Many Russian scholars might indeed have experienced a shock of the war, and some left the country as a gesture of protest, yet those who stayed quite fast adapted to the conditions of the war and even supported it, thus strengthening Putin's bellicose narrative.

Under these conditions the exterioriszation of war by the Russian academic community may look like an attempt to exempt scholars and other public intellectuals from their—at least moral—responsibility for the war and its consequences.

All this is important for two discussions shaped by the Russo–Ukrainian war. One concerns sources and forms of politicisation of Russian studies as an academic discipline. A good illustration of the scale of the problem in the Western academia is migration of the ill-defined idea of "Russophobia" from political discourse to research language. However, this issue seems to be much more acute for Russian scholarship in political science and international relations than for Western expertise on Russia. Another discussion relates to the importance of ethical issues for political analysis: for example, would it be acceptable to criticise the victims of foreign aggression during the ongoing war (chapter 7)? And wouldn't the rejection of such a critical attitude on ethical grounds create new zones of silencing? These are serious questions for Western epistemic communities, but—again—they appear to be much more challenging for scholars working in Russia where ethical considerations are obviously marginalised and even erased from the academic discourse. Politicisation and silencing are two points of contention that were only tangentially touched upon in this book, but in the future we may expect a much more intense debate on each of them.

Bibliography

InfoNapalm. 2025. OSINT analiz: yak rakhuvaty vtraty rosiyskoi armii po nomerakh nagorod 'Orden Muzhestva'. https://informnapalm.org/ua/osint-analiz-yak-rakhuvaty-vtraty-rosiisk/.

The Insider. 2022. Ubiytsy na udalionke. Kto i kak navodir upravliaemye rakety na ukrainskie grazhadanskie obyekty. https://theins.ru/politika/255158.

Rose R. 1991. Comparing Forms of Comparative Analysis, *Political Studies* XXXIX, 454.

Teschke, B. 2017. After the Tilly Thesis Social Conflict, Differential State-formation and Geopolitics in the Construction of the European System of States, in Kasperse L.B. and Strandsbjerg J. (eds.), Does War Make States? Cambridge University Press, 25–51.

Volya. 2025. Chto proizoshlo v Kurskoi oblasti, chto vsio eto znachit i chto budet dal'she. https://teletype.in/@volyamedia/LFOZq9asA05.

Index

www.ingramcontent.com/pod-product-compliance
Lightning Source LLC
Chambersburg PA
CBHW071333230926
51849CB00002B/386
* 9 7 8 9 0 0 4 7 4 7 3 5 7 *